THE BRIGHTEST JEWEL

An historic Lumière Autochrome colour photograph of the herbaceous border
at the National Botanic Gardens, Glasnevin (see page iv)

...Our Botanical Establishment [is] the brightest jewel I am proud to say in the [Dublin] Socyty's cap, admired by all who have visited it — foreigners who have from time to time paid attention to it's different compartments on learning, and hearing their tendency explain'd, seem'd to feel, indeed declare, that there was no similar Garden, founded on the same nationally useful principles.

Dr Walter Wade to the Right Honourable John Foster, 3 May 1819

THE BRIGHTEST JEWEL

A History of the National Botanic Gardens
Glasnevin, Dublin

E. Charles Nelson
Eileen M. McCracken

with original watercolours by Wendy F. Walsh

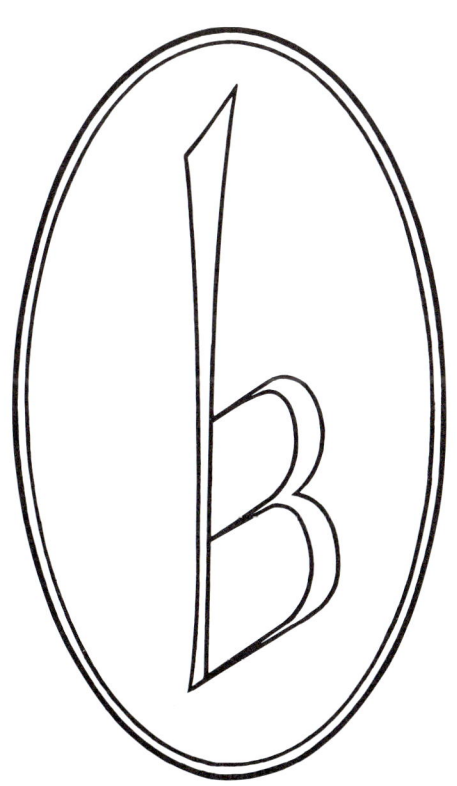

BOETHIUS PRESS
Kilkenny
Ireland

© 1987 E. Charles Nelson
Eileen M. McCracken

All rights reserved. No part of this work may be reproduced or used in any form or by any means, including photocopying or information storage and retrieval systems without the written permission of the publisher.

British Library Cataloguing in Publication Data
Nelson, E. C.
The brightest jewel : a history of the National Botanic Gardens, Glasnevin, Dublin.
1. National Botanic Gardens—History
I. Title II. McCracken, Eileen
580'.74'441835 QK73.I732N3/
ISBN 0 86314 083 1

The colour photograph used for the frontispiece was taken at the Botanic Gardens, Glasnevin, before January 1912, most probably during the summer of 1911. The photographer's name is not known. It is an early and very fine example of a screen plate colour photograph.

All systems of colour photography and colour printing are based upon the principle that any colour sensation can be produced by a suitable mixture, in addition or subtraction, of three colours: coloured light in the case of addition, colour pigment in the case of subtraction. There is no other commercially practical way to reproduce patterns in 'full' colour. Additive techniques preceded subtractive.

In additive technique, white light is analysed optically (by colour filtering) into three wavebands of colours red, green and blue, the scientific primaries. To make an additive colour photograph, three photographs of the same object are made, from the same viewpoint, at the same time, one through a red filter, one through a green filter and one through a blue. These photographs can be projected in register on a white screen through the same three filters, using three different lanterns. The resulting colour picture by projection will resemble the original object more or less accurately.

In the screen plate process, one exposure only is made using a photographic plate covered with tiny transparent cells coloured red, green and blue. These cells are tiny colour filters and act just as the separate filters described above. An additive colour photograph can thus be made in one step and can be viewed by holding it up to white light.

Screen plates were first made, circa 1905, by dyeing potato starch grains in the three colours, red, green or blue and thoroughly mixing them before squeezing them flat between glasses. Squeezed thus, they are more or less transparent and will act as colour filters. The original frontispiece photograph is a starch grain screen plate colour photograph, here reproduced lithographically by subtraction technique.

For details of these and other methods of colour reproduction see any good textbook on photography, under Trichromatic Theory and colour reproduction: Helmholtz, Maxwell and Young, du Hauron and others.

Colour Separations
GRAPHIC REPRODUCTIONS·DUBLIN

Designed and Typeset in Bembo
Process Photographs· Colour Plates· Printing
BOETHIUS
Kilkenny

In memory of my grandparents
ECN

In memory of Ruby Brown
EMMcC

The Department of Agriculture gave a grant to support the publication of this history.

Michael McNamara (Forestry Consultant, Cork) generously made a private gift towards costs.

The Stanley Smith Horticultural Trust made an interest free loan during the period of greatest expense.

The publishers gratefully acknowledge this help.

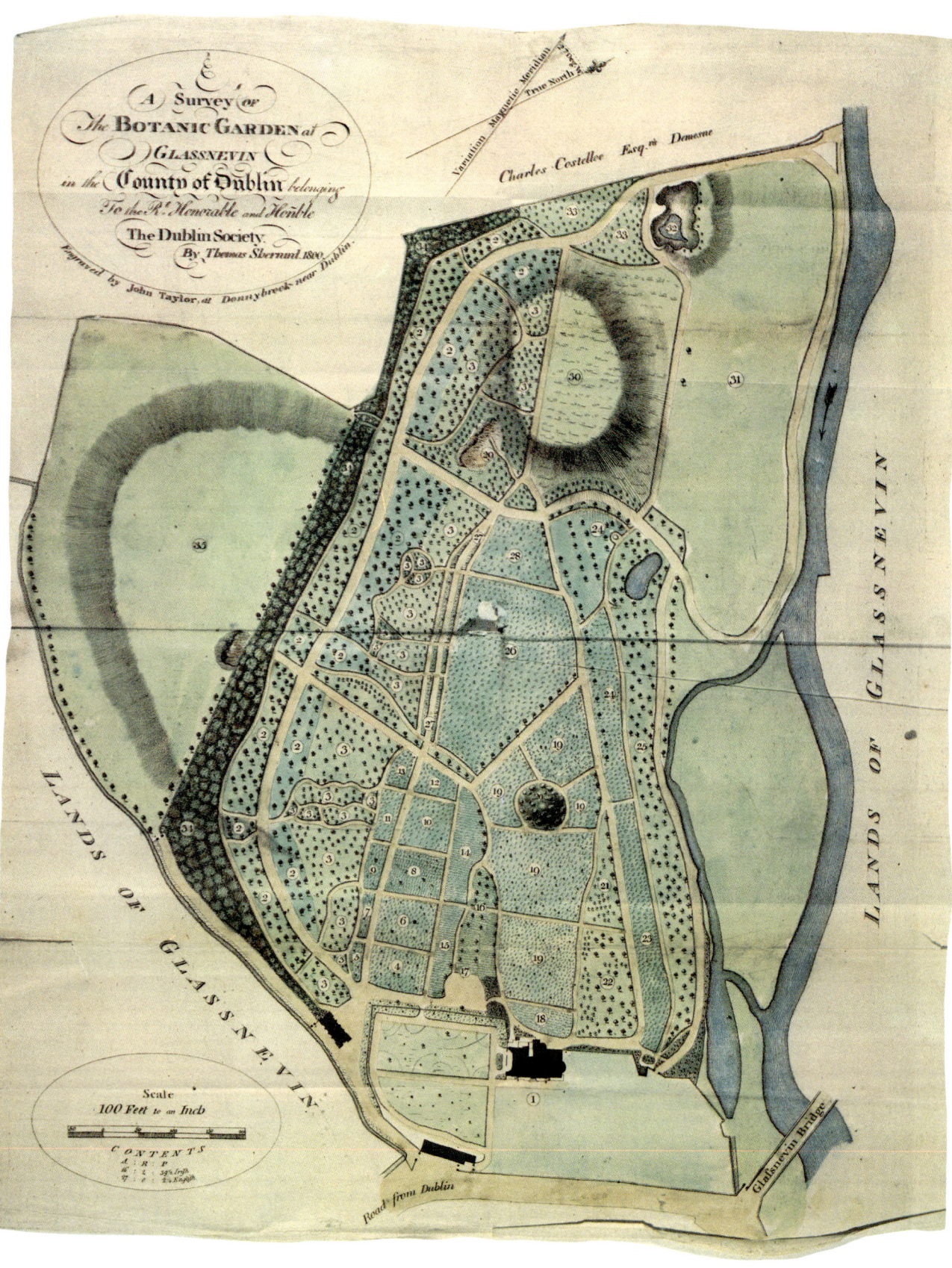

Map Colour Plate 1: Sherrard's map of the Botanic Gardens, 1800 — hand-coloured copy from *Transactions of the Dublin Society* (by courtesy of the National Library, Dublin) (see Fig. 27, p. 54)

Map Colour Plate 2: Mutlow's map of the Botanic Gardens, 1818—hand-coloured copy
(Dr E. C. Nelson) (see Fig. 28, p. 55)

As this book was being printed, on Wednesday, 14 January 1987 the Government made a decision to transfer authority for the National Botanic Gardens from the Department of Agriculture to the Department of Finance. The National Botanic Gardens thus is linked within the Office of Public Works to Ireland's National Parks which include the gardens of Muckross in Killarney, and Glenveagh in County Donegal, and with Ilnacullin at Glengarriff, County Cork.

ECN
19 January 1987

FOREWORD

Gardening is a delightful pursuit, absorbing to practise and rewarding in its results. The popular enthusiasm for gardening is very evident in the display of attractive shrubs and flowers surrounding many an Irish home today. Many of us who garden for pleasure are accustomed to making an occasional visit to Glasnevin to enjoy the charm and appeal of the Botanic Gardens and to admire and learn from the ornamental horticulture there. Glasnevin's role in contributing to the improvement of the private garden is well known but its primary function as an important scientific institution is not so widely recognised.

The Botanic Gardens is undeniably an outstanding part of our heritage. Yet, surprising as it must seem, until now the full story of the Gardens and of the many personalities who contributed to its foundation and development has not been adequately chronicled. This splendid book makes good the deficiency. It gives us a comprehensive history of the Botanic Gardens from its inception towards the end of the eighteenth century to the present day.

From it we learn about the origin and purposes of botanical gardens in general and the fascinating course of events leading to the formation of the Glasnevin Gardens in 1795. The vital roles played in all this by Walter Wade, the Dublin physician, and John Foster MP, the County Louth landowner "with a rage for planting", are vividly described. Chapters are devoted to those whose work contributed to the development of Glasnevin's international reputation. Outstanding among these were the father and son, David and Frederick Moore, whose direction spanned a remarkable 84 years extending from 1838 to 1922, a period during which Glasnevin's international reputation was firmly established. Indeed we read that in 1922 a distinguished authority described Glasnevin among the leading botanical gardens of Europe as 'surpassed by none and equalled by very few'. We are told the story of the elegant curvilinear range of glasshouses erected in the 1840s mainly by the innovative Dubliner, Richard Turner, the finest builder of glasshouses and conservatories of his day.

There is much in this study of particular interest to the horticulturist, botanist, naturalist and social historian. But above all, this absorbing and beautifully illustrated book is a delight to read and will enrich any visit to the Gardens for the general reader.

The authors' enthusiasm and feeling for their subject breathe through every page of this stylishly written, thoroughly researched and scholarly work. They deserve our gratitude.

Austin Deasy TD
Minister for Agriculture
23 May 1986

COLOUR PLATES

Map Colour Plate 1 Glasnevin in 1800
Map Colour Plate 2 Glasnevin in 1818

Plate 1   *Leptospermum laevigatum* (J. Gaertner) F. Mueller
          Myrtaceae

Plate 2   *Verbena phlogiflora* Chamisso
          Verbenaceae

Plate 3   *Alopecurus pratensis* Linnaeus
          Poaceae

Plate 4   *Ulex europaeus* Linnaeus 'Strictus'
          Papilionaceae

Plate 5   *Carex buxbaumii* Wahlenberg
          Cyperaceae

Plate 6   *Inula salicina* Linnaeus
          Asteraceae

Plate 7   *Abelia triflora* R. Brown
          Caprifoliaceae

Plate 8   *Lankesteria barteri* J. D. Hooker
          Acanthaceae

Plate 9   *Helleborus orientalis* Linnaeus 'Dr Moore'
          Ranunculaceae

Plate 10 *Lachenalia bulbifera* (Cyrillo) Ascherson & Graebner
          Liliaceae

Plate 11 *Sarracenia rubra* Walter var. *acuminata* de Candolle
          Sarraceniaceae

Plate 12 *Paphiopedilum dayanum* (Lindley) Pfitzer
          Orchidaceae

Plate 13 *Anguloa virginalis* Schlechter
          Orchidaceae

Plate 14 *Escallonia* 'C. F. Ball'
          Escalloniaceae

Plate 15 *Tulipa sylvestris* Linnaeus
          Liliaceae

(Plates 4, 6, 7, 14 are original watercolours by Wendy Walsh, specially commissioned for this volume. Plates 9, 10, 11, 12 are original watercolours by Lydia Shackleton, from the National Botanic Gardens. Plate 13 is an original watercolour by Alice Jacob, from the National Botanic Gardens. The other plates are from published sources as acknowledged.)

CONTENTS

	Prologue	1
1	The origins of botanical gardens	3
2	The early history of Irish botany and the first botanical gardens	6
3	Botany and politics: Walter Wade and John Foster 1790–1795	21
4	Before 1795: The romantic Glasnevin of Tickell and Delany	37
5	The formation and early years of the Botanic Gardens	46
6	The fostering mantle 1800–1833	64
7	Ninian Niven 1834–1838	83
8	David Moore: the first decade 1838–1848	97
9	'An era in gardening' David Moore's last three decades 1848–1879	117
10	Botanical explorers and the supply of plants 1800–1920	152
11	Frederick William Moore 1879–1922	173
12	National Botanic Gardens 1922–1968	214
	Postscript	235
	Appendix 1: 'Erinensis' on Dr Litton	240
	Appendix 2: *Botanical Magazine* plates (a) associated with Glasnevin (b) by W. E. Trevithick	245
	Appendix 3: Personnel associated with Glasnevin	250
	Bibliography, Notes and references	252
	Index	269

AUTHORS' ACKNOWLEDGEMENTS

This history has taken over twenty years to research and write. At the beginning, Eileen McCracken worked on the general history of Irish botanical gardens, paying special attention to the Belfast Botanic Gardens and its fine palm house which has been superbly restored since the publication of her book in 1971. Dr McCracken read and transcribed the manuscript minute books of the Royal Dublin Society, and also studied the correspondence archive in the National Botanic Gardens. In the early years of this undertaking constant encouragement was given by Cearbhall Ó Dálaigh, President of Ireland (1975-1976). We greatly appreciated the interest that he showed and we know that he would have been pleased to see the history of the National Botanic Gardens, in which he was so interested, finished.

In the late 1970s, shortly after his appointment to the post of taxonomist in the Botanic Gardens at Glasnevin, Charles Nelson expanded his interests in the history of botany, and commenced study of the history of Irish botany and horticulture. His work has included research in the correspondence archives of the Linnean Society, London, and the Royal Botanic Gardens, Kew, as well as the papers of John Foster in the Public Record Office of Northern Ireland in Belfast.

Immediately after the publication in December 1979 of the Royal Horticultural Society of Ireland's volume *Irish Gardening and Horticulture*, at the instigation of Dr McCracken we agreed to co-operate on the writing of the history of the National Botanic Gardens, combining the research which had been carried out independently before. The final major archival discovery, that is of the papers belonging to the Moore family, happened after this through the good offices of Dr M. R. D. Seaward and Miss Evelyn Booth.

Eileen McCracken has for the past few years spent a great deal of her time overseas and in consequence much of the burden of co-ordinating the script and conducting the negotiations relating to publication fell on the shoulders of Charles Nelson. He was also responsible for collecting most of the illustrations and preparing the bibliography.

Over more than twenty years many institutions and individuals have assisted our research. The late Desmond Clarke and his successor Alan Eager, librarians to the Royal Dublin Society, Ballsbridge, gave us unrestricted access to the archives of the Society. The same may be said for the librarians and archivists in many other places: Trinity College, Dublin (W. O'Sullivan and Miss M. P. Pollard), Royal College of Physicians of Ireland, Dublin (R. Mills), Royal College of Surgeons in Ireland, Dublin (Mrs K. Bishop), Department of Agriculture, Dublin (Mary Doyle), Archbishop Marsh's Library, Dublin (Mrs M. McCarthy), Linnean Society, London (G. Bridson, Miss G. Douglas), Royal Botanic Gardens, Kew (V. Parry, Miss S. Fitzgerald), Botany Library, British Museum (Natural History), London (Mrs J. Diment), University of Melbourne and Hunt Institute for Botanical Documentation, Pittsburgh, USA (Dr R. Kiger, Dr M. Steiber). Our thanks are also due to the Director, National Library of Ireland, Dublin, the Assistant Keeper, Public Record Office of Northern Ireland, Belfast, and the librarians of the Royal Society, London, and the Religious Society of Friends, Dublin, for their assistance. The Rector of Glasnevin, Reverend D. Harmon, kindly facilitated our work.

Many individual botanists, historians, gardeners and librarians have given their time to assist us and we particularly wish to record our gratitude to Dr Mark Seaward, Evelyn Booth, Brian Mulligan, Graham Stuart Thomas, David Shackleton, Mike Snowden, Susyn Andrews, Dr Keith Ferguson, Paul Wilson, Michael Walpole, Mrs Margaret Garner, the late Miss Blanche Henrey, Miss E. J. Willson, Ray Desmond, Paddy and Jennifer Woods, Mr and Mrs Sean O'Shea, Ninian C. Niven and Mrs J. Scott, F. E. Dixon, Jim White, Mrs Cowie, Valerie Ingram (for carefully reading the manuscript and for bibliographic assistance), Mary Davies (for drawing the maps) and David Davison (for his photographic assistance and expertise).

This history could not have been completed without the kind co-operation of Major-General M. E. Tickell who provided much valuable information on his forebears and allowed us to examine the deeds which belonged to Thomas Tickell. Lord Massereene and Ferrard granted permission to quote from the invaluable archive of John Foster (Baron Oriel) and also sent us additional information about Glasnevin's founder. Of equal value was the help given by Mrs Margaret Mansfield, daughter of John Besant, who gave us photographs and reminiscences of the Botanic Gardens in the 1920s. Dr Maureen Walsh provided information and photographs of her late husband, Dr Thomas Walsh. And, we have received the most courteous help, including unrestricted access to all his family papers, from Major-General Frederick Moore; his co-operation has been of outstanding importance and has allowed us to form a fuller view of the contributions made by Dr David Moore and Sir Frederick Moore. We thank him most sincerely.

Finally, as this is a history of the National Botanic Gardens, we must record our profound appreciation for the encouragement and help of its present and former staff; the late Dr Walsh and John Fanning, Miss Elsie Miller who shared with us her vivid memories of the days of Sir Frederick Moore and John Besant, Dr Brian Morley, Tom Crawford and Donal Synnott, and the present Director Aidan Brady.

Publication of this work has been supported by a grant from the Department of Agriculture, and the assistance of Michael McNamara and of the Stanley Smith Horticultural Trust (UK) is also gratefully acknowledged.

E. Charles Nelson
Eileen M. McCracken
November 1985

New Range of Conservatories.
(see Fig. 54, p. 111)

Prologue

THE NATIONAL Botanic Gardens[1] at Glasnevin, a few miles north of the centre of Dublin, has a distinguished record in the progress of Irish botany and in the history of horticulture world-wide. Established in 1795 by the Dublin Society[2] at the behest of the Irish Parliament, it was conceived as a great national institution to serve the scientific community, and especially the agricultural interests of Ireland. In time, the Botanic Gardens came to serve a much wider section of the community and its priorities were altered. Exotic plants were introduced from distant continents and new varieties were bred to the enrichment of humbler gardens. Knowledge of the flora of Ireland, and of the plant kingdom in general, increased through field studies and experiments and the resulting publications gained for the Botanic Gardens an international reputation. Simply by being open every day, for more than a century free of charge, to the citizens of Dublin and visitors, the Gardens provided pleasant surroundings for recreation and quiet relaxation. Above all, it became a place for learning.

Nearly two centuries have passed since the first plants were set into the newly-fashioned beds and original glasshouses. The Botanic Gardens has expanded and now covers nearly fifty acres. About 20,000 different plant species and varieties are cultivated out of doors and in greenhouses, representing habitats from arctic tundra to tropical rainforest, parched desert to Irish blanket bog. The reason why so many different plants are grown is that a botanical garden is primarily a place where botany is taught and research into the biology and distribution of plants is carried out. Many of the plants are cultivated because they are of interest to scientists—they may not have pretty flowers. This teaching role, which arose from the ancient link between medicine and botany, distinguishes botanical gardens from other gardens which contain large collections of different plants.

The earliest botanical gardens were called physic gardens—*hortus medicus, hortus sanitatus*—and were attached to the medical faculties of universities. Students of physic (medicine) had to learn to identify the plants used in primitive medicine; they achieved this by studying the plants grown in the university gardens. The first botanical garden was started in Italy a quarter of a millenium before the foundation of the Botanic Gardens at Glasnevin. However, the traditions of the early physic gardens were inherent in the plans drawn up in 1790 for a botanical garden in Dublin, and are embodied in some of the sections of the National Botanic Gardens today.

This is a history of the National Botanic Gardens from its formation in the closing years of the eighteenth century, through the peak of its international fame in the early years of the twentieth century, to the present day. It examines the political and scientific background to the foundation. The land on which the Botanic Gardens was established has 'romantic' associations for it was owned by the poet Thomas Tickell before it was acquired by the Dublin Society. Nearby had lived Patrick and Mary Delany whose own garden made the village of Glasnevin famous throughout Ireland and Britain in the Age of Enlightenment. The 'romantic' past is also recorded in this story of the people and plants that have lived within the Gardens' bounds.

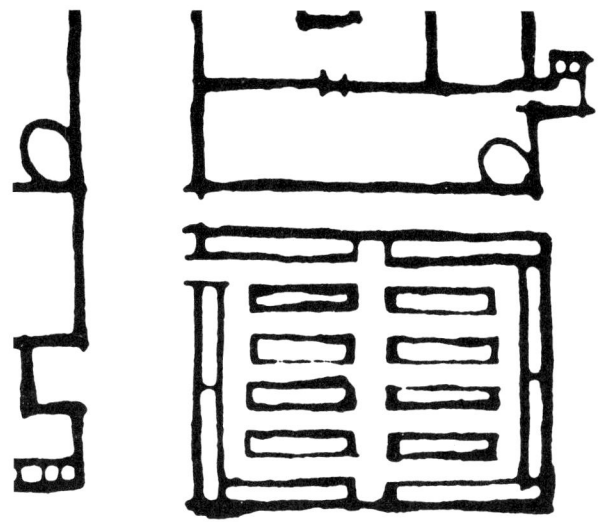

1. The herb garden or *herbularius* situated beside the infirmary and the doctor's quarters (*domus medicorum*) as shown on the St Gall plan (opposite) (below; redrawn after A. W. Hill, 1915[3])

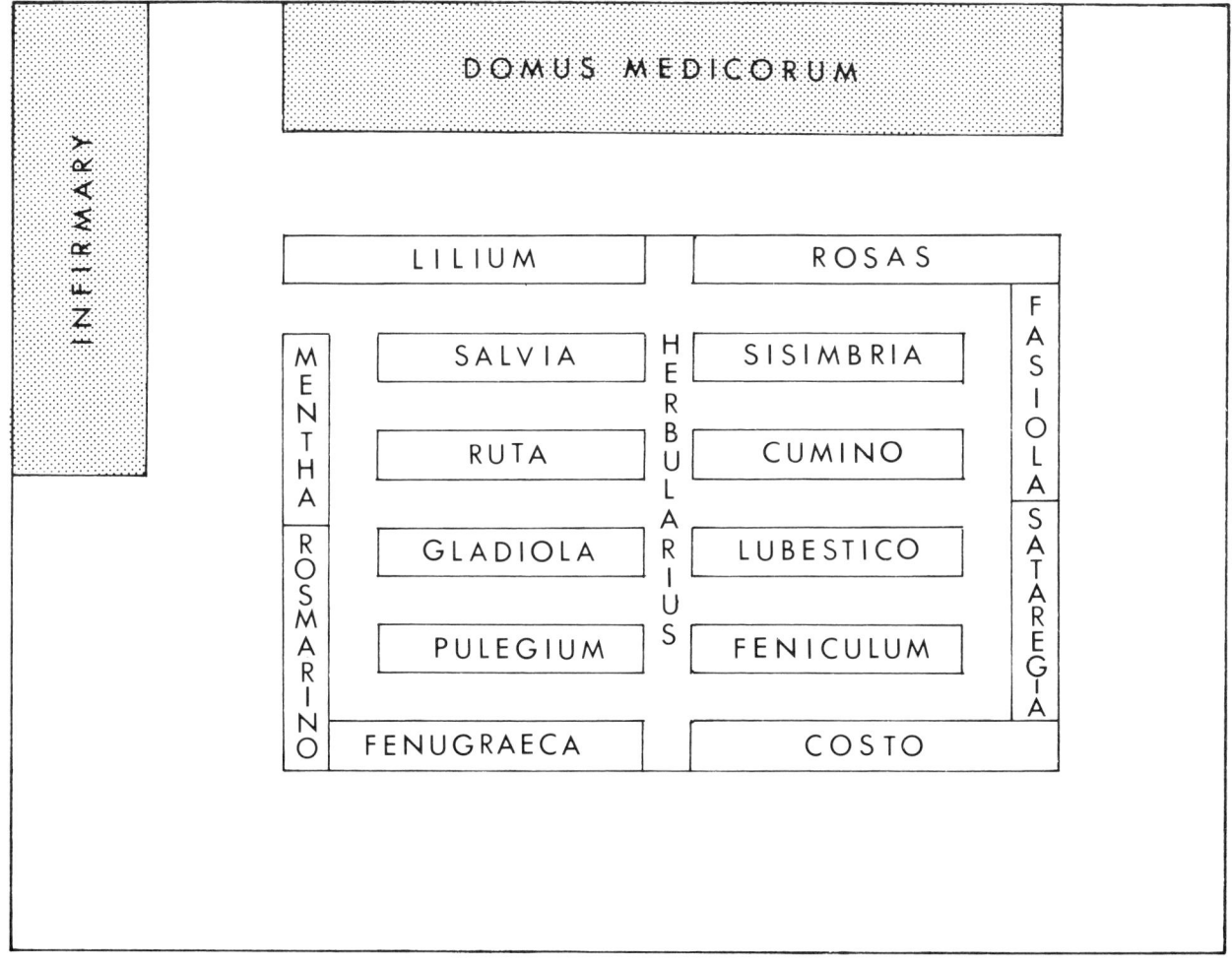

1

The origins of botanical gardens

THE SCIENCE of botany had its beginning in ancient Greece. Theophrastus of Eresos (370-285 BC), the tutor of Alexander the Great and disciple of Aristotle, made careful observations of plants and pioneered the art of describing them. He gave botanical lectures in the Lyceum and had a garden in which he grew native Greek plants as well as vegetables. During Theophrastus' lifetime, Alexander's military campaigns reached into India and Greek ships sailed in the Persian Gulf. The well-trained and observant officers in the army and navy sent reports back to Greece of plants such as bamboo, banyan, banana and mango, as well as descriptions of mangrove swamps on the shores of the Indian Ocean. Theophrastus incorporated these observations and accounts into his botanical teachings which were recorded in two works, *Historia plantarum* (Enquiry into plants) and *Causae plantarum* (Causes of plants). Theophrastus noted the fundamental differences between plant organs such as the roots and shoots, and discussed the relationship of plants to the natural environment. Like all subsequent authors he used information collected by others; in Theophrastus' works there are facts garnered from the *rhizotomoi*, the herb gatherers, who collected plants for use in medicines from their native habitats and knew a considerable amount of folk botany.[1]

Four centuries later Pedianos Dioscorides (*fl.* 50-70 AD), a Greek from Asia Minor, compiled an encyclopaedia of medicinal plants. He was probably a learned herbalist, perhaps a military doctor who was widely travelled. Dioscorides' work contained information about the real and imagined medicinal properties of the plants of the Mediterranean region. He also copied and used the writing of earlier authors. Some of his sources are no longer known, but according to his Roman contemporary, Caius Pliny, these included herbalists who had painted plants and written their properties under the illustrations.

The significance of Dioscorides' encyclopaedia lies in the fact that it was copied many times. The earliest copy which survives is the *Codex Vindobonensis* (Vienna Codex), prepared about 512 AD in Constantinople (Istanbul) for a princess. It contains nearly five hundred illustrations, many of which may be copies of earlier drawings.[2] Throughout mediaeval Europe, Dioscorides' books were widely used, especially by the monks who ministered to the sick. They regarded the texts as authoritative and used the handwritten books to pass on knowledge of medicinal herbs to their successors. But, in the process of copying, the information in the encyclopaedia was revised and expanded.

In some mediaeval monasteries, medicinal herbs were cultivated in special gardens so that there could be a ready supply of these vital plants. The plan for an ideal foundation, drawn at the monastery founded by St Gall, an Irish missionary from Bangor, shows several gardens, including a small *herbularius* (physic or medicinal garden) besides the physicians' quarters (Fig. 1). In this, the herbs which were not available in the local fields and woodlands were grown, including rosemary, rue, cummin and fennel, as well as roses and lilies. Each plant was cultivated in a separate rectangular plot within the enclosure.[3]

In the twelfth and thirteenth centuries the role of monastic foundations as centres of learning began to diminish. New communities of teachers and students were formed apart from the monastic and cathedral schools. They soon found it necessary to create organisations to defend their privileges and order their lives. The result was the formation of the forerunners of modern universities. Initially these embryo universities trained young men for vocations in the Church, but later they helped to instruct students for the two other great mediaeval professions, Law and Medicine.

In the following centuries, as faculties of medicine developed, especially in the universities of southern Europe, they attracted students from distant places. In the early fourteenth century at least, Irish scholars studied medicine at Montpellier in the south of France. Some of these students returned to Ireland

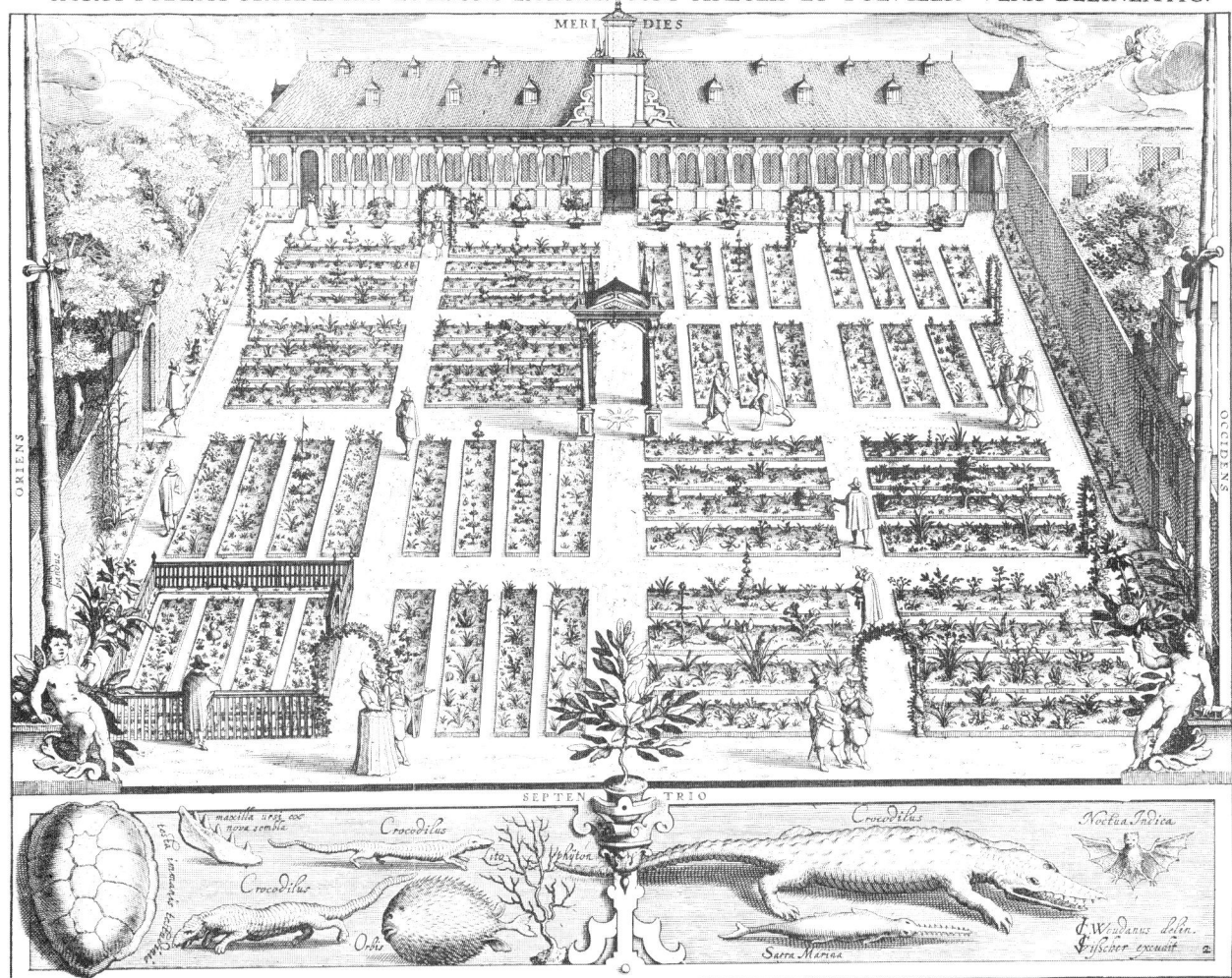

2. A plan and view of the University of Leiden Physic Garden about 1610
(reproduced by courtesy of the Director, Hortus Botanicus, University of Leiden)

and translated medical texts into Irish, including the encyclopaedia of Dioscorides. At this time there was no Irish university and no indigenous school of medicine.[4]

In the medical faculties it was important to train students to distinguish correctly the numerous herbs used in medicines — mistakes were easily made which were quite unacceptable in the mission of healing. As it was not possible to find within the immediate environs of a single university all the useful plants growing wild, the idea arose of growing them side by side in a systematic way. The obvious model was the physic garden of the monasteries with its separate, ordered beds. The first botanical garden, in a modern sense, for which records survive was established at the University of Pisa in northern Italy in 1543 under Luca Ghini (1490-1556). Ghini was a professor of botany at the University of Bologna from about 1534 until 1544, and had had his own private botanical garden there. He was a great teacher and his influence was far-reaching.[5]

Luca Ghini is also credited with the invention of the herbarium — a collection of pressed and dried plant specimens fixed on to sheets of paper and labelled with their scientific names and the places and dates of collection. The early herbarium usually was formed of bound volumes, and was called an *hortus siccus*, literally a dried garden. The volumes were portable and enabled botanists to compare plants from one region with those of another. It also permitted plants to be studied at all stages of growth, as specimens could be collected at different seasons.

The formation of botanical gardens by universities followed the example of Pisa. In 1545, two years after the foundation of the garden in Pisa, others were established at the universities of Florence and Padua. The latter garden still exists on its original site and retains the character of these early botanical gardens—formal beds arranged in a formal pattern with the plants grown separately in labelled plots. By the end of the sixteenth century there were gardens in Luca Ghini's old university at Bologna, and at Montpellier, famous for its medical school. In north-western Europe, botanical gardens had been formed at Heidelberg and in the Netherlands at Leiden. The University of Leiden later became one of the most important centres for botany in Europe and a leading horticultural establishment, pioneering the introduction of plants from subtropical and tropical regions and their cultivation in a primitive greenhouse (Fig. 2).

In England botanical studies were stimulated in the mid-sixteenth century by the publication of the first indigenous books on plants. The Father of English Botany, William Turner (c. 1508-1568), who was one of Luca Ghini's students, published a number of works of which *The names of herbes in Greek, Latin, Englishe, Duche and Frenche wyth the commune name that herbaries and apothecaries use* (1548) was the first to contain localities where plants had been found. Turner also published *A new herball* (1551) which was written in English and included woodcuts of plants. Following William Turner's herbal, the most significant book was John Gerard's *The herball or generall historie of plantes* (1597), a monumental work derived from many earlier sources. However, Gerard did provide a substantial amount of original information, including the first equation of shamrock with clover—'the common Meadow Trefoiles...and called in Irish Shamrockes'.[6] He had a garden at Holborn near London and published a catalogue of its contents in 1596. A contemporary wrote that John Gerard grew 'all the rare simples [i.e. plants] which by any meanes he could attaine unto...all manner of strange trees, herbes, rootes, plants, flowers, and other such rare thinges, that it would make a man woonder...'.[7]

The majority of early English botanical books were slanted towards medical botany, giving the medicinal properties of the various plants mentioned. This link between medicine and botany remained strong for another two centuries and was underlined in the formation of the first botanical gardens in the British Isles. In 1621, through the munificence of Henry, Lord Danvers (1573-1649), later Earl of Danby, a physic garden was established at Oxford. Danvers wanted to give the university 'a place whereby learning, especially the faculty of medicine, might be improved'.[8] He purchased ground near Magdalen College and the physic garden was laid out beside the River Cherwell. In 1633 the surrounding walls and fine gateway designed by Nicholas Stone were completed. The garden was 'primarily founded for a nursery of simples, and that a Professor of botanicey should read there, and shew the use and virtues of them to his auditors', but a professor or keeper was not appointed until 1642 when Jacob Bobart (1600-1680), a native of Brunswick, was given the post of curator. It was said that Bobart was 'ye man yt first gave life and beauty to this famous place'. He established a network of contacts who sent him seeds and plants so that he was able to 'bedeck the earth with great variety of trees, plants and exotick flowers, dayly augmented by botanists, who bring them hither from ye remote quarters of ye world'.[9]

For half a century there was no other botanical garden in Britain. In 1670 Dr Robert Sibbald (1641-1722) and Andrew Balfour (1630-1694) leased an enclosed piece of land near Holyrood House in Edinburgh and quickly established a collection of between eight and nine hundred plants. The garden soon became too small for the number of plants, and the collection was transferred in 1675 to a new site. A number of other moves were made later until the present site of the Royal Botanic Garden was occupied in 1820.[10] Three years after Robert Sibbald and Andrew Balfour started their garden the Society of Apothecaries in London established a physic garden on land at Chelsea leased from Charles Cheyne (1625-1698), later Viscount Newhaven.[11] Both the University Botanic Garden in Oxford and the Chelsea Physic Garden still occupy their original sites.

By the latter decades of the seventeenth century, a botanical garden was recognised as a useful adjunct to university faculties of medicine for the training of physicians. Exploration of remote regions resulted in an influx of seeds of strange plants which were eagerly cultivated. Horticultural technology had become sufficiently advanced to provide the means of growing plants from warmer countries. The stage was now set for the rapid development of botany in European universities in the eighteenth century.

2

The early history of Irish botany and the first botanical gardens

BOTH the science of botany and the art of gardening developed outside Ireland, and the skills needed for their pursuit were brought to Ireland at different times.

Until the arrival of the Normans in the twelfth century, gardening in Ireland was almost solely concerned with the production of food. The first people to till the land in the Neolithic period grew primitive wheats, barley and oats. Later they also cultivated flax. The monks of the Early Christian period probably introduced to Ireland such vegetables as cabbages, onions and leeks, as these were important items in their strictly regulated diet. Apples were grown by the monks and they may have understood the techniques of grafting as early as the eighth century. It is probable that the monks cultivated some medicinal herbs, but they could have gathered many that grew in the wild. From the wild they collected plants which yielded dyes, but woad, at least, was deliberately grown in gardens[1].

As in most of the rest of Europe during the Dark Ages there was no systematic study of plants in Ireland. Irish scribes working in Continental monasteries were engaged in copying, and in the process altering, the works of the Greek and Latin natural philosophers, but they added no new observations. Yet the writers of Ireland possessed an unrivalled ability to express in words the beauty of nature. They saw all creation as a gift from God, and recorded the plants and animals they knew in poetry and prose. From these delightful works may be gleaned the first records of Ireland's native flora.

The period after the Norman invasion is marked by an increased interest in gardening. It is likely that some medicinal and pot herbs such as parsley, mallow, hemlock, and annual nettle, which are often found growing wild today about Norman castles and abbeys in Ireland, were introduced by the Normans[2]. In the late thirteenth and early fourteenth century ornamental plants were being grown for chaplets of roses and single red rose blossoms were required as rents for land[3]. Towards the end of the fourteenth century there was sufficient interest in both kitchen and flower gardens for the poem of Master Jon Gardener to be adapted for Irish readers. One of the earliest surviving manuscripts of this versified set of gardening instructions is contained in a book of poems copied by a Kildare scribe[4]. By the sixteenth century, the large well-regulated monasteries had substantial, productive gardens and orchards, but many of these declined after the dissolution of the abbeys and monasteries in the middle of the 1500s. However, the skills needed for the cultivation of plants were not lost.

While knowledge of medicinal herbs was not all imported—there was a substantial corpus of indigenous lore about the properties and use of plants—Irish scholars who travelled to Europe came into contact with many plants not found in Irish woods and meadows. Those students who went to universities in Italy and France to study medicine were taught about the herbs of southern Europe and would have read the works of Dioscorides and others. Undoubtedly such new knowledge and indigenous information were welded together eventually, but no original books on medicinal plants were prepared in Ireland before the eighteenth century.

Apart from the poetry of Early Christian Ireland, the earliest known records of native Irish plants are contained in letters written by English administrators in the last decades of the sixteenth century. From these people and occasional travellers, John Gerard and others probably gained their scanty information on Irish botany. Two exiles, Dr Peter Lombard (1554-1625), a brilliant philosopher and theologian who was created Archbishop of Armagh, and Don Philip O'Sullivan Beare (c. 1590-c. 1660), wrote general accounts of the geography of Ireland in the early 1600s. They both mentioned briefly the native flora and the state of gardening and agriculture in Ireland, but their isolated notes do not represent the results of concerted attempts to study the Irish plants or to promote gardening in this island; they are simply personal impressions of a homeland.

The Reverend Richard Heaton (c. 1601-1666) was the first person known to have collected botanical specimens in Ireland. He was born in Yorkshire and graduated from the University of Cambridge. About July 1633 he came to Ireland as a chaplain to Sir Thomas Wentworth's Troop of Horse. During his stay he visited the Burren in County Clare, where he observed juniper, mountain avens and spring gentian. Elsewhere Heaton collected the small insectivorous sundew (*Drosera*) which he gave to a Dublin apothecary, Zanchie Silliard, who promptly despatched the specimens to the English botanist John Parkinson (1567-1650). Parkinson mentioned the sundew in *Theatrum Botanicum* (1640), a rambling and poorly written book which, nevertheless, contains the first published records of native Irish plants. He noted that the strawberry tree (*Arbutus unedo*) grew in the 'west part of Ireland' and that the *ros solis* (sundew) 'was sent me by Mr Zanchie Silliard'. Richard Heaton was annoyed that his discovery had been attributed to someone else, and later took the opportunity to correct Parkinson.

Richard Heaton passed his plant records to another English botanist, Dr William How (1620-1656), a graduate of Oxford, who published them in 1650, along with many others, in an anonymous flora. Eight Irish plants collected by Heaton are recorded in How's *Phytologia Britannica*, including the sundew of which Heaton wrote: 'I gave [plants] to Zanchie Syllliard...which he sent to Mr Parkinson, who in his description mentions the saide Zanchie as if he had found it'.

Heaton left Ireland at the outbreak of the 1641 Rebellion, but as he was rector of Birr in County Offaly, and owner of land at Roscrea, he returned in 1660. He was awarded a doctorate of divinity by the University of Dublin and succeeded to various other appointments in the Church of Ireland but he did not resume his botanical wanderings.[5]

The year after Heaton left Ireland the English Civil War erupted and it ended in the defeat and execution of King Charles I. The Commonwealth followed, with Oliver Cromwell as the head of government. In Ireland the rule of Parliament was not accepted universally, and in the summer of 1649 Cromwell came over from England with troops to assert the authority of the new regime. The severity of his campaign in Ireland is a bitter memory, but the consequences included an enforced peace and social stability. Some of Cromwell's soldiers and supporters who acquired confiscated lands set about improving their estates and embellishing them with gardens, and among those who came to Ireland with the Commonwealth administration were William Petty and Benjamin Worsley.

William Petty (1623-1687), who has been called 'that remarkable genius', was a graduate of the school of medicine in Oxford and had also studied at the universities of Leiden, Utrecht, Amsterdam and Paris.[6] As a physician he must have learnt botany and he would have been familiar with the botanical gardens attached to these great universities. It is probable that this experience led him to plan a physic garden in Dublin in 1653, in association with the Surveyor-General for Ireland, Benjamin Worsley.[7] Worsley was also a physician and had a strong interest in gardening—he imported thousands of roses to Dublin for grafting and reported on the cultivation of madder which had been collected for him in the wild.

Nothing is known about the progress of the plans laid by Petty and Worsley. The only mention of them occurs in a letter written by another person to the horticultural publisher, Samuel Hartlib, in England.[7] It is possible that this garden was associated with the Fraternity of Physicians, formed in Dublin in 1654 by Dr John Stearne (1624-1669). It may have been in the garden of Trinity Hall, a building situated near the University, which was leased to Stearne 'during his natural life'. The lease bound him 'to convert the remainder to what should be unto him allotted for his own accommodation, unto the sole and proper use of Physicians'. A garden was attached to this house, but there is unfortunately no report of its use as a physic garden.[8]

The Fraternity eventually became formalised in 1665 as the College of Physicians in Ireland. Occasionally throughout the first century-and-a-half of its existence, the possibility of establishing a physic garden was discussed by the Fellows. As will be seen later, it was as an indirect result of such arguments in the late eighteenth century that a botanical garden was established outside Dublin city with no direct function in medical teaching.

After the restoration of the Monarchy, in the person of King Charles II, peaceful conditions prevailed in Ireland for nearly thirty years. Dublin prospered and soon outstretched its mediaeval limits. The University, the College of the Holy and Undivided Trinity, founded by Queen Elizabeth in 1592, also flourished and there was a general resurgence of interest in the arts and sciences. In October 1683, a group of Dublin *virtuosi* met privately in a coffee-house 'merely to discourse of philosophy, mathematics and other polite literature'.[9] They soon attracted others and the group expanded so that in January 1684 the Provost of Trinity College, the Reverend Dr Robert Huntingdon, invited the members to meet in his lodgings. At this meeting they formed themselves into a society and 'took on us the name of the Dublin Society', later called the Dublin Philosophical Society (not to be confused with the Dublin Society formed in 1731). The prime mover of the Dublin Philosophical Society was William Molyneux (1656-1698), a Dublin lawyer who was a correspondent of certain Fellows of the Royal Society in London and other leading figures of the awakening Age of Enlightenment, including the philosopher Joseph Locke and the astronomer Edmund Halley. In April 1684 William Molyneux wrote to his brother Thomas (1661-1733) who was at the University of Leiden studying medicine and botany, informing him that the new society had just acquired premises off Dame Street near Trinity College, where it had a well-equipped laboratory and 'a fair garden'.

There was an interest in botany among members of the Dublin Philosophical Society so perhaps the garden flourished for a while. Dr Alan Mullen (c. 1654-1690), a Dublin physician best known for his extraordinary dissection of an elephant burnt to death in a local circus, collected plants. After leaving Dublin because of an 'indelicate love-affair', Mullen went to the West Indies accompanied by James Harlow, who was commissioned by Sir Arthur Rawdon, of Moira in County Down, to collect living plants in Jamaica for his garden.[10] Mullen died in Barbados from 'a surfeit of alcohol'. Another member, Samuel Foley, presented a paper to the Society in 1684 on the anatomy of a 'large garden bean', and in 1686, Edward Smyth spoke fancifully on the spontaneous generation of poppies in a field in County Donegal. Thus the facility of a botanical garden, or even just an experimental vegetable plot, was in accord with the interests of members. Alas, as in the case of the garden of Petty and Worsley, nothing is known of the progress or contents of this plot, but it has to be assumed that it had only a fleeting existence, being abandoned in 1687 when the Society itself went into rapid decline.

The last few years of the 1680s were troubled. King James II and the Earl of Tyrconnel were replacing Protestants with Roman Catholics in the Irish Establishment and Army. Fearing repressive measures against them, many Protestant intellectuals and landowners left for England. The Molyneux brothers went to Chester. The Dublin Philosophical Society ceased to meet after April 1687, and even the University was eventually closed and troops were billeted on the campus.

However, before these happenings, the Provost, Robert Huntingdon (c. 1640-1701), and the Fellows of Trinity College took a decision that was a signal event in the history of Irish botany. On 25 June 1687, they decided to convert the kitchen garden into a 'Physick garden at the charge of the college'.[11] Huntingdon's role in the foundation of the College Physic Garden is not at all clear, but he had an interest in botany. He may have spent some of his leisure hours while a student at Merton College, Oxford, walking in that university's Physic Garden. After graduating, he accepted a chaplaincy to the Levant Company's factory at Aleppo and stayed there for ten years. From Aleppo he sent herbarium specimens to Jacob Bobart, in Oxford. Having been a field botanist, Huntingdon was most likely sympathetic to the idea of having a botanical garden in Dublin. His association with the short-lived Dublin Philosophical Society probably strengthened his interest in botany.[12]

Trinity College had ornamental and kitchen gardens on the campus from its establishment. In October 1594, plots were let to various Dublin citizens on condition that they created 'fair gardens planted with good and profitable herbs and fruit trees', and that the Fellows were allowed 'to walk therein for their recreation'.[13] In 1605, Harry Holland leased 'five gardens and the great orchard' from the University. He was permitted to harvest 'half of all the herbs that grow, lavender, roses, fruit of the trees', but the other half of the fruit and flowers was to be delivered to the Provost and

Fellows.[14] The College accounts show that the gardens were actively maintained and that, in 1683, the kitchen garden was planted with '5 hundred of Cabbidge plants' as well as young plants raised from seeds supplied by John Cole for a total cost of £1 16s.2d—'parsly, lettos, tungras [town cress], coucumber, corn sallat, nursturssion, 6 quarts of beans, 3 quarts of pease, survigrass [?scurvy grass], nettels, winter cole, purslon, tyme, winter savorey, sweet marjoram [and] half a hundred of hortichots plants'.[15] This was the garden that was taken over for the Physic Garden.

Work on the new Physic Garden began at once. In February 1688, Margaret Armstrong was paid ten pence for one day's work 'mending the ditch of the fisick garding'.[16] In the November following, the Provost and Fellows agreed that the gardener should be granted an extra twenty shillings per quarter 'during pleasure, upon account of the new walk and Physic garden'.[17] As far as is known no botanist was appointed to look after the garden and there is no record of plant collections at the time.

It is uncertain why the Provost and Fellows decided to create a teaching garden in the University in 1687. Although there had been a medical fellowship (*Medicus*) in Trinity College as early as 1618 it was not necessarily occupied by a physician and the first two holders had no medical training. From the University's foundation to the time of the Commonwealth only one medical degree had been awarded and even that is surprising since there was no medical faculty. In 1658, John Stearne, founder of the Fraternity of Trinity Hall, received a doctorate in medicine and in 1662 he became Professor of Medicine. He was appointed for life.[18] In 1687, however, there was no professor of medicine and no active medical faculty, nor was there any botanist teaching or residing in the College.

In September 1689 the campus was seized by Jacobite troops, and the staff and undergraduates were expelled. There was no teaching in the University during the initial stages of the war between the forces of James II and William III, but shortly after the Battle of the Boyne in July 1690, work resumed in the College and most of the exiled intellectuals who had fled to England returned to Dublin.

In 1692, the College of Physicians petitioned for a new charter and this was granted in December. At the same time it was recommended that the College of Physicians should receive one of the houses confiscated from supporters of the vanquished King James, and five or six acres of ground near to the city for a physic garden. On 6 February 1693, the President of the College, Dr Patrick Dun, and Dr Thomas Molyneux, now established in Dublin as a medical practitioner, were instructed to find out 'what to doe for obtaining a plot of ground for a physick garden'.[19] They discovered that the piece of land suggested was not forfeited and the College of Physicians took no further action towards establishing their own physic garden for the time being.

The Physic Garden, started before the Williamite wars, at Trinity College was still intact. It may not have been properly maintained, but at least it was not given over to other uses. By November 1701 work was again being carried out in the garden, as in that month payments were made for fencing. In 1705, the wall was repaired and 'a little canall in ye physick garden' was walled. In the winter of 1705-06 trees were removed from the Physic Garden and ditches levelled. Holly shrubs were planted in April 1707. According to receipts among the Bursar's documents, the Physic Garden was weeded, dug and supplied with dung on a regular basis, from this time onwards.[20]

The University expanded rapidly in the early eighteenth century. On 14 June 1710, the Provost and Fellows ordered that 'ground be laid out at ye south-east corner of the physick garden sufficient for erecting a laboratory and an Anatomie Theatre thereupon'.[21] The Theatre and Laboratory, in a single undistinguished building, were opened on 16 August 1711 with due ceremony. The Provost and Fellows attended to hear several professors give lectures, and, in one case, recite a poem. Two of the lecturers are of special interest—Dr Thomas Molyneux who lectured on physic, and Dr Henry Nicholson who spoke on botany. Molyneux had just been appointed Professor of Physic in the University, an act by which the Faculty of Medicine was effectively established. It seems likely that it was Molyneux, as Professor of Physic, who employed a lecturer to teach botany, and he was fortunate to find a young graduate of the University of Leiden, Henry Nicholson, to fill the post.[22]

Henry Nicholson (c. 1681-c. 1720), a native of Castlerea, County Roscommon, is an interesting character. He was the grandson of another Henry Nicholson who had obtained his bachelor of medicine degree from the University of Dublin in 1674 and so was one of its earliest medical graduates. His

METHODUS
PLANTARUM,
IN
Horto Medico,
COLLEGII DUBLINENSIS,
JAM JAM
Disponendarum;

In duas partes divisa; quarum prima
de Plantis, altera de Fruticibus &
Arboribus agit.

In Usum Studiosorum Academicorum.

Autore *Henrico Nicholson*, M. D.
Botanices Professore.

DVBLINI:
Typis *A. Rhames*, MDCCXII.

3. Title-page of Henry Nicholson's textbook, which was the first original botanical work published in Ireland, 1712

father, Edward, was a clergyman in the Church of Ireland and an acquaintance, if not a friend, of William King, Archbishop of Dublin. Henry was a frustrated law student. He was refused a degree in law no less than three times at Oxford University. After these rebuffs Nicholson went to Holland and enrolled at Leiden, where, a few months later, he obtained a degree in medicine. He seems to have taken an interest in botany at Oxford for he was friendly with Jacob Bobart, junior, the curator of the University Physic Garden, and he used this friendship to good advantage later. At Leiden, he attended the botany classes of Herman Boerhaave, the foremost botanical and medical teacher in Europe at the time who had built up the reputation of Leiden as a botanical institute. Henry Nicholson dedicated his thesis, *Dissertatio physico-chymica de corpore*, to Boerhaave and Archbishop King among others. By 1710 he was back in Ireland and his father wrote to Archbishop King saying 'as to my son's engagement in matrimony, whatever yr Grace shall be pleased to fix him, I shall be most heartily satisfyd with it...I do think it full time he were settled'.[23] On 3 August 1711, Henry Nicholson, 'doctor of physic', married Sarah Baldwin, whose father paid a dowry of two thousand pounds.[24] A fortnight later, Henry delivered his lecture at the inauguration of the Anatomy Theatre.

In October 1711, Nicholson wrote to the English botanist James Petiver saying that he had 'undertaken to furnish the Physick garden here which is about setting up by the College of Dublin with plants and seed of all sorts...and they have obliged me to this task by nominating me their Professor in Botanie'. He also engaged various people to collect plants for the garden in the wild in Ireland. He wrote to Bobart at Oxford and obtained 'a curious supply of seeds' from him. This willingness to give seeds as a gift, or in exchange, has always been a mark of botanical gardens and modern gardens continue this free interchange of plants irrespective of political divisions or national animosities. Nicholson also received a promise of seeds from his 'friend Mr Boerhaave'. It is probable that several *Aloe* and *Pelargonium* species from southern Africa, which are known to have been growing in Trinity College Physic Garden about 1724, were raised from seeds sent by Boerhaave. As well as expanding the plant collections, Nicholson wrote the first botanical text-book published in Ireland. *Methodus plantarum...jam jam disponendarum* (1712) (Fig. 3) is essentially a catalogue of the plants then growing in the Physic Garden, arranged according to the contemporary system of classification. But Nicholson was still a frustrated lawyer. He left Dublin in 1715 with his wife and young son, Edward, and enrolled at the Middle Temple, one of London's famous law schools. In the following year he canvassed Petiver for election to the Royal Society of London—'I should be extremely proud to be honour'd so far as to be admitted a member of the Royal Society'. He was elected on 5 April 1716, but did not live long to enjoy this new status, for he died sometime before 1721.

After Nicholson's departure from Trinity College, the Physic Garden was maintained, apparently under the charge of William Maple (d. 1762). Nicholson described Maple as 'a very Ingenious Gentleman and usfull to us and well received' in a letter written to Petiver in 1711.[25] William Maple was employed in Trinity College that year and, in 1722, was in charge of removing the plants from the original garden site to a new one situated between the Anatomy Theatre and Nassau Street (Fig. 4). He must have acted as curator in the absence of a botany lecturer, for none is known to have been appointed in Nicholson's stead until 1724. Maple had a deep interest in applied botany. In 1727 he was awarded two hundred pounds by the Irish Parliament for discovering that leather could be tanned using the roots of tormentil (*Potentilla erecta*).[26] He published a pamphlet on *A method of tanning without bark* in 1729 which is distinguished by an accurate drawing of tormentil and was the first illustrated botanical book published in Ireland (Fig. 5). It is possible that Maple carried out his work using tormentil grown under his care in the College Physic Garden.

In 1722 and 1723, the plants were carefully removed to the new site. In the following year a new professor took charge of the Physic Garden, and, like Henry Nicholson, he was a graduate of the famous Leiden school. William Stephens (1696-1780) had studied under Herman Boerhaave and, indeed, was promoted for his doctorate in medicine by this venerable teacher. Stephens set about adding to the collections of plants, and began corresponding with overseas botanists, particularly with those who had just founded a botanical society in London. He sent the society a catalogue of the plants growing in the Physic Garden at Trinity College. The manuscript survives[27] and provides a unique record of a remarkable little garden (Fig. 6). Stephens' list shows that the Trinity College Physic Garden contained many commonly cultivated plants and species native to Ireland, but there were also plants from North America and southern Africa, many of them very rare and unlikely to survive the cold winters of Dublin without great care and attention. About five hundred different species were in cultivation as well as countless varieties of garden flowers, such as tulips, daffodils and auriculas.

The University was proud of its botanical garden, which was full of exotic plants that had only recently been discovered and imported into Europe. Public lectures in botany at the garden were advertised in Dublin newspapers in June 1725. It is not known how many people attended, but as there was a keen interest among some Dubliners in botany and gardening at this time, the lectures may have been popular.

The following year William Stephens (Fig. 7) published his lecture notes as a pamphlet to 'avoid the trouble of dictating yearly so many pages to the students of Botany'. *Botanical elements* was more remarkable than Nicholson's book, for Stephens showed in it that he was prepared to accept ideas about plants which were highly controversial, even shocking, to botanists including men like Herman Boerhaave. Stephens noted that the flower contained sexual organs, and that pollen from the stamens represented the male reproductive cells, ideas which were widely published for the first time when Stephens was studying in Leiden.

On 26 October 1726, the first Irish flora was published by Dr Caleb Threlkeld (1676-1728), a dissenting minister and physician. He was assisted in his work by Thomas Molyneux, but there is no indication that Threlkeld had any association, formal or informal, with Trinity College and its Physic Garden. *Synopsis Stirpium Hibernicarum* listed over five hundred native species and included information on the medicinal uses of many.[28]

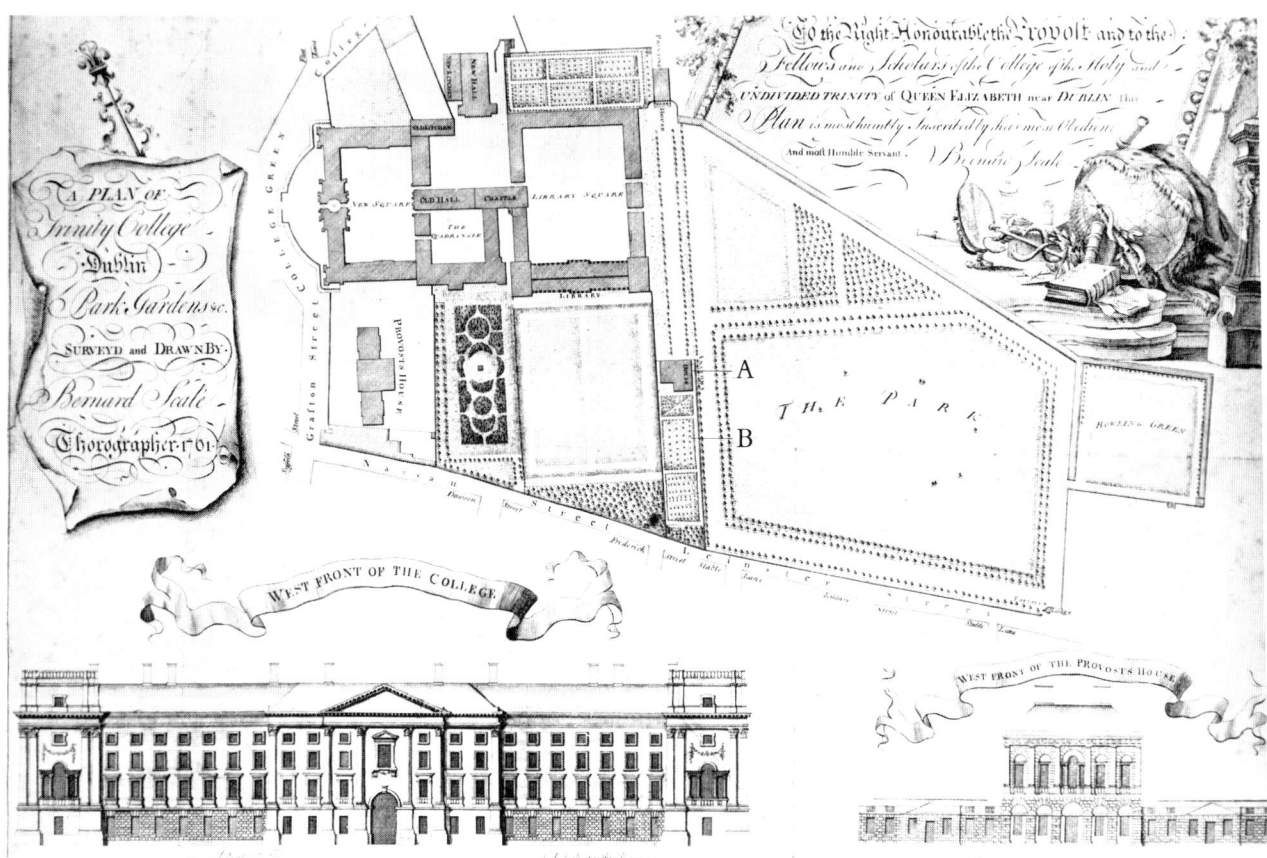

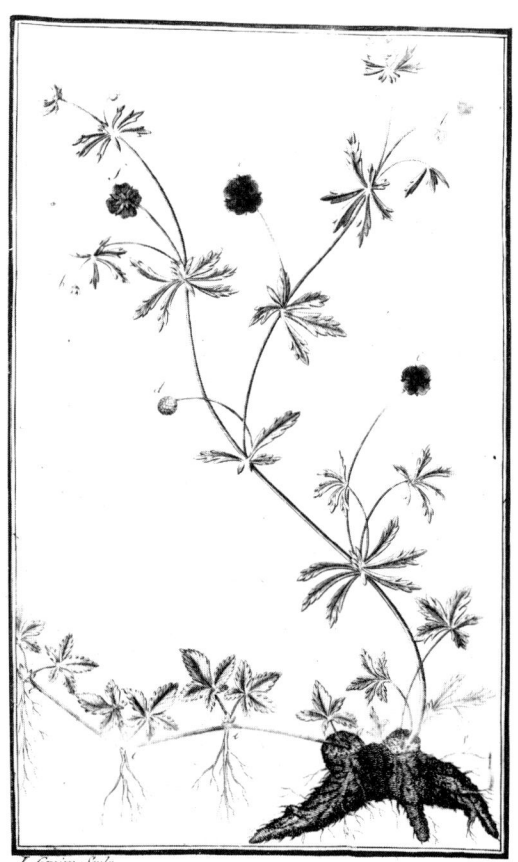

4. Trinity College, Dublin, in 1761, showing the Anatomy Theatre (A) and the Physic Garden (B) that was formed after 1722 (from E. C. Nelson, 1982[7]). (Reproduced by courtesy of Trinity College, Dublin)

5. Hand-coloured engraving of tormentil (*Potentilla erecta*) from William Maple's pamphlet on tanning.[26] Some copies of this plate are not signed by John Gwin. (Reproduced by courtesy of the National Library of Ireland, Dublin)

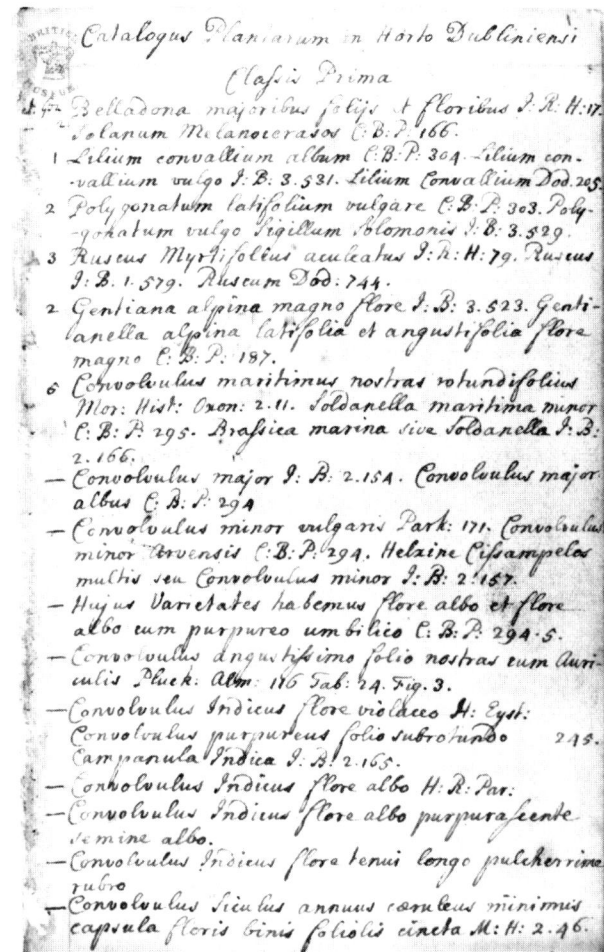

6. First page of William Stephens' manuscript catalogue of Trinity College Physic Garden; this page is not in Stephens' handwriting. (Reproduced by courtesy of the Trustees of the British Museum (Natural History), London, from the original manuscript in the Department of Botany)

William Stephens, Thomas Molyneux and William Maple knew each other intimately, and they probably co-operated in managing the Physic Garden. On 25 June 1731, these three men met with eleven others in the Philosophical Rooms of Trinity College 'in order to promote Improvements of all kinds'. William Stephens took the chair. It was proposed and unanimously agreed to form a society 'by the Name of the Dublin Society, for improving Husbandry, Manufactures and other usefull arts'. Thus, under Stephens' chairmanship the Dublin Society was established and on 8 July 1731, the members agreed to include the improvement of science among its aims. When the first elections for officers took place in December 1731, William Stephens was elected honorary secretary for home affairs and Thomas Prior honorary secretary for foreign affairs. William Maple was elected curator and registrar of the Society.[29]

The Dublin Society was not intended essentially as an intellectual or elitist institution. The founders wanted to promote skills and technologies which would improve the economy of Ireland by assisting home industries, reducing imports, and encouraging modern methods of agriculture. They were concerned to see that new knowledge in the arts and sciences was applied to the practicalities of everyday life and work. Among the initial priorities, the Society sought to increase the acreage under cultivation by reclaiming boggy and marshy land. It also advocated extensive planting of fruit and forest trees. Members were encouraged to undertake experiments and to read widely in a chosen subject in 'Natural History or in Husbandry, Agriculture or Gardening, or some species of Manufacture or other branch of Improvement'.

The reduction of imports was of considerable concern to the Society. William Stephens read a paper to an early meeting in which he outlined some of the 'vegetable substances as are imported here, whose culture or Improvement were either not known or neglected among us, and would very well agree with our Soil and Climate'. He noted that among the dye-yielding plants, woad, woad-wax,

7. William Stephens, a portrait in oils; in the Board Room, Dr Steevens' Hospital, Dublin. (Reproduced by courtesy of Dr Eoin O'Brien; photograph by David H. Davison)

safflower (*Carthamnus tinctoria*), weld and madder were already grown or could be cultivated in Ireland. He spoke at length on the cultivation of woad and weld (*Reseda luteola*).[30]

In 1732, a committee was set up by the Dublin Society 'to look out a piece of ground, about an acre in extent, suitable for a nursery'. In October 1733 two plots were examined and one, belonging to the Reverend Mr Hopkins, at Ballybough Bridge, in north Dublin, was selected. This became known as the Society's garden at Summer Hill, and it served as an experimental plot in which trials were conducted on hops, rye-grass, lucerne, clover and saffron. It was not a botanical garden as it had no teaching role; it was more akin to the experimental grounds of a modern agricultural or horticultural research station. After three years this site was vacated, and a new four-acre garden was established at Martin's Lane in the area east of O'Connell Street, Dublin, now occupied by Tyrone House. Cider apple trees were planted, but the soil was unsuitable, and the garden was abandoned by August 1740.

William Stephens resigned as Professor of Botany at the University in 1733, and was appointed lecturer in chemistry. He was elected President of the College of Physicians for the first time the same year—he eventually served three terms in this distinguished office. At no time in his subsequent career is Stephens known to have returned to lecturing on botany. He became wholly devoted to medicine and chemistry, and was appointed physician to the Royal Hospital, Kilmainham, and to Mercer's and Dr Steevens' Hospitals.

Stephens' successor as Professor of Botany in Trinity College was Charles Chemys (1700-1733); his is the first appointment recorded in the register of the University.[31] He was a graduate of Trinity College and a fellow of the Royal College of Physicians, thus continuing the traditional link between medicine and botany. He held this post for a few months only, and in September 1733, William Clement was chosen to replace Dr Chemys.

While Charles Chemys is an obscure character about whom little is known, William Clement (1702-1782) had a long career within the University. He became Professor of Physic in 1761 and also occupied the posts of Auditor, Librarian and Vice-Provost. He lectured in natural and experimental philosophy and mathematics and even represented the University in the Irish Parliament. With these responsibilities, it is not surprising that Clement left no evidence of active work in botany, no books or botanical manuscripts, and that little is known of his time as Professor of Botany. During his tenure of the position the Physic Garden was at least weeded and manured, but there is no catalogue of the collections.

Clement resigned from the chair of botany in 1763 but remained as Vice-Provost until his death. He was followed in the botany professorship by James Span (c. 1735-1773), who was granted his degrees of bachelor and doctor of physic on the day of his appointment as professor. Span was a popular figure in the University and held two of the teaching positions in the School of Medicine; he was made lecturer in chemistry in 1766. Shortly after his death he was eulogised in a long poem by John Gilbourne, which also makes reference to sixty-two other physicians.

> James Span shakes off the mortuary Gloom
> His bright endowments still retain their Bloom;
> On Earth lamented, and admir'd above,
> His lovely Virtues made him dear to Jove:
> Daisies and Roses spring where'er He treads,
> Tulips and Lillies rear their drooping heads;
> Nor do Plants sensitive his Touch avoid,
> Who for Man's good had all his Thoughts employ'd.[32]

The 'mortuary gloom' which Span shook off, haunted the Physic Garden!

When Edward Hill (1741-1830) (Fig. 8) was appointed to the vacant Chair of Botany in 1773, he took over a dilapidated Physic Garden. Years later, following a bitter row and a law suit against the University, Hill wrote a vitriolic document in which he said that on assuming the professorship he found there were no proper facilities for teaching botany at Trinity College.[33] Although he performed his duties, he was 'destitute of the indispensibly necessary aid of a botanical garden'.

8. Edward Hill, a portrait in oils; in the Royal College of Physicians of Ireland, Dublin. (Reproduced by courtesy of the Royal College of Physicians of Ireland)

What had happened to the Physic Garden behind the Anatomy Theatre? Some writers, unaware of the existence of this garden, have stated that William Clement, due to his position as Vice-Provost, had used as a botanical garden the Vice-Provost's garden situated in the north-eastern corner of the College Park. It is also said that James Span made a botanical garden in the quadrangle known even today as 'Botany Bay'. However, neither Clement nor Span abandoned the Physic Garden behind the Anatomy Theatre. They maintained it, as indicated by accounts paid to gardeners, although the collections seem to have declined, and by 1773 it was in a decrepit state. Hill described luridly what he inherited:

> The cultivated spot, where I am taught to practise a mode of horticulture that would have far outdone the Garden of Alcinous fam'd of old, shall here be described to you. In extent, as I have ascertained it by correct mensuration, it is 250 feet long, and 50 feet broad. It is mostly surrounded by lofty elms which overhang its high walls, and it is continually befogged by the fuliginous vapors of the City. A common path traverses it, towards the door that opens in Nassau-Street, to which there are 500 keys. A Bath, for the use of the Gentlemen of the University, and for every one beside who is the possessor of a key to the aforesaid door, occupies one end of this cultivated spot. It is the only cemetery for the draff and offals of the dissecting room of the Anatomical Theatre, immediately behind which it lies; and is, therefore, burrowed by 10,000 rats, that have mined and countermined the intire soil in every direction, to the absolute prevention of all vegetation.[33]

As well as the rats, Hill was left 'an aged superannuated gardener...whose occupation was reduced merely to the care of a barren Fig-tree'. 'Humanity forbade me', wrote Hill, 'to displace him, but, on his death, I refused to attend to the applications of several who solicited to be appointed in his place'. Hill therefore abandoned 'this rude, unfriendly spot', and suggested that the University should cease to spend money on its upkeep. He sought permission to use a vacant plot of land bordering Townsend Street, but his request was ignored.

Edward Hill was a forceful character who had a singular ability to make enemies. He was a brilliant student at Trinity College, graduating first in arts in 1765 and then in medicine in 1771. In 1781, he became Regius Professor of Physic in the University and, in 1785, his lectureship in botany was formally made a professorship under the School of Physic Act passed by the Irish Parliament. Hill was elected President of the College of Physicians in 1782 and occupied this position four more times, the last occasion being 1813. He was conservative in his attitude to botany, maintaining that botany was an integral part of medicine and that a physic garden (or a botanical garden) was a necessary adjunct to a university or college with a faculty of medicine. In this attitude he was ignoring the trend elsewhere in Europe towards the establishment of botany as a subject worthy of study in its own right, and the foundation of botanical gardens as 'living museums' filled with exotic plants which were of general scientific interest without necessarily having any medicinal value. He may have been unresponsive to new ideas in botanical classification for his library, sold in July 1816, contained none of the many books of the great Swedish botanist Carl Linnaeus whose simplification of biological nomenclature revolutionised botany and zoology in the last half of the eighteenth century.[34]

According to Hill's rather florid history of the events between 1773 and the early years of the next century, the University Physic Garden established in 1722 on the Anatomy Theatre site was abandoned after 1773, and from then on there was no botanical garden on the campus. In January 1775 the Provost was requested, presumably by Hill, to look out for a piece of ground 'proper and convenient' for a botanical garden,[35] but this did not come to fruition due to lack of finance.

Hill however was persistent. In 1783 under his presidency, the College of Physicians decided to enquire into the efficacy of the King's professorships, which had been established under the will of Sir Patrick Dun. The appointments, made by royal charter, were to chairs in physics, chirurgery and midwifery, and materia medica and pharmacy. The professors lectured in Trinity College in Latin. A row between the University and the College of Physicians resulted in placing the responsibility for examinations and exercises for medical degrees on the University, but the 1783 committee decided that it was desirable for the College of Physicians to approach the Provost and Fellows of Trinity College for discussions about examinations. Dr Hill and Dr Hutcheson met with the Provost, John Hely-Hutchinson, who received them enthusiastically. Eventually the committee prepared a report, suggesting six chairs including chemistry, botany and natural history as well as the ones that already existed. Three of the professors were to be elected by the Provost and the Professor of Medicine in the University. It was also suggested that some of the income from Patrick Dun's estate should be devoted to the establishment and maintenance of a botanical garden.

In November 1783, a petition was drawn up asking for alterations to the existing parliamentary Acts. It was presented to the Irish Parliament by the Provost, who was a member of parliament for Cork City and also Secretary of State for Ireland. As a parliamentarian, John Hely-Hutchinson was considered 'profligate [and] very difficult to manage being jealous and tricking and never to be trusted'; he was said to be prepared to 'fight through anything if well paid'.[36] His appointment as Provost created bitter hostility and College members often resented his tyrannical methods. Leave was granted to bring in a new bill and the House of Commons appointed a committee of inquiry which held its first meeting under the Provost's chairmanship. The University opposed points in the College of Physicians' plan. Hely-Hutchinson drafted the bill and this was passed into law as the School of Physic Act of 1785. It conceded all the University's points, while the proposals made by the College of Physicians were extensively modified. However, one part of the bill was not passed; that was the proposal, accepted by all parties including the House of Commons Committee, that a botanical garden 'ought to be supported at the public expense, and that such a garden, for useful medical purposes, may be supported at a moderate expense'.

Hely-Hutchinson's account of the events recorded that the Lord Lieutenant, the Duke of Rutland, supported the scheme, but it was opposed by the Chancellor of the Irish Exchequer in the chamber of the House. The Chancellor, according to the Provost, alleged that the 'expense would be enormous', and thus that part of the bill allowing for a botanical garden lost the support of the Lord Lieutenant's Secretary and was then defeated.

The Chancellor of the Irish Exchequer in the summer of 1785 when the School of Physic Act was passed was the Right Honourable John Foster, who was shortly to be elected Speaker of the House of Commons. He was implacably opposed to John Hely-Hutchinson and this may explain his opposition to the plan for a botanical garden. John Foster later did a remarkable somersault.

Hill's persistence was unquenchable. In 1789, he was elected President of the College of Physicians for a second time and on 25 March that year the College had before it a proposal of the Provost of Trinity College seeking the aid and co-operation of the College of Physicians in forming a botanical garden. The coincidence of discussion and Hill's presidency is hardly accidental. The Provost suggested the appropriation annually by both bodies of money towards the support of a garden. The College of Physicians responded by empowering the President to inform the Provost that the College deemed 'the offer...liberal and that the College of Physicians will be ready to co-operate with [the University] in establishing a botanical garden whenever the necessary estimate can be laid before' the University Board. The annual sum of seventy pounds for the perpetual support of a botanical garden was proposed by the University, but nothing seems to have come of the suggestion. The University did not submit an estimate of costs and the College of Physicians did not discuss the matter further in 1789.

The next development was that Dr Walter Wade, a Dublin man-midwife and surgeon, presented a petition to the Irish Parliament asking for the establishment of a 'Publick Botanical Garden in the city or its environs'. Wade, a Licentiate of the College of Physicians, was a lecturer in botany and the author of *Specimen Florae Dubliniensis*. The petition was tabled on 9 February 1790, and was ably supported in the House especially by John Foster. Such was the general support for Wade's petition that a clause was inserted into the Dublin Society Act of 1790 granting three hundred pounds 'towards maintaining and providing a botanical garden'.

Thus out of the series of events involving the University of Dublin, the College of Physicians and the political leaders of Ireland, the first step was taken towards the formation of a botanical garden under the care of the Dublin Society. That garden is now the National Botanic Gardens in Glasnevin.

Before proceeding to discuss the founding of the garden at Glasnevin, it would be appropriate to conclude the long saga of Edward Hill's attempt to establish a botanical garden for the use of medical students of the University and the College of Physicians.

Shortly after the passing of the Dublin Society Act in 1790 granting that Society monies for a botanical garden, Dr Hill was approached to take over control of the planned 'publick botanical garden', but he declined to resign from the University where he was still Professor of Physic and Professor of Botany. Instead, he proposed that Walter Wade should take charge.[37]

In 1792, Edward Hill suggested that the College of Physicians should appropriate one hundred pounds annually from the legacy of Sir Patrick Dun to support a botanical garden if the University would match that sum. However, a faction in the College resisted this. According to Hill the opposing group was led by 'one of the University Medical Professors, a self-sufficient, vindictive gentleman, singularly obstinate in his own opinions'. This was Dr Robert Perceval, a bitter personal foe of Edward Hill—the two were equally entrenched in their opinions, but Perceval favoured the building of a clinical theatre in which respect he was more abreast of trends in medicine than Hill was of trends in botany.

In 1793, John Hely-Hutchinson again introduced a bill into the House of Commons to provide funds for a botanical garden. The bill was read for the first time on 8 March. In April, Edward Hill proposed for a second time that the College of Physicians should grant one hundred pounds annually from Dun's estate for a garden. This time his proposal was passed, but it had to be passed three times before it became the policy of the College of Physicians. On June 11, the Provost and Fellows of the University petitioned the House of Commons that 'a botany garden is indispensably necessary for the success of a school of physic', but stated that the 'funds of [Trinity] College are totally inadequate'.

Noting that the Dublin Society had been granted money for the same purpose, that the University had agreed to allocate funds, and that the College of Physicians had agreed to do likewise, the University Board suggested that the botanical garden would be most effectively promoted by co-operation between the three institutions. They expressed the hope that whatever regulations were made, the medical and other students would be able to benefit fully from the garden. John Hely-Hutchinson also presented a petition from the President and Fellows of the College of Physicians supporting the University 'that by applications of the...unified fund and co-operation of a certain number of persons, a botanical garden may be established on the most useful principles; that such an establishment would be beneficial to the said university, the School of Physic and the public...'

The two petitions were referred to a committee of the House of Commons. A bill was drafted and the House went into committee on 21 June 1793. The bill was supported by the Secretary of State (the Provost of Trinity College), but was strenuously opposed by the Speaker (John Foster), Samuel Hayes, Thomas Burgh and John Toler. The *Parliamentary Register* records that the 'ground of the opposition was that the Dublin Society should not be compelled thereby to give up 300£, to 112£ of the University and 100£ of the College of Physicians, for a botanical garden for the more particular purpose of only medical and scientific knowledge, when agriculture and arts were the peculiar purposes of the grant of Parliament to the Dublin Society'.[38] Edward Hill and Robert Perceval were both summoned before the House of Commons Committee and John Toler 'gave the amplest testimony of the great abilities and professional science in universal botany of Dr Walter Wade'. The entry in the *Register* concludes: 'Great diversity of opinion also prevailed respecting the construction of the Acts that granted money to the Dublin Society. At length the bill was lost by a majority of six, this being on the motion of the chairman's leaving the chair. Ayes 24. Noes 18'.

John Foster and his cohorts had won. The colleges were defeated and left to sort out their own ideas.

John Hely-Hutchinson died in 1794 and he was succeeded as Provost by Dr Richard Murray. Edward Hill approached the new Provost about the propriety of leasing land for a botanical garden, and even laid a map of the site before him. The Provost approved the plan, and Hill recorded that he sought a lease with the Provost's consent and approbation. The University, for legal reasons, could not engage in a perpetual lease, so Hill undertook it personally in trust for the University. Six acres were obtained, but the Provost and Fellows objected that funds were not sufficient to maintain such an area. Edward Hill suggested that he would personally pay half of the rent and surrender to the University portions of the land as these were developed and demands required. The Provost gave Dr Hill to understand that this was acceptable. Hill signed the lease and took possession of the land from Samuel Stephens. The Board of the University had to approve the scheme, so Hill met the Board and subsequently reported that 'not one dissentient voice' was raised nor 'was the least objection made'. Funds were allocated from 2 November 1795. Continuing to have confidence that the plan would also have the support of the College of Physicians, Edward Hill had the lease legally ratified.

This new University Botanic Garden was at Harold's Cross, then a small village lying south-west of Dublin city. According to Hill the site, the location of which is not known, 'possessed natural advantages, infinitely above all that any Botanic Garden in Europe can have to boast of'.[39] He set about developing this enviable place, and at the same time attempted to overcome the final hurdle, the College of Physicians. The College refused to consider Hill's motion which had been passed twice already but needed assent for a third time. To force matters, in June 1797, it was proposed that a committee of Edward Hill, Robert Perceval and William Harvey should meet the University. They reported back, having agreed to take into consideration a proper arrangement for a jointly funded garden, providing legal counsel confirmed that the College of Physicians could use funds from the Dun legacy. This was the stumbling block. On 15 February 1798, Hill's motion was debated for a third time. Votes were cast equally, but the President of the College cast a deciding vote against the botanical garden. Hill was furious. He wrote to the Provost the following day, telling him of the College of Physicians' decision and hoping, indeed expecting, the University to act unilaterally. 'But my hopes were in vain', he wrote, 'for no manner of notice was taken of me by the University, who looked on...with heedless apathy'. Stubbornly, Hill continued to expend his entire salary on developing the garden at Harold's Cross.

In 1799, yet another committee was set up by the University to investigate matters in the medical faculty. The result of these deliberations was that the University pressed for a new act of parliament to 'regulate the composition of the College of Physicians'. A clause was inserted providing for the two institutions to support in equal measure a garden, under the direction of the University's Professor of Botany. Meanwhile, Robert Perceval approached the Lord Chancellor and as the result of the 'unnecessary private conversation of this ruthless busybody', the Lord Chancellor formed a committee in the House of Lords to inquire into the use of the Dun legacy. Their Lordships formed an unfavourable opinion of the College of Physicians and the use of the bequest, so the School of Physic Act 1800 was passed without any provision for a botanical garden. The Act also made it unlawful for any professor to hold two chairs within the School of Physic. On the day the bill received the Royal Assent, Edward Hill resigned one of the two chairs he held, that in botany which was one of the medical professorships. He felt, perhaps justifiably, that he had been treated unfairly.

Having relinquished the professorship, Hill was left with the personal burden of the botanical garden at Harold's Cross. He tried to settle with the University, and so obtain reimbursement for the money he had spent between November 1795 and October 1800. The University appointed assessors, as did Hill, but they disagreed and Hill took the University to court. In March 1803, the King's Bench awarded Hill £618 19s. 6d and 'the lease of the said garden and other lands therein contained, unto his own possession...and shall keep them to his own use for ever'. Hill was instructed to release the University from the original lease.

The University lost its botanical garden and a sorry tale of argument, personal vindictiveness and petty squabbling ended. The Royal College of Physicians had no botanical garden either, but then botany was no longer an integral part of medicine, it had become an academic discipline in its own right.

Edward Hill was somewhat old-fashioned in his view of botany and medicine. He had had a vision that had faded. In a strange pamphlet addressed to the students of physic, he wrote about indulging

> ...that idle propensity of fond imaginings, called building castles in the air...I gave to airy nothing a local habitation and a name. A Paradise rose like an exhaltation to my view, and charm'd my senses with its varied beauties: I stood upon the border of the Lake, that to the fring'd bank with Myrtles crown'd, its crystal mirror held, and viewed its tranquil surface spangled with the snow-white blossoms of all the aquatic Vegetable tribes. I ascended the moss-clad, rocky mound, and from its summit contemplated the subjacent plain, on which were spread in wide encampment ten thousand trees and shrubs and flowers, by care and art made, denizens of a climate not their own. But, whilst I inhal'd their grateful odors, and exulted in admiration of their bloom, and varied tints of glowing colours, the baseless fabrick, like the aereal pictures of the Fata Morgana, dissolved in air, and was succeeded by the noisome streams and sick'ning vapours of a Lazar-house.[40]

Meanwhile, the Irish Parliament had granted the Dublin Society funds and charged it with forming a separate botanical garden.

Botany and politics:
Walter Wade and John Foster, 1790-1795

NONE of the botanists who worked in Trinity College during the eighteenth century actively studied the native flora. Work on indigenous plants was almost the sole preserve of those amateur botanists whose backgrounds were strongly connected with medicine. Following the publication of Dr Caleb Threlkeld's *Synopsis Stirpium Hibernicarum* in October 1726, only two other works on the plants of Ireland were produced before 1790. The Earl of Kingston's chaplain, John Keogh (c. 1681-1754) wrote *Botanalogia universalis Hibernica* which was published in Cork in 1735. It was a herbal with comments on the supposed medicinal properties of native species and plants which Keogh saw growing in Lord Kingston's garden at Mitchelstown.

In the middle of the 1700s, a Quaker physician, John Rutty (1697-1775) did promote study of native plants through the short-lived Physico-Historical Society of Dublin. He engaged a naturalist, Isaac Butler, to collect plants, but the results of Butler's researches were never published. John Rutty tried to interest other physicians and apothecaries in botany, but he was not successful. Perhaps his excessive religious zeal made him too distant from his fellows. In his diary he sometimes recorded botanical excursions: 'Yesterday's botanic walk gave peace and delight upon reflection, even as a testimony against the idle sons of Aesculapius in this city, immersed in sloth and sensuality'.[1] Dr Rutty gathered some of his observations into his *Natural History of Dublin*, but these mainly concerned cultivated plants. He mentioned that another physician, Dr Abraham Lionel Jenkins, was preparing a flora of Ireland arranged according to Carl Linnaeus' new binomial system of nomenclature, but it was never completed.

Another of the small company of botanists at work in Ireland before 1790 was Dr Patrick Browne (c. 1720-1790) a native of County Mayo and a graduate of the University of Leiden. He spent many years working in the West Indies as a doctor. On his return to Ireland, Browne settled on family estates near Ballinrobe in Mayo and before his death in 1790 compiled a manuscript flora of counties Mayo and Galway. On the advice of the English botanical patron, Sir Joseph Banks, Patrick Browne sent a copy of the flora to the Provost of Trinity College.[2] The flora was never published. However, a number of copies of the manuscript were circulating in Dublin about 1790; one was in the possession of Walter Wade and another, which survives today, was given to Aylmer Lambert and by him to the Linnean Society in London.

Patrick Browne was friendly with Dr Edward Hill and he gave Hill his herbarium of West Indian and Irish plants for Trinity College. He may also have known Walter Wade. Indeed, Browne could have served as a catalyst to Wade and Hill in their separate attempts to form botanical gardens in Dublin. Significantly, Dr Browne wrote the following in the preface to his manuscript flora (Fig. 9):

> Hitherto natural history seems to have been greatly neglected among us. No repository but ye incipient one at ye College of Dublin, no publick gardens of any use, and but one only Professor of natural History in ye Kingdom. Kilkenny may probably raise to be a considerable seminary of learning, and ye open air, ye central situation and ye great convenience of pure water and fuel must render it an eligible situation for a botanical garden and greenhouse, and without such how can the apothecary know ye plants he employs in his shop and compositions, how can ye physician know ye plants he prescribes or examine ye nature or virtues of them, or ye cook or confectioner avoid those of a deliterious nature which have been too often mistaken for wholesome food and ye use of them often attended with death. The greenhouse should have brought us acquaintance with medicines of very extraordinary vertues hitherto but little or not at all known to us and time may bring to our knowledge many more of these if there was but proper places to raise them...Is not ye Beauty of ye flower garden of some account? What source of curiosity have we not in ye greenhouse? I could ad many other reasons for encoraging good publick gardens, and finding a proper intendant for them, such would be (in my opinion) some of ye most usefull members of society.[3]

9. Part of Patrick Browne's preface to his manuscript flora of Ireland, making reference to the desirability of forming a botanical garden. (Reproduced by courtesy of the Linnean Society, London; photograph by Dr Edward Diestelkamp)

Dr Browne wrote that in the summer of 1788, before Edward Hill made one of his approaches to the College of Physicians, and before Walter Wade petitioned Parliament. Hill and Wade may not have read Browne's words, but they probably knew his opinions about botanical gardens. Wade eventually achieved his goal, but, as we have seen, Hill's dream evaporated.

Little is known about Walter Wade's background and early life. His father was a Dublin apothecary, John Wade,[4] who had proposed in 1767 the erection of a 'Pharmacopoeia pauperum' in the city, to dispense medicines to the poor. The Irish Parliament empowered the Dublin Society to spend not more than two hundred and fifty pounds on it, Wade's 'Chymical Elaboratory and Dispensary for the Poor' was established in Capel Street, and its proprietor is said to have 'devoted his whole attention and industry to chemistry, to which science pharmacy is not only indebted but owes its chief support'.[5] We do not know when Walter was born, but it was about 1740. He probably attended a school in Dublin. Wade credited himself with a doctorate in medicine, but it is not known at which university he studied.[6] There was a long family tradition in medicine: his grandfather, John (died c. 1738) was a physician, as was his uncle, also called Walter, who graduated from the University of Rheims in 1735.[7] Dr Walter Wade senior emigrated to Portugal where he published a medical treatise in 1768;[8] two of his sons were doctors of medicine, including Charles Wade who was physician to the Court of Portugal. In view of the strong association with medicine, it is not surprising that the younger Walter should take up the profession too.

10. Advertisement published in *The Dublin Chronicle* for Walter Wade's *Flora Dubliniensis* (see Fig. 11). (Reproduced by courtesy of the National Library of Ireland, Dublin)

Propofal for printing by Subfcription,

FLORA DUBLINIENSIS,

CONTAINING correct Delineations and accurate Defcriptions, in Latin and English, of the feveral PLANTS which grow wild in the County of Dublin, arranged according to the Sexual Syftem of the late celebrated Profeffor Linnæus; and their Ufes in MEDICINE, AGRICULTURE and ARTS, pointed out. To which are added, fuch IRISH NAMES, as can with Confidence be depended upon, or as they are generally known, and received by the Inhabitants. And an APPENDIX, containing PLATES and DESCRIPTIONS of fuch officinal, and ufeful Excticks, as are to be feen in the feveral Gardens of the County of Dublin.

By WALTER WADE, M. D.

Licentiate of the King and Queen's College of Phyficians.

*** Encouragers of this Undertaking are requefted to leave their Directions at Mr. Sleater's, No. 28, Dame-ftreet, where Conditions at large may be feen.

Walter Wade obtained his experience and qualifications some time before 1776 when he started to practise in Dublin as a surgeon and man-midwife.[6] He lived for a few years in Bolton Street but in 1781 he moved to Capel Street where his father had the pharmacy. In the same year he married Mary Chambers, a Quaker, who was 47 years old. It was a rule of the Society of Friends that its members could not marry non-Quakers, so she was expelled. The records of the Society are explicit:

> [Mary Chambers] hath lately joined herself in marriage to a man (not of our Society) (as she saith) contrary to the established rules of the Society...Therefore we are under the disagreeable necessity of disowning her the said Mary Chambers (otherwise Wade) to be a member of our Society; yet [hope] that she may in future be enabled to walk in the path of true rectitude.[9]

Walter and Mary Wade had no children. Mary survived her husband by six years; she died on 10 April 1831 and was buried in the Quaker cemetery in Cork Street.[10]

Walter Wade was an active member of the Experimental Society of Dublin for Promoting Natural Knowledge which was formed in 1771 and lasted for about a decade. Little is known about its activities. However, at this time Wade's main occupation was medicine. On 30 October 1786 he was examined for the first time by the College of Physicians; he was approved and paid a fee of £5 2s. 8½d. On 23 April following, he was examined again and was admitted as a Licentiate of the College on signing the declaration and paying the other half of the fee.[11] He could not become a Fellow of the College

11. Engraving from the proposed *Flora Dubliniensis* depicting henbane (*Hyoscyamus niger*); the names of the artist and engraver are not known (see E. C. Nelson, 1981[12]) (Reproduced by courtesy of the National Library of Ireland, Dublin)

because he was not a graduate of the University of Dublin. In 1789 Dr Wade advertised that he was giving public lectures on midwifery. He also gave botanical lectures at his house in Capel Street in the same year.[12]

For the remaining thirty-five years of his life, Walter Wade combined careers in medicine and botany. His interest in the latter subject may have arisen while he was a student, and by 1787 his involvement in Irish botany was considerable. In that December, he advertised in the *Dublin Chronicle* for subscribers to support the publication of an elaborate book on the plants of Dublin (Fig. 10). At his own expense he had sample sheets printed of which only the superbly engraved illustration of henbane (*Hyoscyamus niger*: Fig. 11) and its letterpress survive. Dr Wade's *Flora Dubliniensis* was intended to rival the famous *Flora Londinensis* being published by William Curtis. Like the London book, *Flora Dubliniensis* was to have coloured plates and a large format. Everything about Wade's ambitious work was modelled on that of Curtis, even the title. Like Curtis, Wade experienced financial problems. Alas, *Flora Dubliniensis* was not published, as too few people were prepared to subscribe.[12] But the proposal established Wade as a prominent figure in Irish botany, and as there were few others active at the time, he soon became the leading Irish botanist.

It is not apparent why Dr Wade initiated the moves to form a botanical garden in Ireland. However, it is unlikely that he would have petitioned Parliament without discussing his petition with some of the powerful figures in the Irish Establishment who could ensure a satisfactory result. He could not have acted alone. Indeed, Walter Wade may have been skilfully used by some individual or lobby group to bring into effect a cherished plan.

Wade's petition was presented to the House of Commons in Dublin on 9 February 1790, by the Solicitor-General, John Toler, who later became Lord Norbury. Toler was a keen planter of trees — he planted at least 30,000 trees on his estate in County Tipperary — and a good speaker in the House. In the 1790s, he was a member of the so-called Irish Cabinet which advised the Lord Lieutenant, the Earl of Westmoreland. Significantly, John Toler was a friend and supporter of John Foster. As Speaker of the House of Commons, Foster could not have presented a petition, nor could he speak on bills unless the House went into committee. Although little is explicit in available contemporary documents, it is probable that John Foster was the person who cherished the idea of a public garden, and who manoeuvred the various parliamentary stages to a successful conclusion.

John Foster (1740-1828) was a scion of a County Louth landowning family whose ancestors were 'mowers of hay', perhaps immigrants from Cumberland in the late seventeenth century.[13] John's father, Anthony Foster (1705-1779), entered the Irish House of Commons in 1737 as a member for Dunleer in County Louth. Anthony was a barrister with an extensive practice which included many wealthy and august clients. In 1766, Foster retired from the House of Commons on his appointment as Lord Chief Baron of the Court of Exchequer. Dr Anthony Malcomson, in his study of Irish politics and the Foster family, remarked that Anthony Foster was 'an able man whose misfortune it has been to be overshadowed by an even abler son'. He was an 'improving landlord', lauded by the English agricultural writer Arthur Young as 'this Great Improver'.[14] Young was impressed by Foster's estate at Collon in County Louth and exclaimed that Anthony Foster lived 'now to overlook a country flourishing only from his own exertions'. Dr Malcomson estimated that this had been achieved by Foster putting his entire professional earnings, which were probably about three thousand guineas a year, into agricultural improvements. Arthur Young wrote glowingly that Anthony Foster 'has made a barren wilderness smile with cultivation, planted it with people and made those people happy. Such are the men to whom monarchs should decree their honours, and nations erect their statues'.[14]

His overshadowing son entered the Irish House of Commons as a member for Dunleer in 1761, when he was just below the age of majority. John Foster had been educated by Richard Norris at Drogheda Grammar School and on 1 February 1757, at the age of seventeen, he entered Trinity College, Dublin. Following his graduation in 1760, he went to pursue his legal training at the Middle Temple in London. But John Foster did not take kindly to law as a career; he was more interested in politics and as early as 1766 his alleged neglect of the legal profession led to a strained relationship with his father. His other passion was collecting and planting trees.

The Foster family home at Collon had a fine garden. Anthony Foster had built a range of glasshouses in 1763 for the cultivation of grapes and pineapples. He attempted to get seeds of new, interesting plants from his friends and acquaintances. He was a member of the Linen Board, a body which commanded a substantial amount of patronage. The London agent for the Board was John Ellis, whose origins are obscure but who was probably born about 1707 in the north of Ireland. Ellis was one of the most outstanding amateur naturalists of the last half of the eighteenth century, and conducted a considerable correspondence with scientists including Carl Linnaeus. He sent Anthony Foster seeds for his garden at Collon.[15]

In May 1764, while John Foster was studying at the Middle Temple, his father wrote a letter to John Ellis introducing his son and expressing the wish that they should meet. They did, and Ellis became one of John Foster's main sources of exotic plants. In October 1768 Foster wrote from Collon to John Ellis:

12. *Cupressus lusitanica* (Portuguese cypress) at Oriel Temple about 1910; this tree is recorded as having been raised from seed collected by John Foster's son in 1809. The photograph was originally reproduced in H. J. Elwes and A. Henry, *Trees of Great Britain and Ireland* (vol. 5, pl. 301), 1910.
(National Botanic Gardens, Dublin)

Dear Sir

I return many thanks for the particular favour you have done me in sending the Print and the Account of the new Sensitive Plant which is an acquisition not only of curiosity but of real use. When a plant of it can be with ease procured, my father would be happy in obtaining it.

Will you give me leave to tell you how you could oblige me most highly? I am a great Planter and as many very beautiful and hardy trees are only to be had by seed from America, your sending me any when you have it in your power would make me very happy. I will not mention any particular species for all would be welcome; but at present I would trouble you only for such as will live in the open air of our Climate.

I should not take the liberty of making this request to you, but that I am sure, you will excuse it, if impertinent.

I am dear Sir, with great esteem, your very obliged and obedient servant
John Foster.[16]

That Ellis sent the younger Foster a drawing and account of the Venus fly-trap ('the Sensitive Plant') which had only just been brought to the attention of botanists in Europe, was a singular mark of favour from the renowned naturalist to the young politician. However, it probably had as much to do with the politics of the Linen Board as with any scientific curiosity on the part of the Fosters. John Ellis knew how to please those whose patronage he relied upon.

John Foster became most enthusiastic about planting. He continued to correspond with Ellis, who responded by sending seeds to Ireland. Foster shared these with other gardeners. Among the first batch of seeds was a 'Convolvulus from Pensacola [Florida]'. Foster boldly asked Ellis to purchase 'seeds of such Forest trees, & shrubs, as will thrive here, particularly the various species of Oak, Pines, Firrs and cedars' from some of his American contacts.[17] In 1769, Foster told Ellis of his previous year's activities: 'I have planted among others last autumn & spring many thousand beech; they, larch, Firrs & Pines of all sorts grow wonderfully well on our highest hills'.[17] He said that he had heard much about the beauty of the copper-leaved beech. As he was unable to get seeds in Ireland John Foster begged Ellis to purchase plants and send them to Collon. Frustrated by the poor selection of plants available in Irish nurseries, Foster pleaded on several occasions for copper-leaved beech and various other trees, and was ultimately rewarded with seeds and plants. John Ellis purchased a copper beech in London in March 1770 and shipped it to Foster along with pines, cedars, maples and firs. Thus John Foster became one of the first to plant copper beech in Ireland, and it was only one of many different trees that he grew and probably introduced into Ireland (Fig. 12). In September 1770, Foster thanked Ellis for his assistance; 'All the trees you sent me...have flourished as if they never changed their situation'.[18] He wrote again in the following month, commenting on some of the plants he had been sent, as they did not fit the descriptions he had read. He told Ellis that he would try

> ...to list all the Trees & shrubs I have according to Miller; they will make but a small figure on paper for it is but two years since I began my Rage of planting, and if you will give me leave I will point out to you those sorts which I have not & am most anxious to get...for my Ambition is to Plant every tree & shrub in plenty that will stand our climate, & in a few years when I can reside more in the country to prepare for Greenhouse plants.[19]

At that time Foster had planted eleven acres, but his 'rage' continued until he had amassed a collection of about seventeen hundred different trees and shrubs at Collon. It was undoubtedly the finest collection of its kind in Ireland in the late eighteenth and early nineteenth century. It fulfilled Foster's ambition 'to have every tree and shrub in plenty that will stand our climate'.[19]

With his 'rage of planting' John Foster had an interest in, and a need for, botanical and horticultural books. He bought works such as Carl Linnaeus's *Genera Plantarum*, a standard work of the period, Colin Milne's *A botanical dictionary*, and William Curtis' sumptuous *Flora Londinensis*. The comments he made to John Ellis on the identity of the plants he grew were often based on Philip Miller's great *Gardener's dictionary* and the French arboriculturist, Duhamel du Monceau's *Traité des arbres et arbustes*. These books were also in his library.[20]

John Foster (Fig. 13) was not only interested in the esoteric delights of planting an arboretum but like his father he was deeply concerned with promoting agricultural improvement in Ireland. He was a leading member of the Dublin Society and in 1800 was a co-founder, with the Marquis of Sligo, of the Farming Society of Ireland. John Foster was not too pleased with individuals or societies, such as the Dublin Society in the late 1700s, which were over-exercised with theory or with pure science. Foster preferred a practical approach. He wanted to get things done in the field and to demonstrate what improvements could be made by applying theory and science to the everyday management of Irish farms. This was not a wholly disinterested concern on Foster's part, since improved agricultural methods led to greater income for landlords and ultimately greater wealth and power for Ireland's ruling class, the Anglo-Irish Ascendancy.

Foster's work within the Dublin Society led to his election as a vice-president in 1775 and he continued to serve in this position for over 50 years. He was diligent in attending committee meetings and actively promoted those parts of the Society's work which he felt were important. In 1781, at the request of the Society, he brought to Ireland an English farmer, Thomas Dawson, to instruct 'such as might desire to improve the mode of agriculture in the Kingdom'.[21]

John Foster's political views undoubtedly coloured his own attitudes, and had a considerable bearing on his support for bills debated in the Irish Parliament before its dissolution in 1800, on the passing of the Act of Union. He was preoccupied with the economic advancement of Ireland. The actions he supported or promoted in the House of Commons, particularly while he was Chancellor of the Irish Exchequer (1784-1785), were intended to strengthen the economy and make Ireland more prosperous. His celebrated Corn Law of 1784 is an example. Foster saw in increased prosperity the cure for political discontent, for he believed that unrest and revolution had social and economic roots. Writing about his own home county of Louth, he stated that he 'never saw more content nor less disposition to think of politics or Popery or Protestantism. Every man gets full employment, and

13. John Foster with his wife, Margaretta (née Burgh), and family outside the portico of Oriel Temple; a watercolour by John J. Barralet, 1786. (Reproduced by courtesy of the Viscount Massereene and Ferrard. (see A. Crookshank and the Knight of Glin, *Irish Portraits 1660-1860*, 1969))

there is plenty of work even for women and children'.[22] This statement also reveals another attitude. Foster was a Protestant and regarded the majority of Roman Catholics as incompetent, especially in political matters. He preferred to have Protestant tenants on his own lands and, in fact, advertised for people of that persuasion. His policy had the benefit of building up the number of Protestant freeholders—in other words, voters—who would be reliable and unlikely, on pain of losing their farms, to vote against the Foster family in their parliamentary constituencies.

As a Protestant and a landowner, John Foster was a member and supporter of the Anglo-Irish Ascendancy, a social elite which held a monopoly of political power in Ireland before the Act of Union. That power and position Foster fought to maintain. Foster, like other members of his class, believed that the Ascendancy was the Irish Nation and that Ireland was an equal partner with Great Britain 'the two great members of the Empire'. He did not consider Great Britain to be the mother country but the sister of Ireland. His opposition to the Act of Union was based, therefore, on his view of the financial and economic repercussions of Ireland's loss of direct control over its own affairs. He also opposed the Union because he feared that the Anglo-Irish Ascendancy might lose its political monopoly.

With social and political views such as these, it may be suggested that the forming of a botanical garden in Dublin was seen by John Foster as a good idea in two ways. In viewing Ireland as the equal partner of Great Britain, it would surely be pleasing to have a botanical garden in the Irish capital to rival, if not excel, any garden which existed in London. At the close of the eighteenth century, there was no publicly supported botanical garden in London, although the Royal Garden at Kew was rapidly becoming Britain's premier botanical institution. Secondly, in furthering Ireland's economic progress, any facility which improved agriculture, by educating the public (particularly the landed gentry) and which demonstrated the virtues of new and improved plants, would be a great benefit. Although Foster had a strong personal interest in collecting exotic trees and shrubs, it was his position as a politician, above all his status as Speaker of the Irish House of Commons, which was of paramount significance in his actions towards the formation of a botanical garden. As Speaker, Foster was a powerful figure in the Irish Establishment. His influence was described as 'amazing' and he skilfully used his power, often to promote his own ideas. Dr Malcomson has remarked that

> ...one thing can be said of Foster: his love of power and office may have been unhealthy, but it derived from his ambition to do good; his conception of good may have been narrow, but at least good, as he conceived it, was the object.[23]

In view of this background it would be naïve to argue that John Foster had nothing to do with the petition which was presented by Walter Wade. The event surely was no accident or isolated incident. Although Wade was not a member of parliament he must have been well-known to Foster and his intimates in the House of Commons. The success of Wade's proposal and his later preferment to important positions within the botanical establishment deny any contrary suggestion.

In considering the actions of Walter Wade and especially of John Foster, it is interesting to recall some events associated with the last abortive attempt in the Irish Parliament to establish a physic garden under the control of the two colleges. John Foster, as Chancellor of the Irish Exchequer, opposed the idea on the grounds of cost. However, it is likely that this stated position was only the public expression of his opposition. It is probable that he fought the plan because of his animosity towards the Provost, John Hely-Hutchinson, who had devised and proposed the scheme, and his own anti-intellectual bias. Edward Hill recorded that Foster told him he would have nothing to do with the University[24] as regards a botanical garden, and, as has been noted, Foster was often impatient with those who were devoted to pure science.

Dr Wade's petition (Fig. 14) was ordered to lie on the table, and it was 'ably supported in the House'. It resulted in a clause being inserted into the Dublin Society Bill of 1790, granting three hundred pounds 'towards providing and maintaining a botanic garden'. This bill was passed on 5 April 1790.[25]

This was the first occasion on which the involvement of the Dublin Society was even mentioned. Wade's petition was not discussed by the Society before it was presented. Indeed, Walter Wade was

TO THE

RIGHT HONOURABLE AND HONOURABLE

THE

KNIGHTS, CITIZENS, and BURGESSES,

IN PARLIAMENT ASSEMBLED,

THE

MEMORIAL of WALTER WADE, M. D.

Licentiate of the College of Physicians, Lecturer on Botany, and Author of the Specimen Floræ Dubliniensis,

HUMBLY SHEWETH,

THAT the Establishment of a PUBLICK BOTANICAL GARDEN in this City, or its Environs, on an enlarged Plan, would be of the greatest NATIONAL ADVANTAGE.—As an intimate Knowledge of the various Vegetable Productions and Processes of Vegetation must be allowed to have a near Relation to many of the Arts and Sciences, especially to AGRICULTURE.

THAT your Memorialist will not presume to offer at present any Scheme for the carrying into Effect so desirable and much wanted AN INSTITUTION, unless honoured by the special Direction of your Lordships and Honours, firmly relying on your Wisdom and Information, for the Accomplishment and Execution of what your Memorialist confidently apprehends will appear on Investigation A GREAT NATIONAL OBJECT.

THEREFORE it is with deference hoped that your Lordships and Honours will be pleased to take your Memorialist's Observations into Consideration, and he, as in Duty bound, will pray.

14. Walter Wade's petition. (Reproduced from the only extant copy, by courtesy of the Public Record Office of Northern Ireland, Belfast, and the Viscount Massereene and Ferrard. (Foster/Massereene Papers D562/7823))

not a member of the Dublin Society at the time. The Society clearly did not instigate the petition, and from the lack of action which followed the granting of the money, it can be concluded that the Society had no plans to establish a botanical garden—there is nothing about a botanical garden in the Society's minutes before July 1790.

The Dublin Society had had experimental or demonstration gardens in the first few years of its existence but these were not botanical gardens. As one of its aims was to promote husbandry, the Society employed John Wynne Baker in 1763 to set up a model farm. In the previous year, Baker had presented a pamphlet to the Society, titled *Hints on husbandry*, which had impressed members. He was given various grants for conducting experiments on the cultivation of carrots, turnips and other crops. In 1763 he rented a farm near Loughlinstown in County Dublin, established a small manufactory for agricultural implements and proposed a scheme for educating agricultural apprentices which the Society approved. The implements were made by the apprentices, and sold by Baker who got a premium on the sales. In December 1769, at a packed meeting of the Dublin Society, Baker was granted three hundred pounds as a salary for conducting his experiments and for giving advice. He was also granted a tenth of the proceeds of the sales of implements up to a maximum of two hundred pounds. H. F. Berry, in his fine history of the Dublin Society, detailed Baker's work and commented that 'it seems fitting that prominence should be afforded to the enlightened policy of the Society...the story of John Wynne Baker shows in a remarkable manner what care and discrimination were evinced in carrying out its plans for the encouragement of more scientific methods in agriculture'.[26] Arthur Young was not so wholehearted in his praise. He said that despite Baker's exertions he had 'not answered the expectations formed of him'.[27] Baker died in 1775, but the memory of his employment lingered, and worried some members of the Society as it set about planning a botanical garden.

In July 1790, the Dublin Society held a meeting to start planning the garden but because it had not been able 'hitherto to obtain such sufficient information as might enable [it] to decide the most efficacious mode of applying' the money granted for the botanical garden, the task was postponed. It was resolved 'in the month of November next to proceed to consider such plans...as shall be submitted to [the Society] by gentlemen of the Medical Faculty'.[28] This decision demonstrates just how unprepared the members of the Dublin Society were for the task that Parliament had assigned to them. The Society wrote to the Faculty, and according to Wade, a plan and observations were 'drawn up agreeable to the general plan of a highly respected and exalted character, when he condescended to communicate his sentiments on this business'.[29] The character's name was not given, but certainly was John Foster.

Details of the plan survive in a manuscript, and in a *Statement of the Progress which has been made for the Purpose of instituting a Public Botanic Garden...* which Wade submitted to both Houses of Parliament in 1793.[30] Wade himself probably had a considerable influence on the plan. A garden of eight or ten Irish acres was suggested, situated south-west of the Circular Road and about a quarter of a mile beyond it, perhaps somewhere in the Harold's Cross area. The major part was to be set aside for experimental agriculture. It was proposed to appoint a lecturer in botany and another in agriculture 'of the most acknowledged and distinguished abilities as to theoretical and practical agriculture, and no matter of what nation so he speak the English language intelligibly'. Foster's opinions are evident in these proposals. A house would be provided containing lecturing facilities and a library of botanical and agricultural textbooks. The garden would be furnished with glasshouses, and in the 'open garden' the plants would be arranged systematically. The plan envisaged sections containing poisonous plants, grasses, native Irish plants, native and exotic mushrooms, native and foreign ferns and mosses, foreign and Irish leguminous plants, and medicinal plants. Ponds and artificial rock-gardens were also proposed. The plants would all be labelled with their Latin and vernacular English names as well as genuine Irish names if available. Though not entirely devoted to agriculture this garden clearly aimed to display plants useful in improving agriculture in Ireland. The proposal placed before the Dublin Society noted that the greatest obstacle 'ultimately frustrating the intentions of the legislature' was the fact that no land was available. This continued to thwart the plan for a further four years.

We the Undersigned Professors of the Several Branches of Medicine in the University of Dublin, Constituting the Faculty of Medicine in the Kingdom of Ireland, Do hereby Certify to all whom it may Concern, That we know **WALTER WADE** of the City of Dublin **M:D:**, a Licentiate of the College of Physicians, and from that Knowledge are Satisfied, that he is perfectly well Skilled in the Theory and Practice of the Science of **BOTANY**, and that he is otherwise extremely fit to be entrusted with the Superintendance, Direction and Management of a Public Botanical Garden. Witness our Hands this 24th day of July 1791, Ninety one.

(A Copy)

Edwd. Hill, Professor of Botany
Edwd. Cullen, Profr. of Mat. Med. & Pharm.
Edwd. Brereton, Profr. of Practice
Stephr. Dickson, Profr. of Institutes
James Cleghorn, Profr. of Anatomy

15. Testimonial for Walter Wade; the original document cannot be traced but this contemporary copy is in the Foster/Massereene Papers (D562/7824). (Reproduced by courtesy of the Public Record Office of Northern Ireland, Belfast, and the Viscount Massereene and Ferrard)

Before the Dublin Society had a chance to consider the plan—it did not have it in November 1790 as hoped—Parliament granted another three hundred pounds towards the foundation and maintenance of a botanic garden. Thus six hundred pounds was available on 26 May 1791 when the Society decided to discuss the matter on the following Thursday. The medical gentlemen Drs Perceval, Wade and Hill were invited to attend and communicate 'such scheme as shall appear to them most eligible'. On Thursday, 2 June, Perceval and Wade attended the Society's meeting, but Hill was unable to be present. Hill and Perceval were bitter opponents so inviting both was somewhat tactless. The Dublin Society decided to write to the College of Physicians and the University telling them that money had been allocated by Parliament and requesting their advice and assistance in forming a botanical garden. This was done because 'a gentleman...supposed much in the confidence of both bodies...asserted that he believed and had some reason to know, that each...would contribute towards the support of a botanic garden if it were once established'.[31] On 20 June, the College of Physicians appointed Hill and Perceval to confer with the University and the Dublin Society on this matter. The University also considered the Society's letter and appointed a deputation to meet the other bodies.

About this time John Foster invited Edward Hill to discuss the establishment of a botanical garden. Hill recorded that 'when the Right Hon. The Speaker...was projecting the foundation of an agricultural botanical garden, at his desire I had some confidence with him relative to it. He proposed to me a scheme of a course of lectures annually on botany...and showed a desire that I would form the arrangement of his garden, and attach myself to his establishment as professor and lecturer'. Although Edward Hill assured Foster of his support 'in promoting so patriotic a design', he declined the offer of the lectureship. He felt that his commitment to the University was such that he could not accept another position. Hearing that Dr Hill had declined John Foster's invitation, Samuel Hayes of Avondale, a close friend of both men, a member of the Dublin Society and a keen forester, went to see Hill and urged him to resign from the University and join the Dublin Society as a lecturer 'with a salary of £300 a year which he engaged I [Hill] should have, through the Speaker's and his influence on the Dublin Society'. Edward Hill again declined and Hayes left 'with evident indications of concern'.[32]

Shortly afterwards, according to Edward Hill's account of these events, 'the Gentleman, who at present [1803] so properly fills the employment' as lecturer in botany to the Dublin Society, sought from Hill 'an attestation of my knowledge of his acquaintance with the science of botany, which I gave him without reluctance'. This was Walter Wade. Although he was not appointed Professor of Botany to the Society for a further five years, the testimonial was given to him on 24 July 1791; significantly the only known copy (Fig. 15) is among John Foster's papers.[33] The five signatories were members of the Faculty of Medicine and of the College of Physicians.

Edward Hill said he had suggested to John Foster that a botanical garden should be jointly managed and funded by the Dublin Society, the University and the College of Physicians. Hill did this 'with some exultation', confident of 'infallible success' if the funds could be allocated. Foster, according to Hill, was 'struck with this idea, but after some pause, declared peremptorily that he would have nothing to do with the Colleges'.[34]

By 24 November 1791, the University and College of Physicians had both replied to the Dublin Society's invitation. The Society appointed Sir William Gleadowe Newcomen, Andrew Caldwell and Patrick Bride as their committee to meet the others. Progress remained very slow. On 2 February 1793, the Dublin Society again considered the matter and, on a motion from Colonel Eustace, seven members, including the Colonel, were appointed to form a committee 'to take ground proper for a botanic garden'. The other members were John Foster, Sir William Newcomen, Samuel Hayes, Thomas Burgh, Andrew Caldwell and the Bishop of Kilmore. Burgh was John Foster's brother-in-law and the Bishop was Foster's younger brother. Shortly after this, Parliament granted another three hundred pounds to the Society.

On 6 March 1793, the Provost of Trinity College sought leave to introduce his bill 'directing the application of funds to provide and maintain a botanic garden'. Two days later the Provost presented the bill. On 11 June, he presented the petition of the University and the College of Physicians, suggesting a botanical garden under the control of the three institutions, and on 21 June the House of Commons went into committee on the bill. This enabled the Speaker to participate in the debate, and probably Foster used the opportunity to oppose the plan vigorously. Hill and Perceval were called before the committee, and the Solicitor-General, John Toler, 'gave the amplest testimony of the great abilities and professional science in universal botany of Doctor Walter Wade, by whose aid a complete plan and arrangement of an universal botanic garden for every useful purpose of both art and science, has been made'.[35] Wade was unable to attend the session of the committee, but he had had printed a nine page document in which he set out the actions already taken by the Dublin Society and the plan submitted to it late in 1790. The document was sent to both Houses of Parliament. In its final paragraphs, Wade emphasised the patriotic need for a botanical garden, and he also pointed out that a school of medicine could not be complete without a teaching garden. Botanical gardens, Wade asserted, were not merely confined to colleges and universities, but

> every wise political and polished government has seen the absolute necessity of such establishments…What a pre-eminent instance is the Royal Garden at Kew condescendingly superintended with a noble degree of enthusiastic regard, by our much beloved and most gracious King and Queen, whose exemplary partiality for Botany is such as to make it respected as a science and as an amusement fashionable and agreeable to persons of all ranks and situations.

Wade commented that in England horticulture had advanced greatly, and that contacts with distant parts of the world gave British gardens superiority over those of other nations. By implication, he was asking how could Ireland be so unpatriotic, so unfashionable as to lag behind. Walter Wade was probably articulating the thoughts of John Foster. 'It is sincerely to be hoped and ardently to be wished that the love of botany and the cultivation of plants may at length strike root and shed its salutary influence among the people of rank and opulence in this kingdom'. Wade appealed to the instincts of the parliamentarians as members of the Anglo-Irish Ascendancy, noting that a botanic garden was a national necessity. His final thrust was to ask that the government did nothing less than to institute a

> *Royal Botanic Garden*…[on land] being in the immediate gift of our most gracious sovereign…in his Majesties deer park adjoining Mrs Talbot's Lodge. And here if a Garden were established, the professor of Botany might have occasional access for the instruction of his pupils.[36]

The committee debate was lengthy. The construction of the acts granting money to the Dublin Society for a botanical garden was discussed. In the end, the bill presented by the Provost was lost on a motion that the committee chairman should leave the chair; the motion was carried by twenty-four votes to eighteen. John Foster had won and the way was now clear for his plan to proceed, for the Dublin Society to form a garden independent of the constraints of academic botany and medicine, a garden slanted towards the needs of practical husbandry.

The Dublin Society now acted with somewhat greater urgency. Its committee started seeking a suitable site. Wade had suggested an area within Phoenix Park, but that seems to have been ignored. Among documents in John Foster's papers is a map dated 1793 of 'part of the lands of Priesthouse' (Fig. 16), on the southern outskirts of Dublin near Roebuck and Goatstown, which may have been considered.[37] Nothing else is known about this proposal.

By September 1793, a site had been selected. In a letter to the English botanist, James Edward Smith, the founder of the Linnean Society, one of the committee members, Andrew Caldwell wrote:

> You will have pleasure in being inform'd that a garden for indigenous botany under the patronage of the Dublin Society is a measure determined upon. A committee, at the head of which is Mr Foster, Speaker of the House of Commons, and of which I have the honour to be a member, is vested with proper powers and a sum of money for purchasing Ground. We have actually agreed for the House and Garden formerly belonging to Doctor Delany, it is within a short mile of town, the soil excellent, the ground well varied with high & low, a small stream running thru' it seems well adapted in every respect.[38]

This desirable property lay in the village of Glasnevin, three miles north of Dublin City. Caldwell noted that Foster wanted the garden to be for agricultural experiments, but he did not agree and obviously other committee members had similar feelings. One of the reasons for these misgivings was

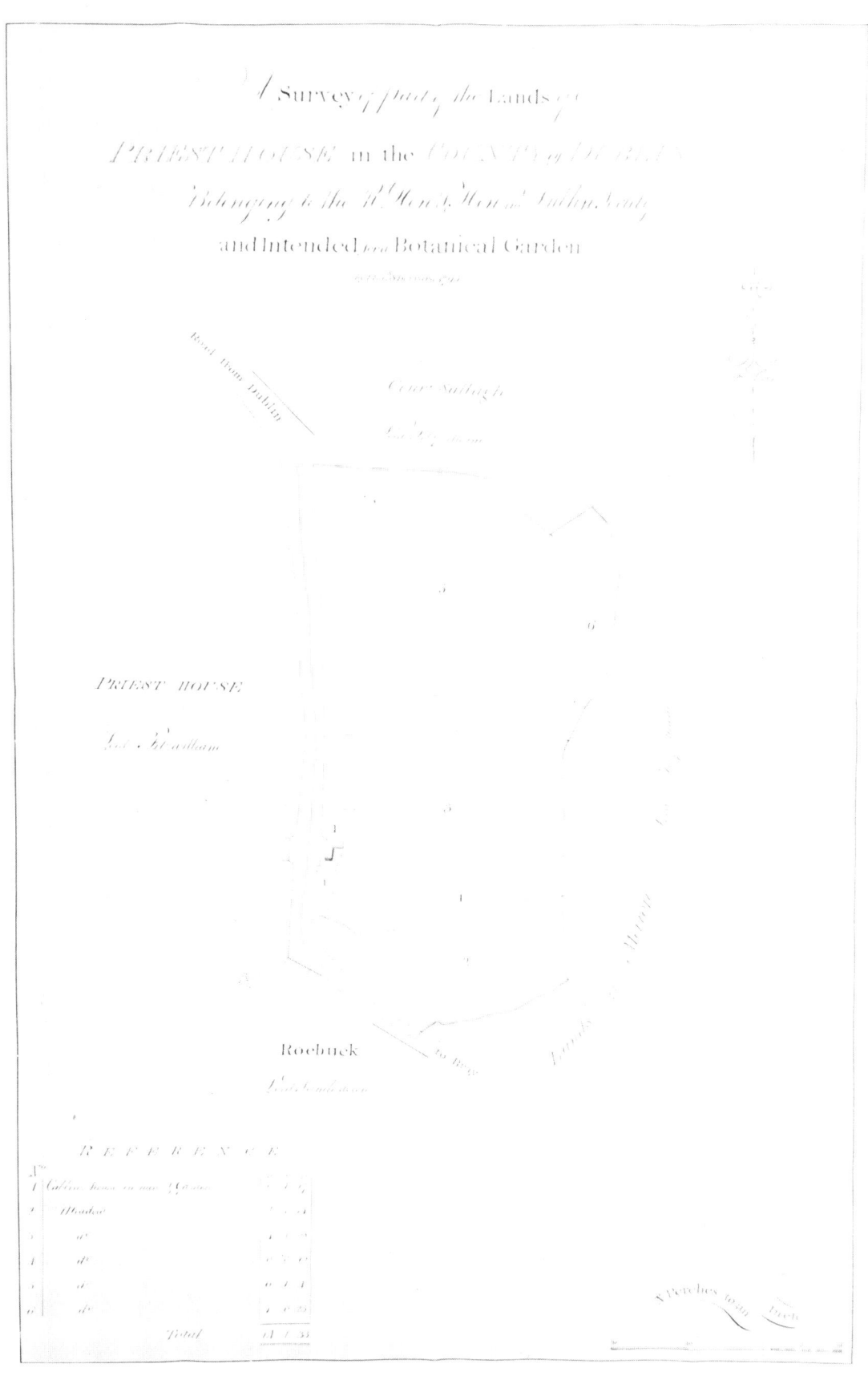

16. A survey of land at Roebuck, south of Dublin city, 'Intended for a Botanical Garden'; the plan was prepared by Thomas Sherrard in 1793. (Reproduced by courtesy of the Public Record Office of Northern Ireland, Belfast, and the Viscount Massereene and Ferrard. (Foster/Massereene Papers, D562/7810))

the problem previously associated with John Wynne Baker's model farm. Caldwell candidly remarked that Baker's project had turned out very unfortunately: Baker had been brought from England and

> allow'd...a sallary of £300 pr Anno, but this did not satisfye him, he was for ever bringing bills of contingencies & carrying them by his influence. He was a man of Talents, agreeable conversation & convivial, an excellent farmer at his desk & over the table, but execrable in the field.

Caldwell told Smith that when he first joined the Dublin Society there were

> pitch'd battles twice a year between him [Baker] & the real publick spirited Gentlemen but Baker always brought down a crowd of Jovial Squires and members of Parliament, and outvoted us, a fever at last carried him off and relieved the Society, & had he liv'd on he would long before this have swallow'd up the Dublin Society and their funds altogether.

In conclusion Caldwell confided that after this experience he was 'dreadfully afraid of any thing like a renewal'.[39]

For another year, negotiations progressed slowly. On 20 September 1794 Walter Wade wrote to Smith enquiring if he had met Andrew Caldwell. Wade repeated to James Smith that 'Parliament has already granted £1,700 and £500 a year' for the botanical garden. He also said that he had every reason to suppose that he would be appointed superintendent of the garden.[40] Sometime between September 1794 and February 1795, the negotiation for the Delany property failed, because the lease included a clause preventing the breaking of the ground, and the removal of trees and shrubs. Caldwell had noted that the garden was 'too much covered with trees, but I shall be for cutting them down sparingly and with caution.'[41]

The Dublin Society committee turned its sights across the River Tolka to a nearby property then under lease to John Kiernan. On 1 February 1795, Kiernan wrote to John Foster saying that for a down-payment of twelve hundred pounds, or one hundred and twenty pounds immediately, and one hundred and twenty pounds each year for the rest of his life and that of his wife, he would 'give you up part of the concerns as soon as conveance is made, the remainder I will give up clear of all rent the 25 day of next March'.[42] Kiernan and his wife were near fifty. On 26 February, the committee reported to the Society and was asked to present a full report a few days later. On 5 March, the Bishop of Kilmore, one of the committee members, informed the Society that after examining various properties, they had found none 'so eligible and likely to attend with so little expense in fencing' as that occupied by Kiernan. This site consisted of sixteen Irish acres; 'the condition of the ground, its situation to the north west of town and the buildings on it make it highly desirable'.[43] The committee urged that no time be lost as the planting and preparation of the ground could be started in the spring. The Dublin Society approved the leasing of Kiernan's land and the committee was empowered to appoint a person to take charge of the ground and premises when possession was given up.[44]

John Kiernan's lease was purchased outright and on 25 March 1795 the Dublin Society took possession of the land. A few days earlier, Walter Wade had written with distinct delight to James Smith saying that 'we have at last got ground for a botanic garden, and are to work immediately—its success and progress from time to time I shall have pleasure in transmitting to you'.[45]

Before 1795:
the Romantic Glasnevin of Tickell and Delany

IN 1795, the hamlet of Glasnevin (*Glas Naion*, the stream of the infant(s))[1] contained only a few houses scattered on the gently sloping hillsides overlooking the small River Tolka. There was also an ancient parish church dedicated to Saint Mobhi. It was a 'romantic village...long celebrated for its salubrity and the mildness of its temperature'. T. K. Cromwell, writing in 1820, suggested that the place was entitled 'to some pretension of classic fame', for Sir Richard Steele, Patrick Delany, Jonathan Swift, Thomas Parnell, Joseph Addison and Thomas Tickell had all made 'its vicinity their constant or occasional residence'.[2] This galaxy of genius provided the basis for legends, some of which are repeated as facts, even after two and a half centuries.

According to the *Annals of the Four Masters*, the Abbot of Glasnoidhean, Berchan Clarainech, better known as Saint Mobhi, died in 544.[3] He had founded a monastery beside the river, and there he taught students from many parts of Ireland. The *Book of Lismore* records that fifty young men studied at Glasnevin when Columcille (Columba) of Derry, Comgall of Bangor, Ciaran and Cainnech were students under Mobhi. They lived in huts on the western bank of the river, and the great church that Mobhi built stood on the opposite side.[4] No trace of Mobhi's monastery can be found today. Only ancient tales persist.

The church and scholastic community of Saint Mobhi perhaps did not survive for long, but the association between Glasnevin and the church in Ireland remained. In 1178, the Archbishop of Dublin, Laurence O'Toole, gave land at Glasnevin to the Church of the Holy Trinity, commonly called Christchurch Cathedral. The grant was ratified by Pope Alexander III in 1179. A mill was built on the church's property and the accounts of the Cathedral indicate that by 1306 the manor at Glasnevin was rated (with its tithes) at forty-eight shillings and contained three carucates.[5]

The Cathedral Chapter and the Dean continued to enjoy revenues from Glasnevin after the dissolution of the monasteries and the forfeiture of the monastic properties in the middle of the sixteenth century. In 1551, Christchurch was converted to use as a Protestant cathedral. A 'plott of the Towne and Lands of Glassinevin...belonging to the Dean of the holy Trinity Dublin' which hangs today in the vestry of the parish church of Saint Mobhi, shows the extent of the Cathedral's lands and names the members of the Chapter entitled to the revenues from the numerous fields.[6] It was prepared in 1640 by Richard Francis, although the surviving map is a copy 'now traced out in ffebruary 1719/20 by me John Greene'. In 1640, most of the sixteen Irish acres, acquired by the Dublin Society in 1795, were designated as belonging to 'Seaven Farmes' (Fig. 17). Of the rest, the Dean had just over two acres and the Vicars held about one acre. The Cathedral Treasurer had three roods two perches adjoining the mill-race, for in 1539 the watermill had been granted to the Treasurer. A small portion of the land was part of 'Forster's farme'. The bridge carrying the road from Dublin over the river is shown on the map, and beside it, on the Dublin side, is depicted a mill and water-wheel. In 1640 there was no house on the land later acquired by the Dublin Society, but several buildings, including the parish church, were marked on the northern side of the river.

In 1720, when the map was copied, the division of the property remained as it had been in 1640, as no attempt had been made to revise the field boundaries. Sometime later, considerable alterations were made to these boundaries. The Dean and Chapter agreed to lease the land to Richard Stewart, who in turn leased it to John Putland.[7] Putland then let the property to John Power, but on 16 April 1736, Power handed over the lease to Thomas Tickell. Tickell agreed to lease, for a yearly rent of £39 9s. 0d,

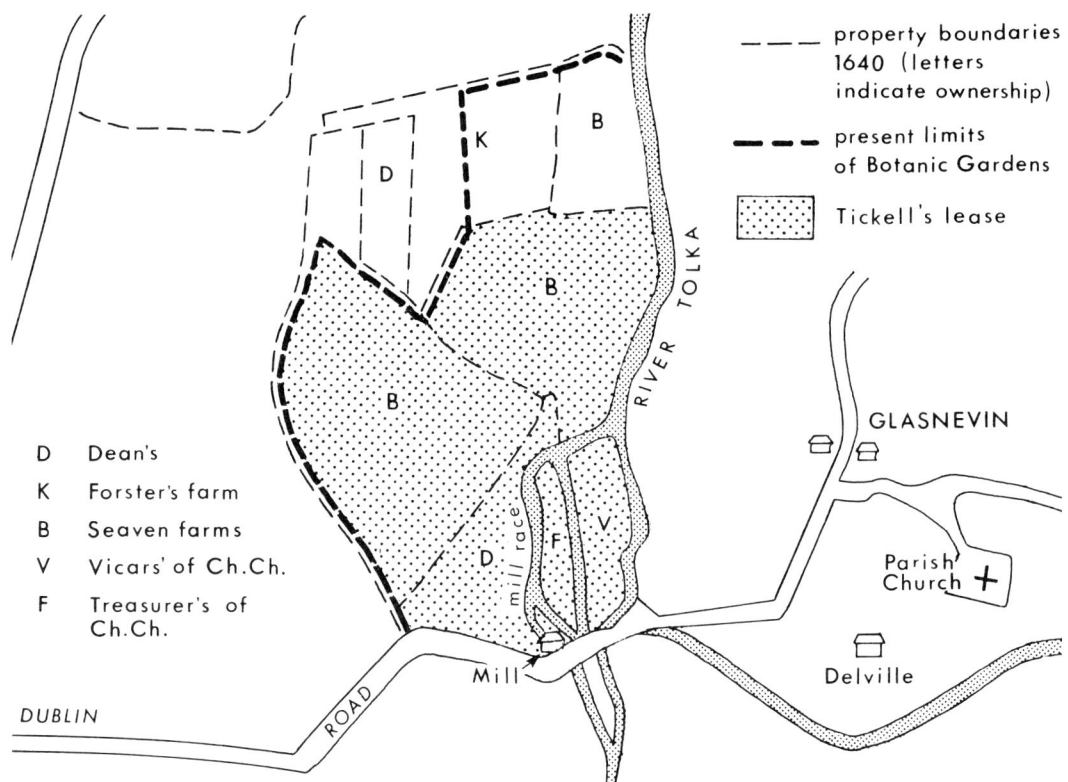

17. Glasnevin in the seventeenth century, showing the ownership of the land on which the Botanic Gardens was eventually established and the area leased by Thomas Tickell in 1736 (see Fig. 18). Redrawn from the original map in Glasnevin Parish Church, which is inscribed 'traced by John Greene in February 1719/20 from a map "performed by Richard Francis 1640".'
(By courtesy of the Rector and Select Vestry, St Mobhi's Parish, Glasnevin)

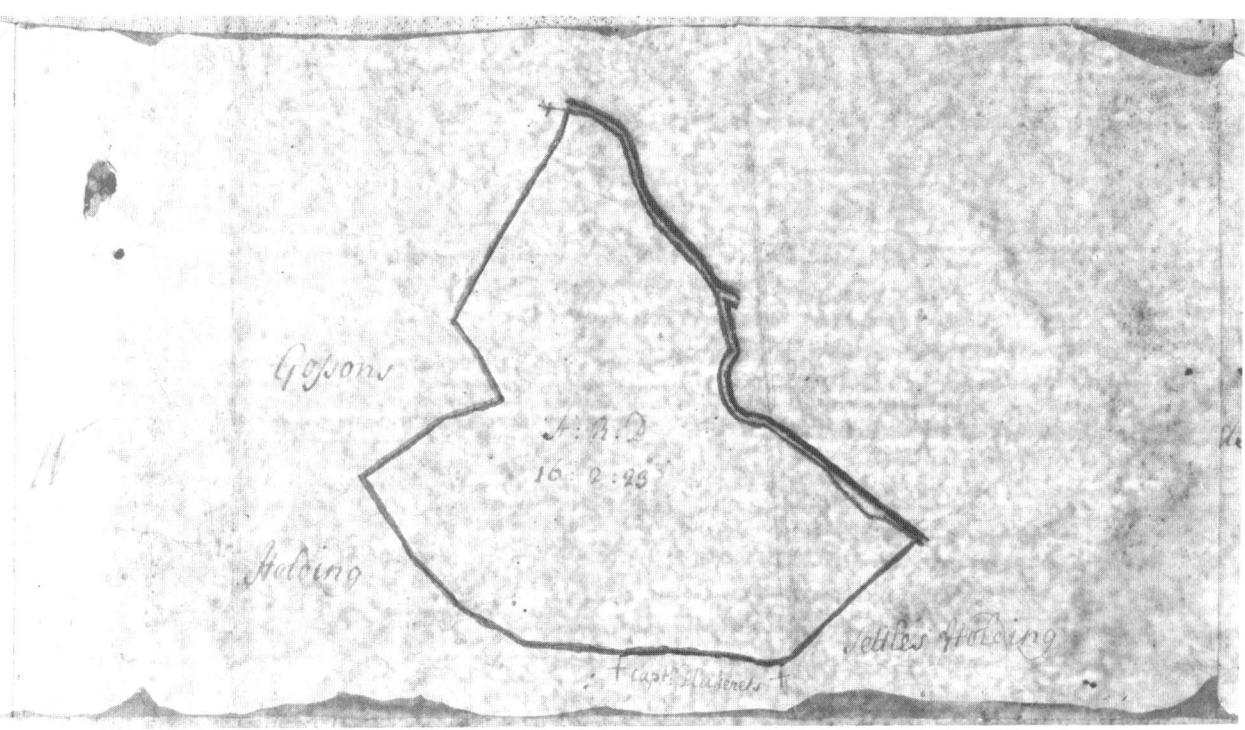

18. Extract showing the extent of Thomas Tickell's lease, from the original deeds which bear his signature. The deeds and map are inscribed on vellum. (Reproduced by courtesy of Major-General M. E. Tickell)

19. Thomas Tickell, a portrait in oils; in Queen's College, Oxford. (Reproduced by courtesy of the Master, Queen's College, Oxford)

all that the House, Garden, Outhouses and Cabbin adjoyning comonly known by the Name of Fielings Holding, together with two acres of meadow, now in the said Thomas Tickell's possession, which said premises are bounded towards the East with the road leading from Dublin to Glasnevin, towards ye north with the Millstream and River of Glasnevin and towards the south with the late Sir John Rogerson's land. (Fig. 18)

Thomas Tickell also agreed to pay an additional fine of £9 7s. 0d every seven years, the amount Putland paid to the Dean and Chapter of Christchurch for his *toties quoties* lease.

Of the early lease-holders the most interesting is John Putland (1709-1773). He was the eldest son of Thomas Putland, a successful London merchant who had settled in Dublin. John Putland went to Trinity College in 1727 and obtained the degree of bachelor of arts in 1731. He remained at the University until 1734 when he graduated as a master of arts and bachelor of divinity.[8] The reason why he leased the property at Glasnevin so early in his life is not known, but one of the executors of Richard Stewart's estate was Dr Richard Helsham, John Putland's stepfather.[9] In February 1735, Putland borrowed fifteen hundred pounds from Jonathan Swift, on which he paid annual interest of eighty pounds. Swift commented that John Putland 'hath a good fortune, and I think it safe'. Many years later, Putland was treasurer of the Dublin Society. In 1760 the Society awarded him a gold medal 'in grateful acknowledgement of the advantage...received from his kind and assiduous attention to its useful purposes'.[10]

20. 'Addison's Yew Walk' in the Botanic Gardens, photographed by W. D. Hemphill, c. 1897. This photograph was originally reproduced in J. Lowe, *The yew-trees of Great Britain and Ireland*, 1897. (National Botanic Gardens, Dublin)

Thomas Tickell was a minor poet (Fig. 19). He was born at Bridekirk in Cumberland on 17 December 1685 and studied at the University of Oxford. He became a Fellow of Queen's College and acted as Professor of Poetry in Oxford during the absence of Joseph Trapp.[11] In May 1714, Joseph Addison, Tickell's patron, was made Secretary to the Lords of the Regency and in September of the same year he became Secretary to the Lord Lieutenant of Ireland, the Earl of Sunderland. Addison chose Thomas Tickell as an Under-Secretary. Although the Irish Administration was being conducted almost entirely from London at this time, plans were made for Tickell to go to Ireland, but these were postponed and later abandoned when Lord Sunderland resigned. Addison and his protégé were left without official positions. It is probable that Tickell was chosen as an Under-Secretary because he was so inoffensive and politically insignificant.[12] Joseph Addison was a statesman, writer and poet.[13] He served as Secretary to the Lord Lieutenant for the first time in 1709. While in Dublin he lived in the official apartments of the Secretary in Dublin Castle. In March 1709, he was told that his 'lodgings in the castle are in a good forwardness, and...will pretty well serve yor occasions during your stay here'.[14] During his short time in Ireland, Addison is not known to have lived elsewhere. There is no evidence that he resided at Glasnevin, although his name is closely associated with the village. It is improbable that Joseph Addison walked in the grounds now occupied by the Botanic Gardens for Tickell did not accompany Addison during his first term of appointment as Secretary to the Lord Lieutenant.[15]

Joseph Addison and Thomas Tickell were close friends. Tickell was Addison's literary executor and biographer and had written many poetical tributes to Addison. When Addison died at Holland House in London on 17 June 1719 'in the weeping presence' of Thomas Tickell, the death-bed scene was

21. Yew Walk in the Botanic Gardens; photographed by David Davison in November 1981

melodramatic. His last words, spoken to his stepson, the young Earl of Warwick, were 'See in what peace a Christian can die'. Addison was interred in Westminster Abbey at dead of night.

Thomas Tickell's association with the Irish Administration continued after his patron's death. Lord Carteret, who became Lord Lieutenant in April 1724, appointed Thomas Tickell as Secretary to the Lords Justices. According to Bishop Downes, a correspondent of his godfather, Dr William Nicolson, Bishop of Derry, Tickell landed in Dublin on 1 June 1724 and was at the time 'entirely unacquainted with the country'.[16] At first Thomas occupied apartments in Dublin Castle. In April 1726, he married Clotilda, daughter of Sir Maurice Eustace of Harristown in County Kildare. She had a share in land at Carnalway and this came to Thomas Tickell on their marriage.

For a few years Tickell was occupied improving the house and property in Kildare by draining bogs and making ditches. The Tickells had five children, and it is possible that he acquired Putland's land and house at Glasnevin to accommodate his growing family and to allow him to live closer to the city. The initial lease signed in April 1736 was for eighteen years, but Thomas Tickell only enjoyed the house and gardens for four years. He died in Bath on 21 April 1740, and was buried in the graveyard of the parish church in Glasnevin. A memorial tablet was erected inside:

> Sacred to the memory of Thomas Tickell Esq...He was sometime Under Secretary in England, and afterwards for many years, Secretary to the Lords Justices of Ireland: but his highest honour was that of having been the friend of Addison.

Legend surrounds Tickell's residence in Glasnevin. It is said, for example, that his best-known ballad, *Colin and Lucy*, which begins with the line, 'In Leinster famed for maidens fair...' was written in Glasnevin.[17] As the first edition was published in Dublin in 1725 this is impossible.

Thomas Tickell is said to have planted the double avenue of yew trees which still survives in the Botanic Gardens (Figs. 20 and 21). It is often called Addison's Walk, and some writers claim that Tickell and Addison walked there. Others say that it was planted by Thomas Tickell in memory of his patron. Certainly the yew trees have been there for a long time, but when they were planted and by whom is not known. As a similar yew walk was in the grounds of Holland House, the London home of Addison's wife, it is not unlikely that Tickell had the avenue planted as a memorial. However, the trees may have been in place when Tickell leased the property, and it is perhaps significant that a similar avenue exists on the opposite bank of the River Tolka, almost parallel with Addison's Walk.

When the Tickells moved to Glasnevin in 1736, they became neighbours of Dr Patrick Delany, a friend of Jonathan Swift, and at the time a widower. Delany had acquired land on the north bank of the Tolka before 1715, in association with Dr Richard Helsham. They originally named the property Hel-Del-Ville, but, perhaps when Helsham relinquished interest in it, it was renamed Delville. During Tickell's absences from Ireland, Dr Delany oversaw the running of Tickell's lands.

Thomas Tickell did not enjoy robust health. When the actor and memoir-writer, John O'Keeffe, was a young boy, he saw Tickell at Glasnevin and described him as 'then old, of the middle size, and [with] a halt in one leg'.[18] In 1736 Thomas Tickell went to Moffat in Scotland to take a cure, and Delany wrote to him there, saying

> I think it is now high time I should give an account of my stewardship. O Sir your meadows did not answer. The low lands were so chilled with the Spring cold, and the uplands so scorched with the summer that they both failed. In short their constitution was greatly hurt as many better constitutions are by heats and colds. The whole crop (tythe included) came to but a hundred & 87 loads, whereas in my estimation it should have amounted to at least 220!.[19]

Delany also informed Tickell that some of the hay 'heated in the field cock and what puzzles me most it was almost all upland hay which they told me was nearly dry the day it was cut, which I own made me suspect some foul play'. Delany's own crop was affected too, and he had to demolish one of his haystacks which overheated.[19]

Clearly Tickell and Delany were friends. They enjoyed each other's company and did not mind the isolation imposed by the winter weather and the undoubtedly bad roads into the city. Lord Orrery, writing to a friend in December 1736, remarked that

> Tickell and Dr Delany perform Quarantine at Glassnevin. We must not hope to see Them till longer Days and milder Weather. The World, I suppose, will reap the Benefit of their Retirement. The Press will groan for it, and Future Ages will revere Delvil [sic] as much as the Present Age honours Praeneste.[20]

Patrick Delany was one of the greatest of the resident Glasnevin *virtuosi*: a brilliant scholar, a Fellow of Trinity College, and a friend of Jonathan Swift, who often visited Delville before illness prevented him venturing far from the Deanery by Saint Patrick's Cathedral. It is said that some of Swift's broadsheets were printed on a press concealed in the basement at Delville. But even Dr Delany was to be overshadowed at Glasnevin by his second wife, Mrs Mary Pendarves, whom he married in 1743. Mary Delany brought to Delville 'singular ingenuity and politeness and...unaffected piety'.[21] It is because of her that Lord Orrery's prediction that 'Future Ages will revere Delvil' is fulfilled.

In 1744, Patrick Delany was appointed Dean of Down, and he and his wife made annual journeys to Downpatrick where they had a house and garden. But Delville was their favourite, and there they created a famous garden, which was lampooned by Jonathan Swift because of its relatively small size:

> Would you that Delville I describe? —
> Believe me, Sir, I will not gibe;
> For who would be satirical
> Upon a thing so very small!
> You scarce upon the borders enter,
> Before you're at the very centre.
> A single crow would make it night,
> If o'er your farm he took his flight;
> Yet in the narrow compass we
> Observe a vast variety;

22. One of Mrs Mary Delany's watercolour sketches showing the view from Delville across Dublin Bay towards the Dublin Mountains. The wall enclosing the garden at Delville is clearly shown, but this is not topographically accurate. (Reproduced by courtesy of the National Gallery of Ireland, Dublin)

> Both walks, walls, meadows and parterres,
> Windows and doors, and rooms and stairs,
> And hills and vales, and woods and fields,
> And hay, and grass, and corn it yields...
> In short, in all your boasted seat
> There's nothing but yourself is — great.[22]

Swift was being unkind for Delville covered eleven Irish acres. However, its size was irrelevant to its reputation. Patrick Delany had followed the newly fashionable style of gardening, more relaxed, less formal than those hitherto in vogue. It was championed by his friend, Alexander Pope, and employed by Jonathan Swift. By a strange coincidence, this gardening style was presaged by Joseph Addison. In one of his lucid philosophical essays in *The Spectator*, published on 25 June 1712, Addison objected to certain contrivances popular in gardens of the time, particularly topiary.[23] He disliked to see 'the marks of Scissars upon every plant and bush'. Addison preferred the 'luxuriancy and diffusion of boughs and branches'. His own estate he described as 'a confusion of kitchen and parterre, orchard and flower garden', a sort of natural wilderness, where, in the words of Jonathan Swift, '...nature is preserv'd in every part, sometimes adorn'd, but nere debauch'd by art'.[24]

Joseph Addison's essays were the first written works in which a natural style of gardening was advocated. He wanted to see the garden extended to include the surrounding countryside, 'to...throw...a whole Estate into a kind of garden', so that 'Man might make a pretty Landskip [sic] of his own Possessions'.[25] These ideas were taken up by Pope who elaborated them in his writings and put them into practice in his garden at Twickenham. Patrick Delany stayed with Alexander Pope in 1728 and probably was influenced by Pope when he was planning and planting Delville.

Delville was situated on the hillside which faces south and east across the River Tolka. It was possible to see the Dublin Mountains and ships in Dublin Bay from the garden (Fig. 22). 'Little wild walks' meandered through the plantations of elms and evergreen oaks. There was a ruined castle with a cave beneath. Temples and grottoes terminated vistas. There was a bowling green, a bath-house, rustic bridges and seats, a parterre around the house and a high perimeter wall which did not obscure the view but served to enclose the estate and keep the herd of deer inside. When Mary Delany came to live there, she found the garden 'a wilderness of sweets...The fields are planted in a wild way with forest-trees as with bushes, that look so naturally you would not imagine it the work of art...'.[26] She added various embellishments, and she decorated the ceiling of the chapel with imitation stucco made chiefly of shells collected from local beaches. In the garden she planted honeysuckles, sweet briar rose and jasmines, violets, primroses and cowslips. In 1750 she considered erecting a greenhouse—'[I] believe I shall build one this spring; my orange trees thrive so well they deserve one. I propose having it 26 feet by 13 and 13 high, and a room under it...that will open into a little back garden, which I intend to make my menagerie'.[27] She was concerned that the fumes from the chimney could harm the orange trees, but, in any case, decided to proceed with the menagerie only. She did however have a greenhouse for 'blowing auriculas'.

From her garden, Mary Delany gathered melons, pears, grapes, filberts and walnuts. She picked baskets of honeysuckle, jasmine, gillyflowers and pippins. Thus Delville contained the same mixture of gardens—kitchen, flower, orchard—that Joseph Addison described, and it was one of the gardens in Ireland, along with those of Jonathan Swift, Lord Orrery and Dr Samuel Madden, that were precursors of the romantic-poetic style. In 1750, Mary Delany described it thus:

> D. D. up to his chin haymaking on the lawn under my closet and the whole house fragrant with the smell...the garden is Paradisiacal.[28]

Dr Delany died in May 1768 and was buried near Thomas Tickell in the graveyard of Glasnevin Parish Church. His widow did not care to remain in Delville and she returned to England. There she lived full and active elder years as a close friend and companion of the Duchess of Portland. The Duchess kept a famous *salon*, which included naturalists such as Daniel Solander, and botanical artists like Georg Ehret and James Bolton. Between 1772 and 1782, Mary Delany executed her most lasting monument, a series of nearly one thousand pictures of wild flowers and garden plants. They are paper mosaics, the leaves, stems, petals and even rose thorns carefully and delicately cut from coloured tissue paper and then glued down. These strange works of art stimulated Dr Erasmus Darwin to eulogise Mrs Delany in his long poem, *The Loves of Plants*:

> So now Delany forms her mimic bowers
> Her paper foliage, and her silken flowers;...
> Cold Winter views amid his realms of snow
> Delany's vegetable statues blow;
> Smooths his stern brow, delays his hoary wing,
> And eyes with wonder all the blooms of spring.[29]

Mary Delany was adopted by King George III and Queen Charlotte after the death of the Duchess in 1785; they gave her a house at Windsor where she spent her final years.

It is pleasant to speculate that the committee of the Dublin Society was influenced in trying to procure Delville for a botanical garden by the lingering reputation of the garden created by the Delanys, perhaps literally by the 'genius of the place'. It is sad that the final remnants of the garden, the temple, in which there was a medallion with Jonathan Swift's Stella portrayed on it, and the punning motto *Fastigia despicit urbis*, should have been demolished in 1951. A modern hospital now occupies the site that was

> ...happy Delville, blissful seat
> The Muses's best belov'd retreat
> With prospects large and unconfined
> Blest emblems of their master's mind.
> Where fragrant gardens, painted meads
> Wide opening walks and twilight shades—
> Inspiring scenes!—elate the heart!
> Nature improved, and raised by Art,
> So Paradise delightful smil'd
> Blooming, and beautifully wild.[30]

But this paradise did not remain fashionable. By the end of the eighteenth century, it seemed staid and old-fashioned. Although in 1791, J. C. Walker described Delville as the first attempt at modern gardening in Ireland,[31] Andrew Caldwell commented to James Smith two years later that 'much of the old fashion'd Garden remains and is poetry in its kind, it is a sort of Classic Ground, having been admir'd and frequented by Swift and all the People of Literature of that day'.[32] Walker had commented that at Delville straight lines had been 'softened into a curve, the terrace melted into a swelling bank and walks opened to catch the vicinal country'. To T. K. Cromwell in 1820, however, there was still 'much of the stiffness of the old style; the walks in straight lines terminating in little porticoes, and the valleys...crossed by regular artificial mounds'.[33]

Unlike Mrs Delany, Mrs Tickell continued to live at Glasnevin after her husband's death in 1740. In 1754, the lease with John Putland was renewed and extended for a further eighty-one years, making it a ninety-nine year lease. On 24 October that year, Clotilda Tickell handed the property to her eldest son, John, for a yearly rent of a peppercorn and £37 9s. 0d. However, she arranged, before the deeds were signed, that she would

> reserve and lock up for her own use as she shall think fit any three rooms in [the] house, except the Great Room and the right hand parlour and the study, and...shall also have the vault therein lying, even with the wine cellar and the liberty of having keys to both gardens to resort to and from the same as she shall think fit.[34]

Mrs Tickell then moved to a house on Saint Stephen's Green, and John Tickell and his wife Ester lived at Glasnevin. On 29 January 1761, John Tickell was proposed for membership of the Dublin Society but was not elected; this appears to be the first instance of a candidate being rejected by the Society.[35] In 1765, he was given a post at Windsor Castle and moved to England with his family. The Glasnevin house reverted to his mother. Some time later, perhaps on Clotilda Tickell's death, the house and gardens were leased by John Kiernan.

5
The formation and early years of the Botanic Gardens, 1795-1818

THE SIXTEEN Irish acres (twenty-seven statute acres) at Glasnevin taken over by the Dublin Society on 25 March 1795[1] were gently hilly. From the highest point the land sloped northwards towards the River Tolka and the millstream which formed one part of the boundary. Trees, including the ancient avenue of yews, were scattered throughout the demesne. The soil, a thin loam overlying gravel, was well-drained, and contained lime.

The house was in slight disrepair when acquired. The architect Edward Parke, whose patron was John Foster, superintended necessary work on it during the first few months.[2] No detailed plans of the house survive, but there is a brief description in the Tickell deeds and some sketchy drawings of it on early maps of the Botanic Gardens. It had two storeys and a basement. The main entrance faced the road, and steps led from the door to a sloping lawn within a walled garden. South of the house, the garden contained some ornamental flower beds. The interior of the house included many features associated with Dublin's Georgian villas. The main room on the ground floor was decorated with elegant plasterwork, including a frieze composed of pipes, lyres and other Doric emblems of the Muse.[3]

Between 1790 and 1795 the Irish Parliament had provided a total of £2,200 for the formation of a botanical garden through successive Dublin Society Acts, and by the end of February 1796 the Society had spent over £500 on work in the grounds and on the house.[4] Then the main task of developing and planting the Gardens could be continued. At a meeting on 7 April, a plan for regulating and arranging the botanic ground was laid before the Society, which ordered that this plan should be printed and circulated. The Society's Committee of Agriculture was instructed to request Walter Wade to act as Professor and Lecturer in botany 'as far as a knowledge of the vegetable products and their qualities may tend to promote agriculture, arts and manufactures'. Dr Wade was also invited to undertake the arrangement of the new garden. Of course, he agreed to this not unexpected invitation.

The Committee of Agriculture had a large poster printed (Fig. 23) giving details of the planned garden and even the method of labelling the plants was outlined. It was stated that lectures would be given on the Cattle and Hay gardens 'for the instruction of the common farmers, their servants, or labouring men', all of whom would be admitted to the lectures gratis. Other lectures were proposed on botany at large. A library stocked with botanical and agricultural books was to be provided. Part of the garden was to be set aside for experiments with agricultural machinery, especially the ploughs and drill machines then kept in the Society's Repository. The Dublin Society invited people wishing to encourage so useful an institution by stocking the botanical garden with plants to do so by sending seeds or living plants either to Walter Wade at his home in Capel Street, or to the Botanic Gardens, or to the Society's Repository.[5]

The Botanic Gardens was not intended to be a rarified scientific institution for the sole delight of botanists, but to be a garden in which agriculture had a prominent place. This was clearly and deliberately spelt out at the head of the poster—the Botanic Gardens was 'for promoting a scientific knowledge in the various branches of agriculture'. This reflected the strong influence of John Foster, who was shortly to take a leading role in the formation of the Farming Society of Ireland, and who, from this time onwards, took a close personal interest in the day-to-day running of the Gardens. His permission and approval were frequently sought. He also had a substantial say in the appointment of staff. Walter Wade's professorship was just one of the many actions which displayed John Foster's influence.

THE DUBLIN SOCIETY

HAVING taken Sixteen Acres of Ground, at GLASNEVIN, for the Purpose of forming a BOTANIC GARDEN, pursuant to Act of Parliament, for promoting a scientific Knowledge in the various Branches of Agriculture, have made some Progress in laying them out, pursuant to the following Report from their Committee of Agriculture.

The GARDENS at GLASNEVIN to be laid out as follows:

1.

ITS LINNÆENSIS,
to three Parts

The HERBACEOUS, *(Herbarium.)*
The SHRUBS, *(Fruticetum.)*
The TREES, *(Arboretum.)*

EACH Plant therein to be arranged according to its Class, Order, Genus and Species, beginning with the first Class and proceeding regularly to the last Class of CRYPTOGAMIA, for which a separate Division of Ground is to be allotted.

IN each of these Divisions every Plant is to have a painted Mark affixed to it, which is to shew—the Number in the GLASNEVIN CATALOGUE,—The Class and Order—the Generic and Specific Name, all in Black on a White Ground, and the English Name in Red.

WHEREVER a Genus contains Herbs and Shrubs, or Trees and Shrubs, a Mark will be placed in its proper Order in the HERBARIUM and FRUTICETUM, referring from one to the other, and so in the ARBORETUM, in order to shew the regular Continuation of the System; and in like Manner wherever in the HERBARIUM any Class or Order is omitted, as not containing any Herb, or any Herb not hardy enough for the open Air, a Mark will be fixed in its proper Place, to shew why it is omitted.

IN the ARBORETUM, which is proposed to occupy the West and South Sides of the Ground, and to form a Skreen of about five or six Perches wide, with a broad Gravel Way through the Centre, and the Grass kept as fine as a Bowling-Green, the Trees are to be planted from twenty to thirty Feet apart, and where there is a very delicate or choice Species, two may be planted, lest one should fail; the intermediate Spaces are to be filled with Fir, Larch, Laurel, Elm, &c. for shelter, which are to be cut away when they come to interfere with the LINNÆAN Plants, or are useless as Nurses, always taking Care that the Nurses be as distinct in Appearance as possible from the Species they are planted to protect, as *Deciduous* for Ever-greens, and *Vice versa*.

LINNÆUS, AITON, &c.—do not notice Varieties in general, but in this Garden, every Variety, even those that are merely Seminal, and all Variegations must be arranged in their proper Places.

This Garden is calculated for the scientific Botanist, who studies the Plants systematically.

2.

The CATTLE GARDEN.

THE next Garden is the Cattle Garden or *Pecudarium*, which is to consist of five Divisions, as follows:

1. The SHEEP DIVISION, or *Hortus Ovinus*.
2. The HORNED CATTLE DIVISION, or *Hortus Bovinus*.
3. The HORSE DIVISION, or *Hortus Equinus*.
4. The GOAT DIVISION, or *Hortus Hircinus*.
5. The SWINE DIVISION, or *Hortus Suinus*.

EACH of these is to be laid out in regular Beds, with Alleys three Feet wide between each, and with a Gravel Walk nearly in the Centre, across the Beds; on the one Side of this Walk are to be arranged in LINNÆAN Order, all Plants which the Animal to which the Division is appropriated, is fond of eating, and which are wholesome Food for it, and also all Plants which it is not fond of eating, tho' not unwholesome; on the opposite Side of the Walk are to be arranged all Plants which the same Animal will eat, but are injurious to it, and likewise all which it refuses to eat, whether injurious or not.

THE HERBACEOUS Plants and Shrubs to be kept in each Arrangement distinct, whereby a useful Shelter will be gained in many Parts.

EVERY Plant is to have a like painted Mark to it, as before described, and if a Native, N. is to be painted on the Back of the Mark.

3.

The HAY GARDEN.

THE next Garden will be the Meadow Division, containing all Plants of which Hay can be made, arranged according to their Times of being fit for cutting, placing on one Side of the Walk those that are valuable, and on the other, those that are the least useful, for the Scythe.

THESE Hay and Cattle Gardens are proposed for the Instruction of the Practical Husbandman; he will there see every Plant, Shrub and Weed which grows in Ireland; he will see at once, what are useful, what otherwise, for each Animal; he will learn how to weed his Meadows and Pastures, how to select the Hay Seeds which should be sown together, and what Weeds on his Ditches or Tillage Grounds he should be most anxious to prevent feeding; and the most illiterate Man is capable of Instruction from these, by being told what is the Description of the Division he looks at.

Catalogues are to be prepared for each Division, with a short Account of the Qualities of each Plant, and a reference to all the Authors who treat of it:—A complete Collection of which Authors is proposed to be procured, and kept in the Lecturing-Room, or Library there.

A LARGER Meadow Garden to be laid out in a separate Part of the Ground, where there should be Plots of all the Hay Grasses, quite distinct and sufficiently large to mow, so as to make Experiments for afforting those Grasses together, which require equal length of Time in being made into Hay, and to save the Seeds of each distinct for Curiosity, or Sale.

4.

The ESCULENT GARDEN.

THE next Garden will be the ESCULENT one, or ESCARIUM, which is to contain every Plant that furnishes Food to Man, arranged in Divisions as follows:

1. Those whose *Roots* furnish Food, wholly or principally.
2. Those whose *Stocks* or *Leaves*, ditto.
3. Those whose *Flowers*, ditto.
4. Those whose *Seeds*, ditto.

AND for this Garden, like painted Marks and like Catalogues are to be prepared, and the various Modes and Seasons of Culture noted.

5.

The DYERS' GARDEN.

THE next will be the Dyers' Garden, wherein all Plants, which afford any assistance in dying Colours, will be arranged, according to the Colours they dye, with like Marks and Catalogues.

6.

THE NEXT WILL BE A GARDEN OF
SAXATILE, or ROCK PLANTS.

7.

THE NEXT WILL BE ONE FOR
CREEPERS and CLIMBERS.

8.

THE NEXT FOR
BOG and WATER PLANTS.

9.

THE NEXT FOR
MARINE PLANTS.

10.

THE NEXT WILL CONTAIN A SEPARATE COLLECTION OF ALL
VARIEGATIONS of every Tree, Shrub and Herb.

11.

The NURSERY.

THE next will be the Nursery, where the Propagation of all the choicest Kinds will be attended to, and the different Modes of Layering, Grafting, Innoculating, &c. exhibited for Instruction.

IT is proposed that there shall be a Professor, who shall give Lectures on Botany in general; and also separate Lectures on the Cattle and Hay Gardens, for the Instruction of the common Farmers, their Servants, or Labouring Men, all of whom are to be admitted to the Lectures gratis, on the Order of a Vice President, or the Treasurer, Secretary, or Professor.

THAT like Lectures be given on the Garden for Dyers' use, and that for the purpose of extending Practical Knowledge, particularly in Husbandry, Samples and Seeds be allowed to be given, and even Plants, where they can be had who may wish for them.

THE Lectures on Botany at large, to be given during the Season when the Generality of Plants are in Flower, for the better Demonstration of the Sexual System. And the Professor to be allowed the use of the House and Gardens for demonstrating, and to take Pupils, and receive Subscriptions.

A HORTUS SICCUS to be formed, containing as well Specimens of the Plants in the Garden as of all others throughout the World, which can be procured.

IN Time, it is to be hoped, that the Society may be enabled to send Persons round the Kingdom to explore its Vegetable Products, so as to form a HORTUS, and a FLORA HIBERNICA, and they propose hereafter to extend their Collection of Plants to the Green-House, and afterwards to the Hot-House Assortments; until which time the HORTUS SICCUS, and Drawings must answer as to such Plants, for the Purposes of Curiosity or Instruction.

A MEDICINAL GARDEN has been in Contemplation, but no Plan is yet fixed for it; it is also in Contemplation to furnish their House in GLASNEVIN, with a full Library of all Agricultural and Botanic Books, and to set aside a Part of the Ground for Experiments in ploughing, by trying the Excellence or Defects of the various Ploughs, Drill Machines, &c. in the Repository.

WHEN the Gardens shall be furnished, Regulations must be made for the Admission of Persons, in which it is proposed, to make the Admission as general as it can be with Safety, and to have Persons properly instructed, to attend those who may wish for Information.

THE cultivating a Sufficiency of Medicinal Plants for Sale, has been in Contemplation, particularly of those which it is difficult or the Shops to procure; but this not being so immediately within the Purpose of the Society's Institution, is postponed till the several other Matters are established.

THE SOCIETY having resolved that their COMMITTEE of AGRICULTURE do, in their Name, request Dr. WADE, the Author of the FLORA DUBLINIENSIS, to undertake the Arrangements of the Plants, and to act as their Professor and Lecturer in BOTANY, so far as a Knowledge of the Vegetable Products, and their Qualities may tend to promote Agriculture, Arts or Manufactures; and the Committee having accordingly applied to him, and obtained his Compliance, they have now the Satisfaction of the Aid of his great Knowledge and Abilities, to promote and complete the Undertaking.

AS a great deal of the Ground is already procured, the SOCIETY request the Assistance of all Persons who wish to encourage so useful an Institution, by sending in such Plants and Seeds, as their several Collections, or their Neighbourhood can furnish. It is requested, that all who shall be pleased to send any, will order them to be delivered to Dr. WADE, at his House in Capel-street, or at the GLASNEVIN GARDEN, or to Mr. BRIEN, their Register, at the REPOSITORY, in Poolbeg-street.

DUBLIN: PRINTED BY RICHARD EDWARD MERCIER AND Co. 1796.

23. Large poster printed for the Dublin Society, announcing the formation of the Botanic Gardens. (Reproduced from the only extant copy, which is damaged, preserved in the Foster/Massereene Papers (D562/7829C), by courtesy of the Public Record Office of Northern Ireland, Belfast, and the Viscount Massereene and Ferrard)

Walter Wade (Fig. 24) had not been idle between 1790 and 1796 in securing his right to the professorship. He had coveted the post from the beginning, for he was an ambitious man, determined to make his mark in Ireland especially in botany. Even before the presentation of the petition to the Irish Parliament in 1790, he had established himself as the foremost local botanist with the lavish though stillborn *Flora Dubliniensis*. Both in the proof-pages of this book and in his public lectures, he showed his inclinations towards a practical approach to botany, emphasising the uses of plants in medicine and agriculture. In 1792, Walter Wade was elected an honorary member of the Dublin Society. In December 1793 he sent a letter to the Society informing members that he had been studying the plants of County Dublin, and he enclosed a manuscript of his observations, which so impressed the Council it ordered that it should be published as a book. *Catalogus systematicus plantarum indigenarum in comitatu Dublinensi inventarum* (1794) was a work for the learned botanist, familiar with Latin; not only was the title and the text in Latin but so was the dedication to *Viro Praehonorabili Joanni Foster*. It listed the native plants of Dublin county and gave their Latin, English and Irish names. The Dublin Society had five hundred copies printed.[6]

In March 1795, Wade presented another manuscript to the Society. This one dealt with forty-one species of grasses from County Dublin, including their agricultural, economic and other uses. Inserted into this manuscript were dried and pressed specimens of each species. Again the Society was impressed and applauded its honorary member. The Council ordered that the volume should be placed in the library at the Society's premises in Hawkin Street.[7] On 2 July, Wade gave the Society a set of the proof sheets of his *Flora Dubliniensis*, including the illustrations.[8] Thus he endeared himself to the Dublin Society, made his name well-known, and established himself as the only possible choice for the post which the Society would eventually create at the head of the new botanical garden. His patron, John Foster, simply ensured that nothing went amiss.

Dr Wade undoubtedly took on the task of laying out the grounds at Glasnevin with enthusiasm. No account of the work survives but it is obvious that Wade was in general charge and that he was responsible for the original design. The Council of the Society released substantial sums to the Committee of Agriculture from the government grants for development work at Glasnevin; for example, in May 1796, permission was given for the Committee to withdraw up to £700 for work during the summer recess. On 1 September 1796, £300 was granted for plants and their transport to the Gardens.

The first consignments of plants came from several different sources. Before the end of 1796, trees had been obtained from Cork and other plants arrived from London. John Grimwood, a nurseryman of Charlemont Street, Dublin, supplied sundry trees in May 1797. In the same month, Dr Wade was allowed to spend £100 on travel to find plants, and on employing other people to collect or buy plants directly from nurserymen.

The invitation to the public to support the Gardens by giving plants bore fruit. In November 1797, Charles and Luke Toole, who ran a nursery at Cullenswood, said that they had 'a great number of curious plants' which they believed were not in the Gardens, and they invited the Society to send someone to select specimens. Dr Wade was asked to do this. He was also told to contact Edward Hodgins of Dunganstown in County Wicklow, who had offered plants. The Right Honourable David La Touche, a member of a well-known Dublin banking family who had a fine garden at Marlay in County Dublin, offered to donate several very valuable and scarce plants. Nurserymen who gave plants included Richard Burnet of Richmond, and Messrs Simpson and Sibbert of College Green. Plants were also purchased: Patrick Donegan, who had a seed shop in Christchurch Lane, was paid over twelve pounds for seeds supplied between 1795 and 1799. Those who gave plants were thanked for their patriotism.

John Foster did some of the work of getting plants himself. He was not diffident about asking for gifts from higher authorities and in July 1796 sent a letter to the Lord Lieutenant, Lord Camden:

> My Lord,
>   If I take too great a liberty in mentioning to ye a subject in which I have no other interest than having been the promoter of it for a publick benefit, I trust you will impute it to my presumption.

A botanic garden has been prepared by the Dublin Society chiefly with a view to agricultural knowledge and to an acquaintance with vegetables so far as they are connected with arts, manufactures and rural oeconomy, but in order to promote that knowledge, it will be necessary to have an arrangement for the whole system of botanic science. Their garden is now ready for the reception of plants and if His Majesty would be pleased to order us such plants from his garden at Kew, as can be spared from thence, we should not only be saved a large expence, but be supplied with many, which money could not procure and by obtaining so speedy a collection be enabled to bring our whole plan into use immediately.

My reason for mentioning this to your Excellency now is that we have an intention of sending over in August a skillful person to collect from the nurseries in England as far as our limited funds will allow and if His Majesty should be graciously pleased to promote our undertaking he would attend the gardener to receive the plants and the necessary instruction as to their lives and culture.

If this suggestion should not meet your Excellency's full approbation I entreat you will not think a second time about it.

I have the honour to be with the greatest respect your Excellency's very faithful and obedient servant

John Foster.[9]

The Lord Lieutenant probably did not often get such innocent requests from the Speaker. He forwarded the letter to London with a covering note, saying that he felt he could make no comment upon the request but adding that

The promotion of botanical and agricultural knowledge in a country which, tho' extremely capable stands much in need of improvement, is the only public ground upon which the Dublin Society can presume to request this mark from His Majesty's liberality, and the only excuse I can make in being the channel for this communication.[10]

Alas, it is not known if plants were received from the Royal Garden at Kew, which, under the management of Sir Joseph Banks and William Aiton, contained the finest botanical collections in Europe. Banks had garnered plants from all over the world, especially from the new colonies at Port Jackson in New South Wales, Tasmania and the Cape of Good Hope.

Walter Wade was in sole command at Glasnevin from 1796 until 1798. He had labourers to carry out the manual work but no head gardener was appointed until September 1798. The English horticulturist and publisher, William Curtis, recommended John Underwood for this position after receiving a request from John Foster. Underwood's early life is obscure. It is said that he was Scottish, like several of his successors.[11] The earliest known mention of him is his election on 19 January 1796 as an associate of the Linnean Society of London. On his nomination paper (Fig. 25), Underwood was described as being 'of Brompton Middlesex'. It is possible that he was working at that time in one of the local large gardens, or more likely in William Curtis' own nursery. The gentlemen who proposed him to the Linnean Society stated that John Underwood was 'highly skilled in the knowledge of plants and well versed in their culture'.[12] Had he been otherwise, it is unlikely that he would have sought election to that Society, or, indeed, been elected. His nomination paper was signed by William Curtis, James Fairburn who was curator of the Chelsea Physic Garden, and by James Dickson. All three were among the founders of the Society, and in 1803, James Dickson helped to establish the Horticultural Society of London. He was a prominent nurseryman and botanist, and is commemorated in the Australian genus of tree ferns *Dicksonia*. With such impressive sponsors it is surprising that so little can be learned of Underwood's London years. He arrived at Glasnevin on 25 October.

Undoubtedly Underwood knew the London nurseries well. Two months after his arrival in Glasnevin, the first consignment of plants arrived from James Lee and John Kennedy of Hammersmith.[13] The Dublin Society paid over £370 for the plants, a very considerable sum and greatly exceeding anything paid to Irish nurseries. Lee and Kennedy were famous for their exotic plants, especially the many new species which they raised from seeds collected in Australia. They were among the first to offer Australian plants for sale, and also marketed *Fuchsia* in England for the first time. It is probable that many of the southern hemisphere plants such as *Protea* and *Erica* from southern Africa, and *Banksia* and *Leptospermum* (Plate 1) from Australia, that were grown at Glasnevin in its early years came from Lee and Kennedy. In those years the Dublin Society spent about £1,000 on plants for the Gardens, three-quarters of this going on orders filled by Lee and Kennedy.

24. Dr Walter Wade (a portrait in oils attributed to Allen Ramsay)
(Reproduced by courtesy of G. A. Kenyon)[43]

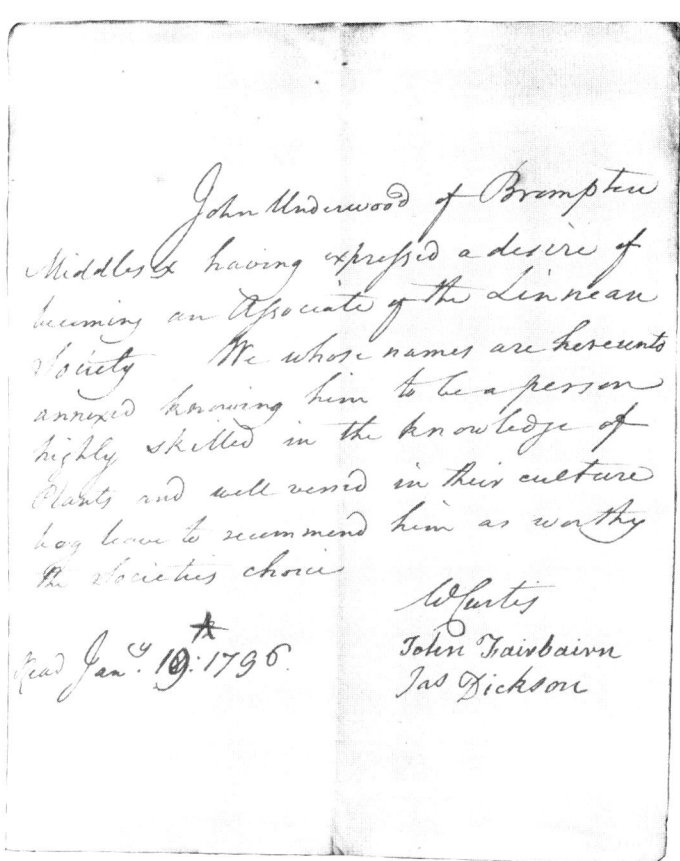

25. Nomination paper for John Underwood, A.L.S., dated 19 January 1796. (Reproduced by courtesy of the Linnean Society, London; photograph by Dr Edward Diestelkamp)

As head gardener, John Underwood relieved Walter Wade of some of the responsibilities of establishing this new institution. Unlike Wade, who lived in the city, Underwood was provided with a house at Glasnevin. His salary was sixty-eight pounds per year, with free coal and candles. The house was furnished and much of the basic equipment was provided by the Dublin Society. Underwood was a bachelor when he came to Ireland. In 1799, he married Mary Newcommen in Dublin.[14] The couple eventually had seven children all of whom were baptised in Glasnevin Parish Church. Some of the infants died young and were buried in the churchyard.[15]

An undergardener had been appointed in 1796 on the recommendation of John Foster. He was John White, a native of County Louth, familiar with the Irish language, and, like Underwood, a Protestant.[16] As he was married and had a young family when he came to live at Glasnevin, White was given a newly built thatched lodge to live in. This was furnished by the Society in December 1799 at a cost of £29 2s. 0d: an account survives for beds, blankets, sheets, quilts and bolsters as well as two mahogany tables and nine chairs. The Whites were also given saucepans, a frying pan, a kettle, a grid-iron, fire-irons, a pair of bellows and two candle-sticks.[17] John White's salary was fifty pounds per year with coal and candles provided free of charge. White's origins are almost as obscure as Underwood's, but it is possible that he was trained at Oriel Temple, John Foster's garden. Later correspondence between John White and his patron suggests the two men were well acquainted.

As well as the construction of lodges for the gardeners, the boundary wall along the main road was rebuilt before 1800, and a shed was erected for parking the horse-cart used in the grounds. In 1799 work was begun on a range of glasshouses, designed by Edward Parke, which would cost an estimated £3,000. The glasshouses occupied the walled garden north-east of the main house, and were linked to the house by a glazed corridor thirty-seven feet in length, with folding doors allowing direct access to the range from the lecture room. The range consisted of a large hothouse, sixty feet long and twenty-five feet wide, flanked by four smaller houses, two on each side. These smaller houses were also sixty feet long but only sixteen feet wide (Figs. 26a and 26b).[18] The houses were built of wood, and joined by a continuous backing shed which contained furnaces supplying heat for each house. The shed was roofed with wood. The final cost of the range was £2,877 19s. 6d.

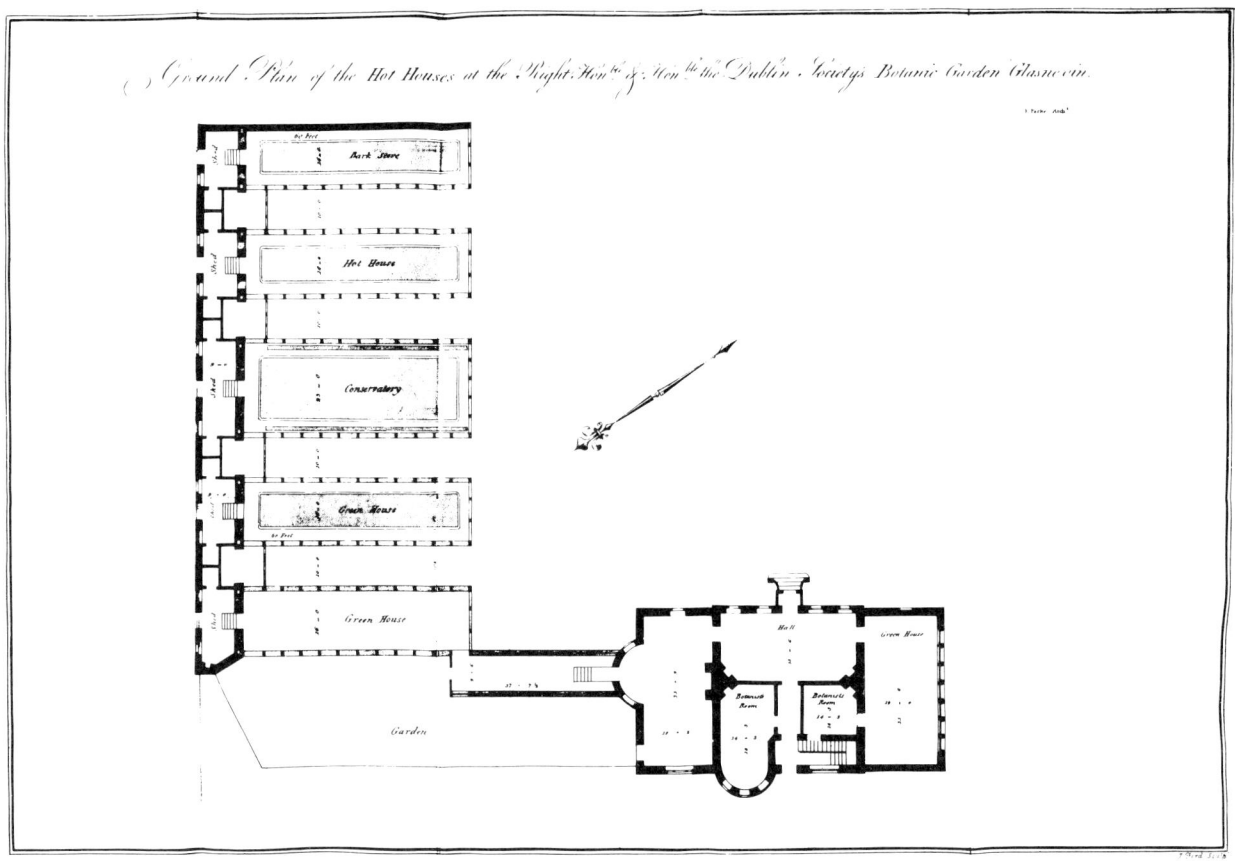
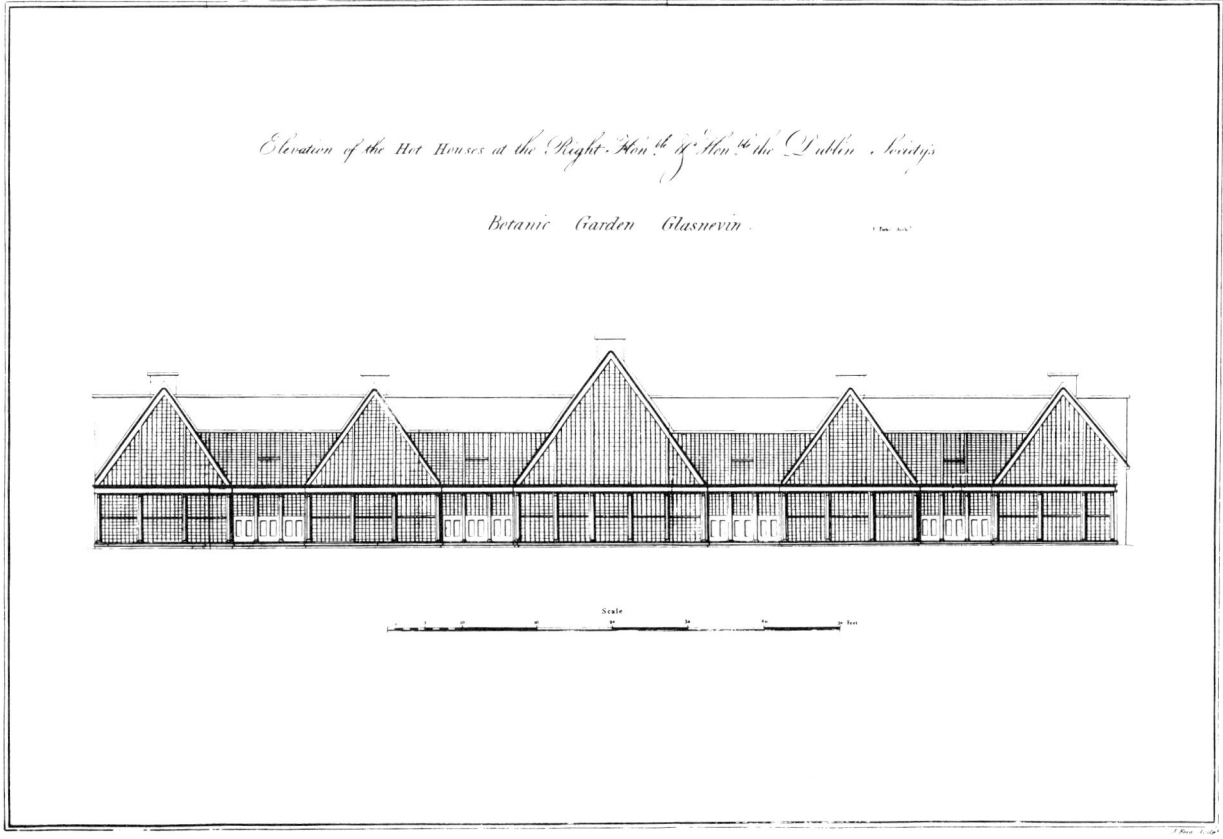

26. Plan and elevation of the first range of greenhouses from *Transactions of the Dublin Society*

Each house was devoted to a different group of plants. The central conservatory contained tender trees. There was a dry stovehouse for succulents including species of *Crassula, Aloe, Cotyledon* and *Mesembryanthemum*. Another stovehouse contained about four hundred species from the humid tropics. Larger plants were grown in the third greenhouse. There were about one hundred and fifty species of *Erica* in the fourth house in 1800, mostly from the Cape of Good Hope.[19] These glasshouses proved most unsatisfactory. They were difficult to heat, requiring fifty tons of coal a year. The houses were too close together and were aligned the wrong way, with their narrow sides facing south and the long sides east-west; thus, each house obstructed the light from its neighbour.

While the tender plants were housed indoors, the hardy species were gradually planted out in the grounds. Some of these were not tolerant of the limey soil, so 258 loads of bog earth were imported to provide peaty conditions. The paths which separated the various sections were gravelled with 457 loads of sea and rock stone. In the centre of the Gardens a rockery was built for the rock, saxatile and alpine plants. This took the form of a mound with spiral paths not visible to the eye winding round the sides. From the top of the mound, visitors were 'gratified with a view of the whole garden'. The rocks for the mound were brought from Howth, having been carefully selected for the varieties of mosses and lichens which were already growing on them. When piled together on the rockery, it was hoped that the minute plants would flourish and therefore add to the botanical collection. Rock plants and ferns were grown in the soil-filled pockets between the rocks.[20]

The Botanic Gardens at Glasnevin was ready to be opened to the public and members of the Dublin Society in May 1800. It was an institution about which Dublin, and Ireland, could boast. It was the largest publicly supported botanical garden in Europe and the largest in the British Empire apart from a seventy-acre garden near Kingston in Jamaica. Glasnevin was 'next to this in size and situation, and not perhaps inferior in beauty'.[21]

At first, members of the Dublin Society and the general public were freely admitted, but the Society was compelled to place a notice in local newspapers for one week in June 1801 stating that in future admission for non-members would be by ticket only. This was precipitated because 'idle persons and particularly children' had got into the Gardens, 'done considerable mischief' to the plants and disturbed other visitors. Daily tickets were obtainable from Dr Wade or from the vice-presidents, who included John Foster. Non-members could also apply for entry to the head gardener, but he was obliged to accompany such visitors around the gardens personally. Henceforth, children were admitted only if they were accompanied by their parents. Dogs were, and still are, totally forbidden. Strangely, visitors could remove specimens with the consent of the head gardener. Copies of the catalogues of the various sections of the Gardens were available at the gate-lodge, free of charge, to any person who requested them. Visitors were forbidden to give money to the garden staff for 'admission, attendance, specimens, flowers or on any other account'. They were invited to put donations in a box kept in the gate-lodge, the proceeds from which were placed in a charitable fund to support either gardeners who had retired from active life due to age or infirmity, or the wives and children of men who had died while employed in the Gardens.

In 1800, only a portion of the sixteen Irish acres was developed (Fig. 27, Map Colour Plate 1). Two large areas remained unplanted and were kept as pasture possibly for grazing the horses which were used for the heavy work. These areas may have been intended for use as demonstration plots for ploughing, but they were soon incorporated into the botanical sections and planted. A screen of trees was planted along the south and west boundaries of the Gardens to shelter the site and hide the rough pasture and the city from view. Within this screen, along both sides of a long gravelled walk which circled the perimeter of the Gardens, were the Arboretum and shrubberies (Fruticetum). These contained, in botanical order, many species and horticultural varieties of trees ranging from pines and spruces, to deciduous oaks, ashes and beeches, including the copper beech. Among the more unusual plants were the maiden-hair tree (*Ginkgo biloba*) from China and *Sophora tetraptera* from New Zealand. There was also a supposed new species of horse chestnut (*Aesculus*). Each class, defined according to Linnaeus' Sexual System, was contained within a separate plot surrounded by a gravel path. The Fruticetum adjacent to the Arboretum contained about a thousand different shrubs although

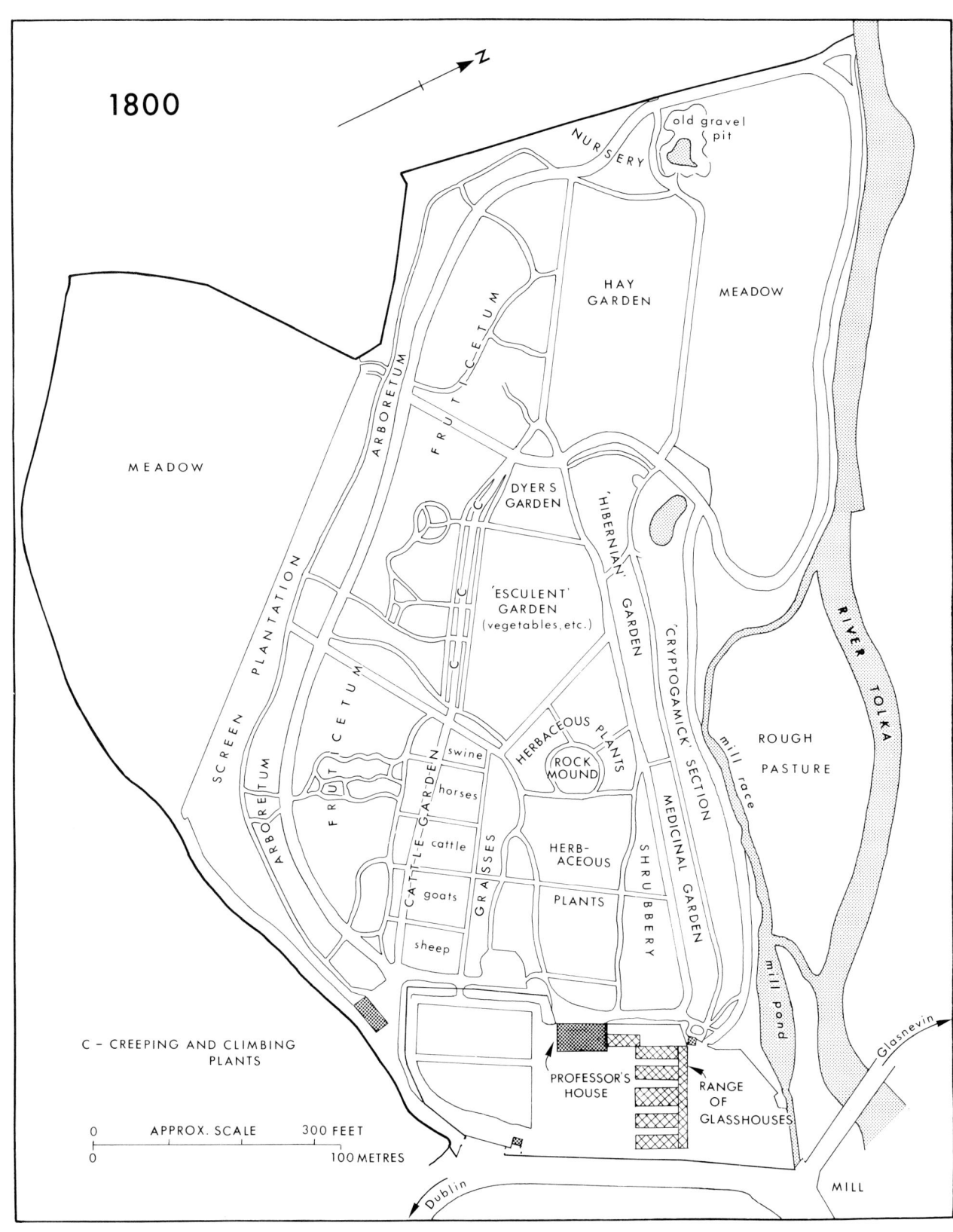

27. Botanic Gardens in 1800 (redrawn by Mary Davies from the original—see Map Colour Plate 1)

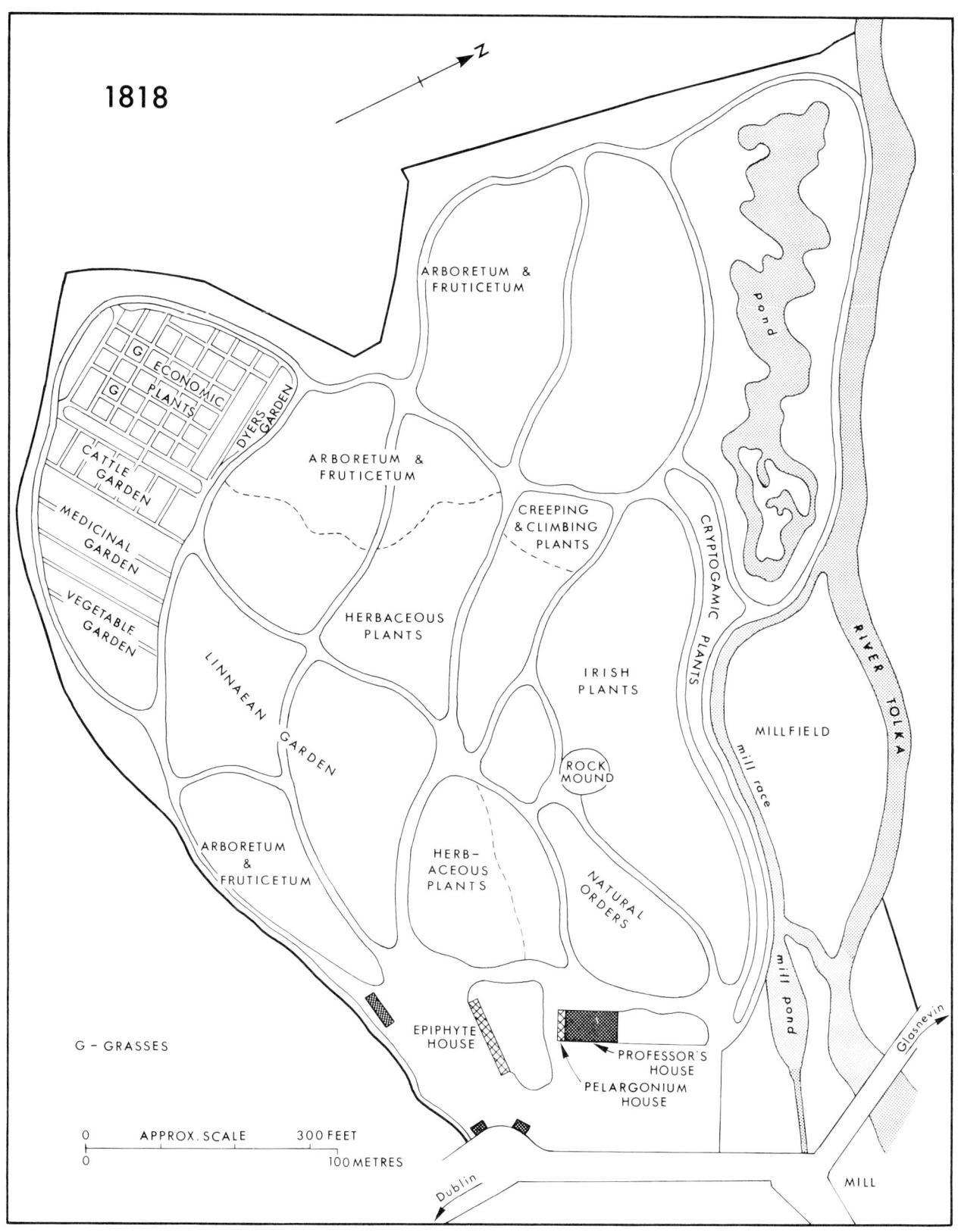

28. Botanic Gardens in 1818 (redrawn by Mary Davies from the original—see Map Colour Plate 2)

29. The pond in the Botanic Gardens, looking eastwards, photographed in the summer by W. D. Hemphill, c. 1897. *Taxodium distichum* is on the left and *Pinus nigra* in the background (centre right) (National Botanic Gardens)

some plants in the shrubberies would eventually become trees, including the various forms of strawberry tree (*Arbutus unedo*) and the willows. About one hundred different roses, both species and garden varieties, were in the Fruticetum, as well as thirteen different hollies. Among the rarities in 1800 — many probably from Lee and Kennedy — were *Fuchsia coccinea* (in fact probably the common fuchsia, *F. magellanica*) and *Buddleja globosa*. Plants which were unsuitable for cultivation in the limey soil abounded: there were many species and varieties of *Rhododendron, Erica, Andromeda* and *Vaccinium*. These were grown in peat beds. At the eastern extremity of the Arboretum was a nursery, situated beside a water-filled gravel pit. On the hill slope and level ground by the river was a meadow and the Hay Garden, which was divided into 150 plots containing plants that were either suitable or unsuitable for making hay. Between the entrance gates and the Hay Garden was the Cattle Garden; variously shaped plots contained collections of edible plants, with a companion plot in which poisonous species were grown. In the first pair, plants edible by sheep were grown alongside those which were injurious to sheep. The succeeding plots contained plants affecting goats, cattle, horses and swine. There were several plots between the Cattle Garden and the herbaceous section which were filled with grasses including cereals such as barley, wheat and oats. The Herbaceous Grounds encircled the rocky mound and contained a botanical arrangement of herbaceous plants including garden varieties. Beyond this, bordering the path to the Hay Garden, there were poles supporting climbing and creeping plants such as *Clematis*. On the river side of these lay the Dyer's Garden with plants from which dyes could be extracted, and the Esculent Garden which contained edible root vegetables like turnips, carrots and parsnips, as well as peas, cabbages and fruit. Along the top of the steep slope which fell to the millstream were long plots containing medicinal herbs and Irish native plants. On the slope was the Cryptogamic Division, the contents of which are not recorded in the 1800 prospectus — it was intended to include species of mosses and ferns, but this section of the Gardens was not a success. Despite the considerable effort made during the first few decades to establish a collection of ferns, by 1818 the Cryptogamic Division was described as 'very incomplete'.[21]

30. The pond, looking westwards, in winter flood. This photograph cannot be dated but was probably taken about 1910. It is printed from a glass-plate negative in the Gardens' archives. In the centre is *Pinus nigra*, planted about 1798 – as this tree collapsed on 20 January 1985 it is no longer a feature of the pond-side planting

According to the original prospectus,[22] near the nursery there was a collection of variegated plants; at that time 'variegated' meant any unusual form, including plants with mottled leaves or prostrate forms of otherwise erect shrubs. The prospectus also indicated that an area was being prepared for marsh and aquatic plants. Most astonishing, however, was the proposed aquarium for marine plants, which would have been unique had it been constructed.

The original layout of the Gardens was modified substantially within the next eighteen years. Comparison of the first map (Map Colour Plate 1 – Fig. 27) and the second (Map Colour Plate 2 – Fig. 28) shows that the paths were realigned especially in the centre of the Gardens near the rocky mound. Of course changes have taken place throughout the whole history of the Gardens. The present layout of paths took form in the 1830s and 1840s under Ninian Niven and David Moore.

The first major development after 1800 was the excavation of the pond (Figs. 29, 30). John Underwood reported progress to John Foster in December 1809, noting that owing to the fine autumn weather he had managed to excavate part of the low ground 'for our piece of water, but within these few days, the back water has drove me from sinking, althow about six days more would have enabled me, for to have had the water stand, about three feet depth'.[23] The pond was two hundred yards long, with an island and promentories from which visitors could get good views of the aquatic plants. The sides were properly dressed. John Underwood told Foster that he considered the pond looked very well. It was situated in the north-western part of the Gardens in an area originally left as meadow. As the pond was excavated below the level of the river, and was in the lowest part of the site, it filled naturally by seepage without requiring any channel leading from the river.[24] Marsh plants were grown around the edges of the pond and aquatic plants flourished in the water.

Another major alteration was the removal of the Vegetable, Hay and Cattle Gardens to the unused area in the south-western corner. This is the site which the Vegetable and Lawn Gardens and the oak collection occupy today. The grasses, medicinal and dye plants were also brought to this area, and the collections were grown in regularly arranged rectangular beds, separated by paths. The site previously occupied by these collections was used to expand the botanical arrangements of trees, shrubs and herbaceous plants. Some of these botanical plots were still organised according to the Linnaean system, but later ones were arranged following the new natural classification of the French botanist, Antoine de Jussieu.[25] His system was the forerunner of the modern classification of plants into families. The present-day order beds in front of the Palm House are arranged according to families.

In 1804, the Dublin Society decided to tidy up the lease arrangements for the site. The terms agreed with Major Thomas Tickell, grandson of the poet, were four hundred pounds plus a further one thousand pounds payable in two years in equal sums.[26] In 1806, it was agreed with Mr Hume, the assignee of the Putland family, to pay £1,250 with interest over three years.[27] This meant that the Dublin Society was then in direct possession of the sixteen Irish acres.

In 1807, the watermill which stood by the bridge on the eastern side of the road and which was fed by the millstream direct from the Tolka was also taken over by the Society. The mill house (see Fig. 36) had a small garden and a meadow—the Millfield—which covered one Irish acre. The head rent was five pounds a year and thirteen years remained of the current lease. The agreed purchase price for the mill and the field was £625, which was paid to Mr Duffin of the Linen Hall. The Society thus added the Millfield to the Gardens and acquired the mill which was to present many problems over the next few decades.[28]

In November 1807, the Bishop of Kildare, who was also Dean of Christchurch Cathedral and owned the land on the northern bank of the Tolka, sent John Foster an advertisement from a Dublin newspaper in which the Royal Canal Company offered a plot of land for sale at the northern end of Glasnevin Bridge. The Bishop told Foster he would like to acquire the land so that he could have an easier and more ornamental approach to his house. Foster obviously suggested that the Dublin Society should purchase it and lease a part to the Bishop for a proportional rent. The Canal Company agreed to lease the land to the Society at twenty-five pounds a year, and Foster informed the Bishop accordingly.[29] Thus a small area on the northern side of the river was added to the Gardens. Later, it was occupied by a cottage and is now the rose garden.

While the grounds were being enlarged, the amount of glasshouse space was also substantially increased. In July 1806, the members of the Committee of Botany visited the Gardens and found that the range of glasshouses needed repairing and painting. It was also obvious that a propagation glasshouse was required with a warmer atmosphere, for the cultivation of certain tender plants. John Underwood was asked for his opinion about the best site for the new glasshouse. By March 1808, it was decided that what was required was the removal of the whole range of glasshouses attached to the house and its re-erection elsewhere. John Foster was not convinced this was necessary, but he was told that the Committee of Botany believed the range was 'badly situated and [had] not sun enough'. Underwood was being blamed for some problems in propagating plants, but in a letter to Foster, General Charles Vallency explained that the Committee of Botany 'speak in high terms of [Underwood's] knowledge and say that the situation of the glasshouses not his ignorance is the cause of all our failure'.[30] Foster insisted that the original range remain intact, so a new house, with its axis lying due east-west was built between the gate and the house. This new glasshouse, the Epiphyte House, was eighty-two feet long and twelve feet wide and cost £258. It was divided into three: the first compartment, entered by a rustic porch, contained cacti; the middle section was used for orchids, and the third division housed other tropical plants and bulbs of genera including *Amaryllis, Brunsvegia* and *Pancratium*.

In January 1802, it had been agreed that Underwood should be entirely responsible for superintending the Gardens and the employees. Indeed, Underwood was often referred to as the Superintendent of the Botanic Gardens. The Professor of Botany was left with the task of lecturing and was no longer directly responsible for the running of the Gardens.

In the early years John Underwood was certainly conscientious. Among other things, he was in charge of the meteorological observations made at Glasnevin. Temperature and barometric readings were taken daily, and the general state of the weather was recorded. These data were published in the *Transactions* of the Dublin Society, along with occasional phenological observations. After a few

31. Letter from John Underwood to General Charles Vallency, 7 September 1807. (Reproduced by courtesy of the Public Record Office of Northern Ireland, Belfast, and the Viscount Massereene and Ferrard. (Foster/Massereene Papers, D207/29/102))

years, however, publication ceased, but records continued to be made, and the weather station at Glasnevin Botanic Gardens has operated more or less continuously since 1800. Underwood also prepared the second catalogue of the collections which was published in *Transactions* in 1804. Unlike Dr Wade's earlier catalogue, and those of the separate divisions of the Gardens, Underwood's revision was arranged in botanical sequence, not according to the divisions of the Gardens. It was also the last complete catalogue published and listed approximately six thousand different species and varieties.

John Underwood was responsible to the Committee of Botany not to Dr Wade. In effect he reported directly to John Foster who had control over all affairs. For example, on 26 May 1803, he wrote to Foster saying that there had been a long period of cold dry winds but the plants were in good health. He continued, 'last December the great Flood did much damage'. Underwood reported that he had just completed making beds for the peat-loving plants and was busy collecting heaps of compost for spreading on the shrubberies in the autumn.[31] Plants were being propagated in the Gardens and surplus ones were sold to members of the Society. Underwood's duties included the listing of the duplicate stock and keeping accounts of the sales. He told Foster that 'there is many things wanted for the Gardens but I suppose that there will be nothing got, this year, but the things that you ordered from Brompton'.[32] In 1803, £100 was spent on plants from an unrecorded source, which may have been William Curtis' Brompton nursery.

In September 1807 the weather again affected the Gardens. Underwood reported that from one Thursday evening to 10 o'clock on Sunday morning 5·07 inches of rain fell 'a wonderfull quantity'. He told General Vallency that 'the River did raise to a great height, and overflowed all the lower

32. Silver trowel used on 15 August 1815 by Joshua Pasley, a Dublin merchant and member of the Dublin Society's Committee of Botany, to lay the foundation stone of the gate-lodges, which are depicted on the trowel (see also Map Colour Plate 2 and p. 63). (Reproduced by courtesy of the Director, National Museum of Ireland)

part of the Garden, and has almost tore it to pieces, and being a very general storm of wind, the trees in the plantations have suffered much; one large elm tree is tore up by the roots, the head of several broken off, and part of the wall at the far end of the Garden down, about 60 feet of it...'[33] (Fig. 31).

But all was not gloomy. In the same year, the Botanic Gardens acquired a plant of the waratah (*Telopea speciosissima*), a remarkable shrub with huge red flower heads from New South Wales that was eagerly sought by collectors in England. It cost ten guineas from the London nurseryman, Conrad Loddiges. John Underwood told Foster that while he worked in London there was only one plant known.[34] In 1807 there were three in England, including one in the Royal collection at Kew. It is not known how this extraordinary plant fared at Glasnevin, but the species was still cultivated in the 1860s.

By about 1810, the Gardens had become well established and all major works affecting the cultivation of plants had been completed, except for frequent alterations to glasshouses. The plant collection had risen to over six thousand species, but progress was scarcely recorded after 1804. In 1815 the last major alteration was carried out; the original Doric cottages at the entrance to the Gardens were removed and replaced by two gate lodges, joined by a pair of gates forming an arc (Figs. 32, 33). The cost of the new gates and gate lodges was met from a gift of £600 made by Thomas Pleasants (1728-1818), a member of the Society. The new gates did not please everyone, for many regretted the removal of the Doric cottages. Pleasants was obviously satisfied with the plans for he wrote to the Society, in the third person, saying

> Mr Pleasants, having his writing-thumb hurted, can hardly hold a pen; but wishes, under his own hand, to present his best compliments to the secretary of the Dublin Society; is heartily glad to find, that an appropriate entrance is determined on for the Botanic Garden, that institution of such immeasurable value to the community, and so worthy of the public's consideration.[35]

The benefactor was an eccentric character. He was deeply devoted to his wife, and after she died he took her slippers to bed with him every night for the remainder of his life. He busily set about spending a considerable fortune which he inherited from her, despite the attempts of her family to restrain him. He bequeathed to the Gardens two large white and blue porcelain jars and instructed that these be filled with something 'lively and showily disposed'. Two china elephants which hitherto accompanied the jars were also left to the Gardens. For the gift of the gates and lodges, the Dublin Society made him an honorary member.[36] The new gate lodges were occupied by John Underwood and his family, and by the family of one of the undergardeners.

It was even contemplated closing the original entrance, and making a new one on the north side of the river. Visitors would have crossed over the river, through an island and the millfield. It was also proposed to demolish the professor's house and erect a new one elsewhere. None of these plans came to pass.

The new Botanic Gardens attracted comments which ranged from polite compliments to trenchant criticism. However, the early progress of the institution can be gauged by noting some of the opinions during the first few years. Shortly after the Dublin Society had acquired the land and work had commenced on establishing a botanical garden, the Belfast naturalist, John Templeton, visited Glasnevin. Templeton was a skilful gardener and grew many plants of botanical interest. Two years later he recalled his impressions of the infant Gardens in a letter addressed to the imprisoned United Irishman, Thomas Russell:

> I saw the Botanic Garden when I was last in Dublin, but at that time I could not form a judgement of it. It is now 2 years since and I suppose great things have since been done. It appeared to me at the time that the plan was so extensive it would in all probability never be completed, and I have heard of nothing since to make me change my opinion. I think the idea of turning every part of science to immediate use is a thing that will constantly retard its progress...A Botanic Garden ought...to aim at getting an extensive collection of plants, for the use of such a place is that we may have at all times plants that are already known to compare with those of which we are doubtfull. One thing which the Dublin Botanic Garden appears to want, and which can alone bring it forward is an intelligent practical gardener, who has also a considerable degree of enthusiasm in pursuit of knowledge in his profession.[37]

33. Gate lodges of the Botanic Gardens, c. 1900, photographed by W. Lawrence. The entrance is still like this today. (Reproduced by courtesy of the National Library of Ireland, Dublin (Lawrence collection 8549)) (see tailpiece opposite)

Templeton's pessimism was not borne out by subsequent developments at Glasnevin, and in later years he eagerly exchanged plants with John Underwood.

Robert Brown, a young Scottish botanist attached as a surgeon's mate to the Fifeshire Regiment of Fencibles, was stationed in Ireland from 1795 and visited Glasnevin on 4 September 1800 in the company of the antiquarian Edward Ledwich. Brown was introduced to Dr Wade, and he examined some of Wade's herbarium specimens, including a rare moss. Of the Gardens, Brown wrote, 'the plan [was] excellent but the execution in its infancy'.[38] He collected several grasses in the Gardens and later prepared detailed botanical descriptions of them. A decade later the British philanthropist and agriculturalist, Edward Wakefield, visited Ireland at the invitation of John Foster. He published a lengthy account of his travels, and was complimentary about Foster's estates. But he wrote that the Botanic Gardens had the appearance of being kept in a slovenly manner.[39] Sir Vere Hunt called at Glasnevin in May 1813 and recorded his impressions in his diary. He was shown a specimen of *Pinus cembra* grafted in the common cleft way on to Scots pine (*P. sylvestris*). This plant was thriving then and was still alive in 1838. Hunt noted that shrubs were trained against walls, especially *Pyracantha* which most beautifully covered one section twelve feet high. He admired the collection of irises, a golden foliaged arbor-vitae (*Thuja plicata*) and numerous varieties of holly, willow and poplar. He was impressed by the greenhouses, which were 'capitally stocked...with a most curious and extensive variety of plants'.[40] In the following year, Anne Plumptre came to spend the summer in Ireland. She found the Botanic Gardens 'most noble...much larger than any other that I have seen either in the British dominions or in France..laid out with great taste and judgement. The conservatories, however, and the collection of exotic plants, are not so good as in the King's garden at Kew'.[41] She remarked that the Gardens were situated on high ground from which there was a fine view over the city and Dublin Bay.

The Reverend Robert Walsh, vicar of the adjacent parish of Finglas, was one of the authors of a history of Dublin published in 1818. This contained a map of the Gardens (Map Colour Plate 2), on which was a vignette of Pleasant's gates and lodges (Figs. 32, 33). Walsh had an interest in botany and was probably responsible for the chapter on the Glasnevin Botanic Gardens which was included in the history. At times he praised the Gardens, but he did not shirk from criticising parts that were unkempt or uninteresting. In general he admired the place where 'the hand of taste embellished the formal face of science'. The flower garden was not to his liking, because the plants 'were neither remarkable for beauty or variety'. 'The Conservatories', wrote Walsh, 'are built without any attention to architectural ornament. They are obtruded into the most conspicuous parts of the garden, where they are coarse and formal objects destroying the picturesque effect which was so particularly attended to in every other arrangement'. He was especially pleased that the planners had retained some of the features of the old demesne, including large clumps of ancient elms, and the 'ivy-clad ruin of some venerable arch'. Walsh approved of the scattered shrubs and the gravel paths and green sward which separated the regular beds of the various sections. He thought that these prevented the regularity of the beds from becoming oppressive and formal. To the Reverend Robert Walsh, the Botanic Gardens at Glasnevin was the successor of the Garden of Eden, where Adam had been the first professor of botany, ordained to tend a garden, part of which 'was laid out with botanic regularity'. Yet, the Dublin Society's Botanical and Agricultural Garden at Glasnevin was only a shadow of Eden for 'the whole [presented] the aspect of unstudied confusion'.[42]

6
The Fostering Mantle; 1800-1833

ON 2 AUGUST 1800 the Irish Parliament sat in its splendid building opposite Trinity College for the last time. One of the final acts passed by this Parliament provided a grant of fifteen hundred pounds for the Botanic Gardens (Act 40 George III c.31) thereby bringing the total sum allocated since 1790 to £7,700. John Foster as Speaker of the House of Commons presided over the last sitting, an event that was highly distasteful to him, an event which also marked the defeat of his rearguard action of trying to maintain the ascendancy of the Anglo-Irish. When the sitting of the House of Commons ended, he refused to surrender the mace and removed it and the Speaker's chair to his own house at Collon.

For almost two years John Foster took no active part in the new Parliament of the United Kingdom although he was a member for County Louth. He made appearances in the House of Commons at Westminster for a short period in the spring of 1802, but did not return again until the beginning of 1804, when he resumed an active role. He served two terms as Chancellor of the Irish Exchequer between 1804 and 1811 after which he retired.[1] But John Foster had many interests to occupy his time after the termination of his speakership—his estates and garden in County Louth, the Farming Society and the Dublin Society, and above all 'his' botanical garden at Glasnevin.

It is clear that John Foster was the person who controlled the Botanic Gardens. During the first three decades the officials of the Dublin Society and the staff of the Gardens all deferred to him, sought his advice, begged his approbation and carried out his orders. He was consulted about major developments and if plants, books or other items had to be purchased, he was the person who gave the instructions. Sometimes there was poor communication between the Society, which held the financial responsibility, and Foster, who gave the commands. For example, John Foster had authorized the ordering of plants from the nurserymen James Lee and John Kennedy in 1804, 1805 and 1810, but he had not informed the Society about the last order. In October 1813, the Assistant Secretary of the Dublin Society, Bucknall McCarthy, considered it necessary to remind Foster that 'according to the present rules of the society the latter must be certified also. Can you do it? If you ordered them be so good as to do it and return it with your signature and I will remit the amount next Thursday'.[2]

Foster used the Gardens as his personal domain and he occasionally acquired plants for his own garden from Glasnevin. On 6 October 1811, he sent a memorandum to John White:

> I will thank you to mention to Underwood my wishes to have some of the Holcus oderatus [sic] and the different Nymphaea and the other herbaceous articles which you took a memorandum of—Is this or Spring the best season for moving them?[3]

John Underwood sometimes received direct orders from Foster and rarely did any major work in the Gardens without first seeking his approval. In 1809 he was particularly keen to have directions. He wrote to John Foster expressing the hope that Foster would be able to come to Glasnevin before returning to London. If this could not be done, Underwood said he would

> goe on with every thing, to the best of my understanding, full of expectation that it will meet your approbation, when time will allow your Honour for to see it, Then I hope that your Honours goodness will be pleased to order a little advance to my salary to enable me to support my family.[4]

The 'Fostering mantle'[5] was appreciated and understood by all connected with the Botanic Gardens. The men employed there owed their positions and salaries to the patronage of John Foster and they dared not forget that fact. Neither could those in authority in the Dublin Society ignore Foster's power and status, and it seems that they were quite content to let him run the Gardens as he wished. Foster's control was complete, at least until his final years when age and infirmity prevented him from continuing to make his presence felt.

The person who was most acutely conscious of John Foster's patronage was Dr Walter Wade. He more than anyone owed his position to John Foster and he frequently tried to raise his own standing both with his patron and within society generally. His letters to Foster appear to the modern reader to be servile in their formality although the style of writing was quite normal at that time. Two examples can be given of Wade's attempts to better himself. In May 1809 he wrote to John Foster saying that he was aware that beer was 'from time to time...[being] infus'd [with] improper and dangerous substances'. He suggested that an ordinary inspector of breweries could not detect such illicit additives because he did not have sufficient knowledge of brewing. Walter Wade asked Foster to appoint him inspector for the city and county of Dublin.[6] In March 1811 the post of physician to Newgate Prison in Dublin became vacant and Wade wrote to Foster asking him to write to the aldermen in his favour, as the prison doctor was elected by the city aldermen. 'I never yet received a favour from Government', complained Dr Wade, 'though my exertions in 1798 and 1803 so fully intitle me to it, as a *loyal protestant*—my exertions in 1798 are well known to Lord Camden, Lord Pelham etc. and as to 1803 it wou'd be unnecessary to call on Lord Hardwicke to vouch for me'.[7] As far as is known, Wade did not obtain the position at Newgate.

One of Wade's duties at the Botanic Gardens was to give lectures on botany. Indeed, this became his main responsibility after John Underwood assumed full charge of the day-to-day running of the Gardens in 1802. The classes were held in the French Room, on the ground floor of the Professor's house; this room, decorated with a plaster frieze depicting emblems of the Muses, was connected to the glasshouses by folding doors. Dr Wade's first course of lectures started early in 1802 and the syllabus was published in the Dublin Society's *Transactions*.[8] He began the lecture series by exclaiming that botany was a captivating subject which added to the 'intellectual enjoyment of the philosophick and contemplative mind', and that exercising one's mind was as necessary to existence as exercising one's body. The study of botany, even as an amusement, was commendable, and was of the greatest advantage from the commercial point of view.

Wade's lectures included discussion of the history of botany. He spoke about sexual reproduction in plants, and explained the functions of the component parts of a flower. The series of lectures concluded with one on Linnaeus' 'Sexual System' of classification. Afterwards he held practical sessions in which each of the various 'Classes' of plants were examined and living examples were dissected by the students. One of those who attended Wade's lectures was John Nuttall of Tithewar (or Titour) near Newtownmountkennedy in County Wicklow. Nuttall's lecture notes have survived and show that the students were encouraged to examine the flowers, press and dry dissected specimens and stick them into their notebooks.[9] Finally, Dr Wade brought his students on a walk through the Botanic Gardens to look at the various sections devoted to plants of economic importance.

Wade delivered his botanical lectures each year from 1802 to 1823 and to accompany them the Dublin Society published a series of lecture notes or syllabuses (Fig. 34), the most detailed being those for the lectures in 1802 and 1823.[10] Apart from the general lectures, Dr Wade gave others devoted entirely to plants of agricultural importance; the notes for these were also published by the Society in 1808 under the titles *Sketch of lectures on artificial or sown grasses...* and *Sketch of lectures on meadow and pasture grasses...*[11]

Wade was never satisfied with the numbers of people who came to his lectures, but the low attendance was not surprising. Glasnevin was three miles north of the city, and, as some of the lectures began at 8 o'clock in the morning, it called for a special effort to reach the Gardens in time. In the summer of 1806, the Lord Lieutenant, the Duke of Bedford, attended Wade's lectures, and as a result they became more popular. The Duke found Wade's exposition of botany simple and clear. His presence was an accolade, which Wade accepted with relish.

Shortly after the Duke arrived in Ireland in 1806, Dr Wade sent him copies of his various publications, including the catalogue of the collections in Glasnevin. The Lord Lieutenant formally acknowledged receiving the books.[12] He was keenly interested in gardening and had a considerable correspondence

BOTANICAL LECTURES,
FIRST COURSE,
THEATRE,
DUBLIN SOCIETY.

By PROFESSOR WADE, F. R. S. &c. &c.

LECTURE I.
Monday, May 4th, 1818, at a Quarter past 3 o'Clock.
INTRODUCTION. A full and comprehensive view of the Dublin Society's grand establishment at the present day—the great benefits derivable to the Public, by their attending the *Open Lectures*, and Instructions, established by the Society---as those of *Chemistry, Botany, Agriculture, Mineralogy, Mining, Natural Philosophy*, the *Veterinary Art*, the *Fine Arts, &c.* fully enlarged on.

LECTURE II.
Wednesday, May 6th.
The different compartments, in the Society's Botanical Garden at Glasnevin explained, and their high importance vindicated.—Specimens of plants handed about, and distributed, illustrative of the facts just mentioned.

LECTURE III.
Friay, May 8th.
The high estimation and great consequence, in which Botanical gardens are held all over the world.—An account of some of the principal ones, with their various intentions.—*Artocarpus*, or bread fruit exhibited,—&c.
Exercise of the mind as necessary to the existence of man, as that of the body—incontrovertible—cannot be more worthily and usefully employed than in contemplating the works of Nature—manifest—arguments adduced, and specimens exhibited in corroboration of several curious facts in this point of view.

LECTURE IV.
Monday, May 11th.
Absolute necessity and high utility of botanical knowledge, and the emoluments that must result from a practical knowledge of it, not only to the nobleman and gentleman, medical practitioner, the merchant and manufacturer, as well as the planter, husbandman, and gardener—many convincing instances adduced.—Advantages arising from joining literary and philosophical inquiries with medical, commercial, and agricultural pursuits.

LECTURE V.
Wednesday, May 13th.
A very brief history of Botany, from the most remote period of antiquity, to the present day.

LECTURE VI.
Friday, May 15th.
A review of most of the best systems of botany, which preceded the Linnæan artificial, or sexual system—their faults and inconveniencies pointed out.—A natural system how desirable—many fruitless attempts to perfect one by Linnæus himself, Haller, Alston, Hope, Gærtner, &c.—Jussieu, a botanist of the first eminence, now living at Paris, has made the very best endeavour at natural classification—his system explained—subject to inconveniencies—such pointed out.—An artificial system becomes necessary—the Linnæan now universally adopted—its beauties, conveniencies, and utility.
A short account of the life of Linnæus, with an analysis of some of his most useful and principal works.

LECTURE VII.
Monday, May 18th.
INCREASE OF PLANTS.—*Sexus Distincti*. Known to the earliest observers of nature—Theophrastus—300 years before the birth of Christ—Arabian and Persian knowledge—*Phœnix dactylifera*; or date bearing palm—*Pistachia*—*Fig, &c.*—Sir Thomas Millington, 1676.—supposed to have given the first hints in England.—Followed by many others—such mentioned.
Nature of vegetable bodies—research from the minutest insect to the vegetable kingdom—the chain or series traced—many familiar instances adduced in proof of the animality of vegetables.

Wednesday, May 20th.
Chemical principles appear to be different in animals and vegetables, when considered *generally*. All the *single* substances found in vegetables by chemical analysis, most of them found in animals.—Such principles enlarged upon, and specimens of plants as afford some of them, exhibited.—The *vital* power in vegetables produces, by mixing those elementary or single substances, *new* formed substances—each fully noticed, and plants handed about as elucidations.
Lower tribes of *Zoophyta*—connecting link between animals and vegetables—*Sertulariæ, Polypi, &c.*

34. Syllabus published in 1818 for Dr Wade's botany lectures. (Reproduced by courtesy of the Public Record Office of Northern Ireland, Belfast and the Viscount Massereene and Ferrard (Foster/Massereene Papers, D562/7848))

LECTURE VIII.
Friday, May 22d.

STAMINA and PISTILLA—*Antheræ*—*Pollen*—essential and important—Doctor Darwin's reasonings on their nature hypothetically beautiful. Tournefort, Alston, Smellie, &c.—their opinions on their uses refuted.—Koelruter's curious and extraordinary experiments with the pollen—changing one species to another!—Remarks on the female date bearing palm *(Phœnix dactylifera)* by Dr. Watson, Hasselquist, Gleditch, Michaux, &c.—Fourcroy and Vauquelin's analysis of the pollen or fertilizing powder of the palm—satisfactorily proves its resemblance to certain animal secretions, and in destining them to the same end Nature has wished to constitute them of the same elements or principles, in order to fulfil the same functions!!

LECTURE IX.
Monday, May 25th.

ANATOMY and PHYSIOLOGY of PLANTS.—Cuticle or epidermis—cellular integument—bark—wood—medulla, or pith.—Sap vessels and course of the sap—sap and insensible perspiration.—Secreted fluids of plants—Grafting,—Heat of the vegetable body, &c.—Malpighi, Grew, Duhamel, Dr. Darwin, Dr. Hope, Sir J. E. Smith, Mons. Mirbel, and above all Mr. Knight, on these interesting inquiries.

LECTURE X.
Wednesday, May 27th.

The Fructification, or flower and fruit—consists of seven principal parts, 1. Calyx. 2. Corolla. 3. Stamen. 4. Pistillum. 5. Pericarpium. 6. Seed, and 7. Receptaculum—their various denominations, uses, anatomy, physiology.—Nectaria—their curious economy.

LECTURE XI.
Friday, May 29th.

Further Observations on Seeds—Darwin, Willdenow, Gærtner, " De Seminibus et Fructibus," Jussieu, Mr. Knight; and lately some curious remarks by that admirable vegetable physiologist Mirbel. Some singular phenomena attending the germination of seeds—Sukkow, Bonnet, Humboldt, &c.—Oxygene acts as a stimulant to the sudden developement of seeds—Simple methods to obtain oxy-mur—acid gas—*(Chlorine* or *Chloric gas* of DAVY, from χλωρος, viridis or green, the gas being of a green colour.) Professor Pohl at Dresden—curious experiment with *Chloric gas*—caused seeds to germinate, of 110 or 120 years standing.

LECTURE XII.
Monday, June 1st.

APHORISMS on VEGETATION.—Plants produced by seeds—germination—require moisture—heat—oxygene gas—light.—Food prepared in the cotyledons—then sent into the radicle—seminal leaves.—Plants require food—water is necessary—absorb food from the atmosphere—water and air not alone sufficient—a third remaining source—the soil on which plants grow.—Soils contain earths—remains of animal and vegetable matter—salts, &c.—Manure necessary for the growth of plants—vegetable manure—acid vegetable mould—animal composts, &c.—GENERAL OBSERVATIONS.

LECTURE XIII.
Wednesday, June 3d.

ROOTS.—Their several species, forms, uses, duration, and best mode of propagation—specimens handed about.—TRUNKS, or stems, ditto—BUDS—general remarks on—LEAVES—their respective situations—insertions—forms—surfaces, &c.—of much use in systematical botany—on the falling of the leaf—some curious practical facts.—The uses, general economy, and physiology of leaves fully entered into.—Hales, Ingenhouz, Saussure, Sennebier, Priestly, Gough, Perceval, Woodward, Henry, Woodhouse, Humboldt, and Mr. Ellis, "ENQUIRY INTO THE CHANGES IN ATMOSPHERIC AIR, &c."—interesting remarks on this subject.—Almost the whole of vegetables composed of carbon, hydrogene, oxygene—azote or nitrogene, forms but a small proportion—into some vegetables do not enter at all—decomposition formed by the *parenchyma*, &c.—carbonic acid absorbed—various in different plants—the *whole* oxygene not emitted—plants refuse to vegetate when confined in nitrogene or azotic gas—they differ in the quantity of oxygene—proofs adduced.—GENERAL REMARKS.

LECTURE XIV.
Friday, June 5th.

Further observations on the economy and physiololy of leaves, &c.—of all parts of plants, they show the most singular irritability—*Dionæa muscipula, Drosera, Mimosa pudica, Nepenthes distillatoria, Hedysarum gyrans, Tylandsia maxima, &c. Fulcra,* or props—inflorescence of plants—various—such explained, and recent specimens handed about as illustrations—elegant specific characters from the inflorescence—*Pyrus communis,* or pear, bears a *corymbus*—*Pyrus malus,* or apple, an *umbel,* &c. &c.

LECTURE XV.
Monday, June 8th.

Linnæan Classes explained, with general observations.

LECTURE XVI.
Wednesday, June 10th.

Linnæan Orders explained—plants handed about as illustrations—their value in various points of view briefly noticed.

CONCLUSION.

Graisberry and Campbell, Printers to the Dublin Society.

with people connected with horticulture and agriculture. The Duke of Bedford was succeeded as Lord Lieutenant by the Duke of Richmond on 19 April 1807. Early in July that year the Duke of Bedford said that he wished to pay a private visit to the Gardens. Dr Wade was delighted. The new Lord Lieutenant joined the Duke and Duchess of Bedford for the occasion, and the party was presumably escorted by Wade. After the visit he wrote to John Foster:

> I now take a...liberty, but merely to inform my rever'd Patron that on Saturday last, their Graces the Duke and Dutchess visited the garden, and with which they were highly pleased, His Grace of Richmond in particular who in the most affable manner entered fully into the general merits of the institution, and listen'd with attention to my descriptions of the different compartments.[13]

But the officers of the Dublin Society were not at all pleased. General Charles Vallency, a vice-president, wrote to Foster a few days later saying that 'Wade (to be sure) has told you with pomp and pride that the Duke of Bedford visited the garden and expressed himself pleased. He desired to be in private (if Wade speaks the truth) and I believe no one of the society, but Wade was there, to whom his intended visit was communicated'.[14]

Dr Wade remembered these visits as the high-point of his professorship and in June 1811 complained to Foster that 'since the time the Duke of Bedford by his presence made it somewhat fashionable to attend [my lectures] the great (except in a very few instances) have paid no attention to this highly interesting subject'.[15] By this time Wade was thoroughly disenchanted with his work as lecturer, complaining that it was 'an extra duty, which neither the Farming [Society] or Dublin Society have ever noticed by any mark of kindness'.[16] Every year, before starting the lecture course, Walter Wade sent the prospectus for the series to John Foster, and he wrote again after the lectures were concluded. On Saturday 21 September 1812, Wade sent this letter:

> I take the liberty of informing my much respected Mr Foster that I have this day, finished a very laborious Course of Lectures on Botany, and it's [sic] connexion with the conveniences and accommodations of life, in the very fullest extent—always considering myself, my rever'd Patron's professor, whose selection of me I was determin'd never to disgrace, or disappoint His views and expectations in the formation of the Botanical Establishment, I am sorry to add, I am in a very declining state of health.
> I have the honour to be, my respected Mr Foster's very grateful humble servt.
> Walter Wade.[17]

Walter Wade frequently managed to slip into his letters to Foster some seemingly inconsequential remark, presumably hoping his patron would take the rather obvious hint and further assist 'his professor'. Dr Wade wanted to go to Paris in September 1814, but could not afford the cost of such a trip.[18] He intended obtaining plants and seed for Glasnevin, and hoped that Foster could persuade the Society to meet the expenses. However, on this occasion he did not receive any support from John Foster, and did not go to Paris.

In a letter sent in October 1819, Wade gave some details of the amount of work involved in his lecture programme. That year he gave sixteen 'theoretical lectures' in the Society's lecture theatre in Leinster House and forty-two 'practical lectures' at Glasnevin. There were also ten lectures on agricultural botany—meadow, pasture and 'artificial' grasses—and fourteen private lectures to the apprentices in the Botanic Gardens, a total of eighty-two lectures. Wade distributed 623 specimens of plants or 'exhibited them by figures' at Kildare Street and used 1,535 species at Glasnevin. All those attending the lectures at the Botanic Gardens were required to sign a book kept in the gate lodge, and this recorded that 2,554 ladies and gentlemen attended the lectures during the months of June, July, August and September. Wade concluded his letter by repeating his belief that he was Foster's personal professor.[19]

Dr Wade's attachment to Foster may have resulted from his impatience with the Dublin Society and its Committee of Botany which was nominally in charge of the Gardens. He complained again and again about the attitude of the members of the Society towards the Gardens and botany. Wade was disheartened by the lack of interest which his lectures received especially from the gentry and nobility—'but I am an Irishman who is never a prophet in his own country!'[20] For example, in October 1811 he wrote to Foster apologising for not being able to visit Collon: he was indisposed although not confined to bed, with a 'vesical complaint'. Wade wanted to speak with Foster about the lecture courses, of which his patron had approved. He was disturbed by 'apathy or envious obstructions which may occasionally be thrown' in his way.[21]

Earlier that year Wade had complained that the Society had become little more than a 'theatre of debate; and the promiscuous admission and weekly harangues are shameful and portend but badly'.[22] He regretted that the Committee of Botany showed no enthusiasm while chemistry was the rage — 'poor botany with every exertion is in the back ground & only occasionally thought of or spoken of!!'.[23] From Wade's point of view things were no better in 1819. In May he protested that the

> majority of...members at present are of a most ungovernable description, more addicted to making speeches than to promoting the real interest of the Society...Some of the members seem to set their faces against our Botanical Establishment, the brightest jewel I am proud to say in the Society's cap, admired by all who have visited it, foreigners who have from time to time paid attention to it's [sic] different compartments on learning and hearing their tendency explain'd, seem'd to feel, indeed declare, that there was no similar garden, founded on the same nationally useful principles. It may be necessary to inform Mr Foster that some individuals in the course of their harangues have express'd their displeasure at the great expense attending its support, ignorant altogether, I must presume, that our Irish Parliament...allocated £1,500 a year for its support.[24]

In the same letter Dr Wade suggested that a clause should be inserted in the next Dublin Society Act compelling the Society to appoint respectable and intelligent people as directors.

By 1821 things had improved at least temporarily. Wade was able to write that the members of the Committee of Botany 'tho' not adept in the science, are most assiduously attentive' and would carry out Foster's instructions about the use of composts.[25] The general state of the Society however continued to displease Wade and the antagonism of some members towards the Gardens annoyed him. In addition, John Foster had recently suggested the establishment of an arboretum in Phoenix Park. While this worried Wade, he thought it a good idea especially if it could be supported by the government. 'How happy would I be', he wrote, 'to see such a beautiful and picturesque assemblage of vegetable creation even in their infant state, at almost the close of a botanical career!!!'.[26]

Up to 1821 Walter Wade's career in botany and medicine had been full. As well as being Professor of Botany to the Dublin Society, he was physician to the Dublin General Dispensary and lecturer in botany at the Royal College of Surgeons. But, there had been some disappointments — in 1809 following the death of Dr Robert Scott, Professor of Botany in Trinity College, Wade had tried unsuccessfully to add that appointment to his others, and, as has been noted, failed to obtain the post at Newgate Prison in 1811. On 13 December 1810, Dr Walter Wade, 'a gentleman eminent for his scientific knowledge and Author of several works on Botany and vegetable Physiology' was proposed for election as a Fellow of the Royal Society in London by Richard Kirwan, the Honourable George Knox, Professor William Higgins and Richard Lovell Edgeworth, who were all prominent members of the Dublin Society, and Aylmer Bourke Lambert. Wade became a Fellow on 14 March 1811.[27] By 1811, Wade had written four important botanical works, as well as catalogues of the Gardens, prospectuses and lecture notes. In 1794, 1801 and 1804 Wade published lists of native Irish plants, many of which he had noted or collected during field excursions carried out as part of his duties. In the first of these lists Wade gave details of plants collected in County Dublin and the second covered Connemara, including the first report of the American pipewort (*Eriocaulon aquaticum*) in Ireland.[28] The 1804 publication listed plants from other parts of the country; a coloured engraving was included showing the rare moss *Buxbaumia aphylla* that Wade discovered near Killarney.[29] He translated François Michaux' treatise on oaks from the original French, and had added notes on the species that he knew growing in the Glasnevin Gardens and in John Foster's collections at Collon.[30] *Quercus or oaks...* was published in 1809, and was followed in 1811 by a book on sallows, willows and osiers (*Salix* spp.) illustrated with two coloured plates (Fig. 35). Wade maintained his interest in grasses, and published several papers and catalogues on them. In 1804 he wrote a short pamphlet on sweet-scented vernal grass (*Anthoxanthum odoratum*) which included a coloured engraving,[32] and in 1818 he produced a catalogue of the grasses growing in Glasnevin which was accompanied by pressed specimens of each of the 125 plants listed.[33]

35. *Salix acutifolia*, hand-coloured engraving from Walter Wade's essay on willows. The illustration was drawn by Richardson and engraved by Ford. (National Botanic Gardens, Dublin)

The paper on sweet-scented vernal grass was coupled with a short account of the orange-flowered *Buddleja globosa*, a native of Chile. This was grown in Glasnevin as a source of nectar, and Wade observed that it '...afforded a very grateful, and...a most diligently sought after repast, to that valuable creature', the domestic honey bee.[34] One of the skills which the Dublin Society promoted during the early years of the Botanic Gardens was that of bee-keeping and in 1800 Christian Schulze was appointed as official bee-keeper. He assisted Dr Wade in establishing an apiary, but left after only three years. Schulze published some notes on bee-keeping and the rearing of silk-worms in the *Transactions* of the Society.[35]

In 1800 the Gardens were opened to the general public and as the visitors had to be conducted round the Gardens by the head gardener, the Society engaged a second undergardener. On 3 November that year John Jones was appointed to this position. Nothing more is known about him and he was replaced before December 1804 by John Wallace. The Dublin Society provided Wallace with a furnished cottage in the Gardens and a salary of fifty pounds per year, raised to sixty guineas in December 1806. In the summer of 1805 Wallace was sent to Scotland to obtain plants for the Gardens. He resigned his post on 23 December 1809 and left Glasnevin.

John Mackie, one of the employees in William Malcolm's nursery at Brompton, in Middlesex, succeeded Wallace as undergardener. Malcolm had recommended him for the job, and John Underwood commented to Foster that if Mackie was 'such a clever man as Malcolm represents, in propagation of plants, I should think with your approbation, he will be a very proper person for your botanic garden'.[36] John Mackie, who was twenty-five years old, travelled from London in January 1810 and received ten guineas for travel and removal expenses. His salary was also set at fifty pounds, and increased to sixty-eight pounds in 1815. Mackie supplemented his salary by collecting native plants: in 1818, for example, he received twenty pounds for one hundred native species. Soon after Mackie arrived in Glasnevin, Dr Wade complained to Foster that 'the young gardener lately come over has taken to the whisky drinking example of the curator'.[37] Mackie's subsequent career was unremarkable. In 1821, the Society became concerned about his poor state of health said to be due to the 'damp and unwholesome situation of his bed-room in the Garden', and allowed him an extra eight shillings per week for two months for lodgings outside the Gardens. Mackie's condition did not improve and he died in May 1821 leaving a widow and six children. Mrs Jane Mackie received twenty pounds from the Society for the plants her late husband had collected during the previous year and was appointed assistant housekeeper in charge of the apartments occupied by the apprentices, but she was dismissed in the following year.

The Botanic Gardens had been a centre for training gardeners since 1812 when the Committee of Botany decided that a plan should be prepared to allow young lads of seventeen to become apprentices. At that time there were eighteen labourers working in the Gardens, and the Committee suggested that apprentices should be taken on in place of an equal number of labourers. Six apprentices were accepted — the first was Thomas Carroll — and the boys lived in the Gardens, each one being paid an allowance of nine shillings a week. They proved to be bothersome lads, idle and unpunctual. As an incentive to hard work, the Dublin Society decided to pay a premium of five pounds for the first year's training, if the performance was satisfactory, and the same sum for a second successful year.

The head gardener, John Underwood, was made responsible for the apprentices' training, and was paid five pounds a year for their tuition. He was required to instruct them in the 'knowledge, use and culture of plants'. In February 1816, two of the six apprentices had completed their courses and left. Instead of taking on two new apprentices, the Society engaged two additional labourers, as the weekly wage for labourers had just been reduced from twelve shillings to nine shillings and six pence. The wages of the remaining apprentices were reduced in the same proportion to six shillings and six pence. In the following February the Society complained that ignorant boys had been admitted as apprentices and the Committee of Botany decided that in future it would be best if apprentices were 'persons educated in gardens and with some proficiency in horticulture'. In May 1818 John Duffy, who had been a gardener for seven years under the Right Honourable Denis Bowes Daly, became an apprentice under this new regime.

About the same time, James White, one of the sons of the undergardener John White, was taken on as an apprentice, and in 1819 was awarded a prize of some valuable books to mark his progress. James worked as an apprentice for two years and was then appointed as a labourer. In 1821, to effect savings in the cost of running Glasnevin, the Society ordered the dismissal of four labourers, and James White was one of the men dismissed. He was a bit of a rascal. Unable to get a job elsewhere, James lived with his parents in their lodge at Glasnevin, but he became bored and decided to do something about it. In January 1822, John Foster received a letter from John White saying:

> It is now nearly twelve months since my son was discharged from the Dublin Society's Garden, during which time he has remained with me although I have repeatedly tried to procure him work in the different nurseries and gentlemen's gardens about Dublin.[38]

Foster was requested to use his influence to get James a job. In May, Foster received a second letter[39] asking for a reference for James, and in June received a third.[40] Foster replied to John White, who was utterly mystified. He had not written to Foster. It must have been with some embarrassment that the father had to tell his patron that his son James had forged the three letters. James was re-employed at Glasnevin, but dismissed in 1832 following an unrecorded offence of an outrageous nature. John Underwood was instructed to call the police if James White attempted to enter the Gardens, and Edward White, who was also working at Glasnevin, was cautioned to avoid his brother's company and to be circumspect in his behaviour if he wished to avoid dismissal too. John White himself committed some misdemeanour two years later for which he was admonished and told that his conduct 'had been highly unsatisfactory and unbecoming of a person in a subordinate position'.

John White's duties as undergardener were many and varied. He was despatched to various counties in Ireland to collect native plants for the Gardens and his travelling expenses were met by the Dublin Society. In 1803 he travelled to the Carlingford and Mourne regions of counties Louth and Down, and in 1804 returned to Down before going on to Antrim. During these excursions White collected native plants and visited some of the larger demesnes. At Mount Stewart on the shores of Strangford Lough, probably in 1804, he discovered an unusual form of gorse (*Ulex europaeus*). He took cuttings and brought these back to Glasnevin where they were successfully rooted. White's gorse was an upright form with smaller, less thorny leaves than the common one (Plate 4). Wade and others regarded it as a new species and Wade even published the name *Ulex downiensis* for it,[41] but it is now considered to be only a horticultural variety (cultivar) and is called *Ulex europaeus* 'Strictus'. It has been in cultivation at Glasnevin for nearly two centuries, the first new plant introduced into cultivation from there.

John White had a particular interest in grasses and prepared a comprehensive book on this subject which was published by the Dublin Society in 1808.[42] It was illustrated with coloured engravings of two species (Plate 3) and contained descriptions of all the known native Irish grasses, the Irish names and habitats being given. As well as collecting plants, White was employed as label painter and was paid at the rate of twenty-five pounds for every three thousand labels which he lettered. In 1817 he was instructed to collect specimens of grasses for which he was paid twenty pounds. It is not known why he was ordered to do this, but the specimens may have been used in the following year to illustrate Walter Wade's catalogue of the grasses cultivated in the Botanic Gardens.[43] White continued to accumulate information on native plants during his travels and, in 1833, in an anonymously published flora of Ireland, many of his records were reported.[44]

The head gardener, John Underwood, and his assistants (including John White) managed the Gardens with skill during its early years. In September 1814 the Dublin Society had declared its approval of Underwood's management of the gardens, and, in April 1818, the Committee of Botany stated its approbation of the 'care and attention evinced in the several departments...as well as in those persons who had the individual direction of the glasshouses and the external departments of the garden, as in the general superintendence of the entire establishment'. However, soon afterwards things turned sour and both Underwood and White fell into disfavour with the Society.

One of the most outstanding plants at Glasnevin in 1818 was a tree of *Araucaria excelsa* (Norfolk Island pine). It had been purchased in October 1798 from Lee and Kennedy at a cost of seven shillings

36. View of the Botanic Gardens from the east bank of the River Tolka showing the old school (left) (nick-named 'The Inkpot') and in the centre distance, the Octagon House; by W. Roe, 1823. The roof of the mill is just visible between the school and Octagon House. (Reproduced by permission of Mr and Mrs S. O'Shea) (See also Fig. 45)

and six pence. Then fifteen inches tall,[45] it was kept in a pot for the next six years. In 1804 it was planted out in the conservatory attached to the house. The tree grew strongly and by November 1819 was sixteen feet tall and the same in diameter. However, as early as July 1814 it had outgrown the conservatory and Underwood recommended to the Committee of Botany that it should be removed to another house. Plans were drawn up then for a new glasshouse, but it was not built.

By February 1817 the original conservatories attached to the house were in danger of total collapse so a new range of houses was planned and built. The Long Range, 153 feet long and nine feet high, was designed by Robert Doyle and cost £400. It stood in front of the artificial rock mound, in the centre of the Gardens, near the site of the present Palm House. A pond with a small island rockery and a fountain was constructed in front of it. The range had a wooden frame with a glass roof and sides. It soon became apparent that the pitch of the roof was too low and allowed rain to drive in between the laps of the panes. This problem was never rectified despite much puttying and painting. In the 1830s, the east wing of the range contained cultivars of *Camellia* and *Rhododendron*. The next section was devoted to species of *Erica* (heathers) from the Cape of Good Hope. The spacious central section, seventy-five feet long and eighteen feet wide and projecting slightly in front of the two end wings, was used as a conservatory for plants from warm, temperate lands, for example *Banksia*, *Grevillea*, *Protea* and *Araucaria*. It also contained a small pool surrounded by a rockery. The next division, a small one, was filled with representatives of Epacridaceae from Australia. The last section was the dry stove and contained cacti.

But, the fine specimen of the Norfolk Island pine was too large for the Long Range. On 5 July 1819 tenders were sought for yet another glasshouse; various estimates were received and that of James Carpenter of Grenville Street, Dublin, was accepted. Robert Doyle was placed in charge of the building work which began in late July. The glasshouse was contracted to be finished within two months as the pine was at this time growing in the open ground, unprotected, the original conservatories having been demolished. The idea was to construct around the tree an octagonal house that could be increased in height as the tree grew. Work went slowly. Although Underwood remonstrated with the contractor, the intended finishing date passed with the house still uncompleted. On 13 November, long after the original completion date, Robert Doyle and John Underwood both met the contractor, and Doyle formally told Carpenter that he was going to complain to the Committee of Botany. Doyle did this in a letter dated 15 November 1819. On 24 November, with the house still not completed, the night temperature fell to 20°F (−7°C). Nothing seemed to be amiss with the pine, and five days later Walter Wade wrote to John Foster suggesting that 'upon the whole from a variety of circumstances wch have occur'd I have every reason to think it wou'd bear the open air with us'.[46]

On 4 December the house was at last finished. It stood forty feet high with a diameter of thirty feet (Fig. 36). All still seemed well with the pine, but in February 1820 it started to show signs of distress. It continued to decline, and as there was general concern about its state, the Committee of Botany went to Glasnevin on 24 April to inspect it. The Committee reported that the tree was in a bad state, 'its health and beauty very much impaired and its recovery very doubtful'. Both Underwood and Mackie seemed to think it would recover. The Committee summoned the two men to a meeting at which several members expressed so much annoyance about the state of this prize specimen that one even proposed the dismissal of Underwood 'for mismanagement and neglect'.

John Foster had been keenly interested in the Norfolk Island pine and in November 1819 he ordered Wade to have it sketched.[47] Although he felt that nobody in the Dublin Society's Drawing School could do it justice, Dr Wade passed on the request and in January 1820 Robert Doyle was instructed to get the tree properly measured and to have a drawing made of both it and the building. S. Kerry prepared the drawing, for which he received a fee of £2 10s. 0d.[48] Walter Wade informed John Foster that the drawing had been completed and Foster replied that he would be 'happy to receive the drawing and plan of the Araucarian Conservatory'. Foster had been aware of the sad state of the tree and had told Wade to advise the artist 'to draw the plant such as it was before the neglect of last winter injured it'. Foster, distressed to hear about the damage, commented that he was certain that Dr Wade's feelings 'are much hurt by the neglect but I am afraid we cannot give sobriety to Underwood or ability to the Hothouse Gardener'[49] — Underwood's fondness for whiskey was well known to Foster and Wade.

Despite the forces ranged against him, John Underwood proceeded to defend himself against the accusation of neglect. He wrote a long and harrowing memorial addressed to the 'Right Honorable and Honorable the President, Vice Presidents and members of the Dublin Society', in which he refuted the allegation that the damage to the Norfolk Island pine was attributable, even in part, to his negligence or incompetence. He stated that he was not informed that the contractor was supposed to have completed the glasshouse by 24 September. Underwood noted that it was most unusual for frost to occur so suddenly and so severely in late November and remarked that

> even supposing your Memorialist could have foreseen the approaching inclemency of the night, your Memorialist was then afflicted with a severe illness, so that your Memorialist's life would have been endangered by removing from his bed, and Memorialist could not, during the night, have given his personal assistance in protecting the plant.
> That your Memorialist was then also without the means of affording any additional protection to the plant either from any resource provided by the Society, or those which his own humble means could have afforded.[50]

John Underwood repeated his conviction that the plant would recover and pleaded with the Society to remember that he had a family to support. He reminded members that he had served the Gardens well since he was brought to Ireland and that 'in no instance ever sought for private gain or emolument beyond the precise salary, which your honours have been pleased to allow him'. On that salary he had raised, educated and presently maintained seven children. Plaintively he wrote that the loss of his employment would be attended with the most afflicting consequences; for, as your Memorialist has been rendered unfit to resume his original profession of a gardener, from the length of time he has devoted solely to botanical pursuits...nothing could follow from his then desolate situation but that his helpless children would be instantly involved in all the miseries of want and starvation.[50]

Pleading for forgiveness for any errors which he may have unintentionally committed, he asked the Society to bear his family in mind.

A special committee was set up to investigate the truth of Underwood's memorial and its report was discussed by the Society. A motion that Underwood should be fined five pounds for his neglect and inattention was defeated but a substitute motion that he should be summoned before the Society and severely censured was passed. The revised motion noted that the injury to the pine was 'not altogether attributable to the head-gardener'. Underwood was summoned to the meeting of 18 May and told that he would be dismissed on the next occasion on which there was a well-founded complaint preferred against him, but he was not fined.

It seems that after this Underwood's relationship with the Committee deteriorated. Certainly he was unjustly accused of failing to supervise the work on the conservatory. Indeed, it is unlikely that he could have protected the sub-tropical plant from damage in the frost, for the Norfolk Island pine could not, as Wade thought, grow out-of-doors in Dublin or anywhere else in Ireland except perhaps in exceptionally well-sheltered areas in the extreme south-west. However, John Underwood continued to carry out his duties and to instruct the apprentices.

John White also came in for censure from the Society at this time. In July 1820 Foster commended White for wishing to visit Liverpool Botanic Garden. Foster instructed Dr Wade to give John White a letter of introduction. White also planned to visit nurseries and Foster asked him to bring back lists and prices for him. Foster would also have liked John White to travel to London at the Society's expense to procure those plants which were not in the Glasnevin collections, but funds would not allow this. Foster asked White to write to him on his return.[51] In December 1820 or perhaps January 1821, White went to Collon and was delayed there. On his return he was summoned before the Committee of Botany. Before White met the Committee Professor Wade presented a letter from John Foster explaining the reasons for the delay. This was received, according to Wade, 'with that exculpatory attention and politeness which it [was] so very justly entitled to'.[52] White was given a friendly reprimand, not for his delay at Collon, but for his absence in England. White had received permission to go to England from members individually, not from the Committee, and in order not to create a precedent he was admonished. The Committee also felt that during White's absence the Gardens had not been kept as they should have been.[53]

One of the members of the Committee of Botany, Sir William Betham, Ulster King of Arms, was particularly vindictive towards White who was again summoned before the Committee in March. He was pilloried by Betham. Walter Wade reported the proceedings at length to Foster, who cannot have been pleased. Sir William brought six apprentices to the meeting hoping that they would be accusers against White, but his plan failed and the only charge which could be levelled against the undergardener was that some years earlier he had obliged two or three members of the Society by laying out and planting several small gardens in the city. John White was questioned by Betham and then asked to leave the room, but the apprentices failed to substantiate the allegations.[54] However, Betham proposed a motion that White should be denied the usual payment of twenty pounds for assisting at lectures.

John White enjoyed Foster's patronage, and it is unlikely that he could have been dismissed from the Botanic Gardens without John Foster's concurrence. He remained at his post and was given leave in later years to spend as long as ten days at Collon selecting plants from Foster's private collections for the Botanic Gardens.

The apathy of the Society towards the Botanic Gardens and the largely unwarranted accusations against the head gardener and undergardener appear to have led to a decline in the state of the Gardens. Walter Wade complained about this to John Foster in the early 1820s and Foster made various attempts to reverse the decline. Wade himself was now elderly and infirm, but his devotion to his patron remained. In 1821 John Foster was created Baron Oriel and on 29 November Wade sent a proof copy of a new catalogue of the Glasnevin Arboretum and Fructicetum to Collon and noted that

> Lord Oriel will perceive our wants to be many and our losses shameful, owing much I fear to great neglect or perhaps to a lack of horticultural knowledge in those entrusted with the permanent care of the garden — there was at one time a rich store of trees and shrubs in the garden.
> I am sorry to inform your Lordship that Lady Norfolk [*Araucaria*] is Pine-ing away very fast and all our hopes of recovery at an end — even Doctor Underwood has given her over. What a pity, such a beauty and once so lovely in form, shou'd be cut off in the prime of life![55]

Laying aside the whimsy and sarcasm, Wade continued by suggesting to Lord Oriel that a special committee should be appointed, composed of noblemen, gentlemen and members of the Society selected by Parliament, to run the Botanic Gardens and that this committee should have sole control of the establishment and disbursement of money granted for its maintenance. 'By such means', Wade suggested, 'all cabaling about the garden may possibly be avoided. As to the Professor, be whom he may in the future, my stay in life being at present very uncertain...implicit confidence should be plac'd in him, even in pecuniary matters, so as to enable him to purchase new plants, replace old ones which have been lost etc'. The prospect of the continuing state of decline grieved Wade for the Botanic Gardens was 'a child of my own rearing, under the auspices, assistance and Fostering care of Lord Oriel'. He concluded the letter in a way which suggested that he thought it might be his last:

> The Botanic Gardens have heretofore flourish'd and I shou'd still hope leave to posterity in an unfading state and glowing with vegetable beauty, one of the brightest gems in the Royal Dublin Society's cap, and ultimately I shou'd hope tend to the advantage of Ireland in all its bearing.[55]

Wade's health continued to deteriorate although he did give lectures in 1823. By the spring of 1825 he was confined to his house in Dublin, and for the most part of the time, to his bed. A short time before he died he was visited by John White, who reported to Lord Oriel that Walter Wade was still very ill (Fig. 37).[56] Dr Wade died in Dublin on 25 July 1825, and was survived by his wife Mary.

Contemporary pen-portraits and assessments of Dr Walter Wade are rare. There is one of considerable interest which was published ten years after Wade's death by *Erinensis*, otherwise Dr Peter Hennis Green, a Dublin doctor who was a member of the staff of the medical journal *The Lancet*.[57] *Erinensis* published a remarkable series of witty pen-pictures of Dublin medical men and Dublin institutions. He sought evidence of malpractice, nepotism and favouritism in Irish medical circles, and was often sarcastic in his articles, so it was necessary for him to maintain the mask of anonymity. In an article on Glasnevin's professors, *Erinensis* painted a devastating picture of Dr Wade. He was described as an old-fashioned prig, a cox-comb, who gave his public lectures attired in a sky-blue waistcoat. 'Tricked out in apparel as tawdry as the pie-bald vestiture of the high priest of Flora herself, and thoroughly imbued with that dancing-master style of manner with which Chesterfield and the last century might have been enraptured. Dr Wade always commenced his didactic duties by a preliminary eulogium on the Society'. The lectures were burlesques, according to *Erinensis*, because Wade '...imagined...[that] counting the pistils and stamens of some showy flowers in the theatre of the Society constituted the whole duties of a professor of botany'. When compared with the praise heaped on Dr Samuel Litton, Walter Wade's successor, *Erinensis*' destruction of Wade's character appears grossly exaggerated, but Dr Green does imply that Wade was less than competent in his latter years.

John Foster was the only other person known to have left a testimonial on Walter Wade. It contrasts sharply with *Erinensis*' invective, and is found in a letter he sent to the Royal Dublin Society on 7 November 1825.

> Your late professor was a man of great worth, zeal and knowledge in that part of the science which is strictly termed Botany, and means merely a nomenclature of all the known plants in existence, arranged in so clear and intelligible a system as to admit within its catalogue, all other plants and vegetables which may at any time thereafter be discussed. I believe no man could excel him in his perfect knowledge thereof nor in the clearness and accuracy with which he explained the simplicity and beauty thereof; but he was very deficient in the whole practical part of the subject...namely the extension and improvement of agriculture and planting, and the production of the various vegetables useful for the food of man and all animals which are the objects of a farmer's care, or are applicable to the uses of manufactures.[58]

37. John White's letter to Lord Oriel concerning Dr Wade, 27 April 1825. (Reproduced by courtesy of the Public Record Office of Northern Ireland, Belfast and the Viscount Massereene and Ferrard (Foster/Massereene Papers, D207/29/191))

While praising Wade's scientific abilities, Foster restated what he had probably said many times during the past thirty years, that the Botanic Gardens at Glasnevin was to be used for the benefit of the practical farmer as well as the academic botanist. In his letter Foster argued that the person in charge of the Gardens should be 'a complete practical Gardener and scientific Botanist lecturer'. He even went as far as suggesting that James Mackay, the curator of the botanical garden attached to Trinity College, should be appointed to look after the Glasnevin Botanic Gardens 'without lessening his care of the charge entrusted to him by the College'. Foster proposed that the University should be approached immediately, and if the authorities there refused to release Mackay, the election of a new professor should be postponed until 'such a distant day as may allow full time for our endeavouring to find in the British empire, particularly in Scotland, a gentleman qualified as I think Mr Mackay is'.[58]

In concluding this letter, John Foster stated clearly his own role in the establishment of the Society's Garden:

> I trust in your kindness to forgive the liberty I have taken in addressing myself so fully and freely to you. I shall only call to your recollection as my apology, that I originally proposed the idea of a Botanic Gardens, that you called on me to procure the present situation of your garden, to lay out the grounds, and furnish them with plants and arrange them; all which under your auspices, have been accomplished, so that I am excusable in feeling a deep and proud interest in their prosperity.[58]

But Foster's opinions no longer held sway in the committees of the Society, and on this occasion his advice was not taken. The Society politely thanked him for his letter which was referred to the Committee of Botany. Before receiving John Foster's letter the Society had advertised the vacant professorship. Candidates were asked to give the Society a statement of their views on the 'utility of Botany, as connected with the useful arts'. The Council also agreed to reduce the salary of the professors of chemistry and botany to two hundred pounds, but to allow them to give private lectures to supplement their incomes. The Committee of Botany received Foster's letter and agreed that if James Mackay applied, John Foster's recommendation would 'have great weight'. Mackay did not put his name forward. There were six candidates by 2 March 1826, including William Allman, who held the botany chair in Trinity College, and Thomas Taylor, who was the Professor of Botany to the Royal Cork Institution. John Frost, lecturer on botany to the Medico-Botanical Society of London, William Steward, lecturer on botany and anatomy in Greenock. Dr Edward Vernon of Dublin also applied, as did the Royal Dublin Society's librarian, Dr Samuel Litton. A Mr Murphy submitted a late application. On 4 May, Dr Samuel Litton was elected with 175 votes; William Allman received two votes and Dr Taylor failed to get any.

Samuel Litton was born in London in 1781. His father, Edward, was an amateur theologian of some repute and an 'ardent lover of literature',[59] who wrote a *Grammatical Instructor* that was widely used in schools. Edward Litton married Rhoda Makon, daughter of an eminent barrister, and by this marriage he acquired a large fortune, which he soon lost in failed business adventures. Mrs Litton died when her son was only three years old. Samuel went to school in Liverpool and while still a lad he attracted the attention of the Reverend Dr Richard Magee. Dr Magee, a Fellow of Trinity College, Dublin, and later Archbishop of Dublin, recommended that young Litton should come to Ireland to attend Trinity College. Samuel Litton matriculated in Trinity College in 1795 when he was only fourteen years old. Although Samuel lived in Dublin during term he returned home to Lancashire during vacations. At the end of one holiday he had to return to Dublin before a certain day so that he could compete for a gold medal. He had sent his luggage on ahead, but when he arrived at Liverpool pier the packetboat VICEROY had already set sail—it never reached Ireland and no trace of it was ever found. Samuel Litton crossed by a later vessel and reached Trinity College just in time to get into the Examination Hall. He won the gold medal. In 1798 he became a scholar and ended a distinguished undergraduate career in 1800 when he obtained his bachelor's degree. In 1804 Samuel Litton graduated as master of arts and intended reading for a Fellowship and entering the Church. He abandoned these ideas and instead chose to study medicine at the University of Edinburgh from where he obtained a doctorate of medicine in 1806. Dr Litton returned to Ireland where he continued his studies, taking a special interest in botany. In 1809 along with Walter Wade, he was one of the unsuccessful candidates for the vacant Chair of Botany in Trinity College. In 1810 when the Dublin Institution was established privately in Sackville Street, Samuel Litton was appointed its professor of Chemistry and Natural Philosophy. He delivered lectures that were very popular and well attended.[59]

In 1815 Dr Litton was elected librarian of the Dublin Society, a post he held until he became Professor of Botany in May 1826. During his time as librarian Litton supervised the transfer of the collections from the Repository in Hawkin Street to the Society's new premises in Leinster House. He also had the pleasure of seeing the considerable expansion of the library, especially through bequests of such members as Thomas Pleasants and in 1817 prepared a catalogue of the Society's library for publication. For his work for the Dublin Society, Samuel Litton was elected an honorary member in 1820.

Litton's appointment to the professorship seems not to have resulted in any improvement in the state of the plant collections at Glasnevin. Like Walter Wade, Samuel Litton was not responsible for the day-to-day management of the Gardens, which remained the duty of John Underwood. The Society required the new Professor of Botany to live in the house at Glasnevin during the lecture season. But, the house, which contained the lecture-room, had not been occupied since 1795 and was in a poor state of repair; urgent work had to be carried out to make it habitable. So Litton did not start his lectures until November 1826 when he gave several on agricultural topics. He undertook a full programme in 1827. Just as his lectures at the Dublin Institution had proved popular, so Litton's botanical lectures at Glasnevin soon attracted large audiences. It is said that his lectures were not as 'dry' as those given by Wade, and it is clear that the emphasis was different.[60] According to *Erinensis*, Litton's classes were masterpieces of popular composition on the study and advantages of botany, and demonstrate[d] how much can be accomplished by a man of correct taste and general acquirements, when his energies and resources are concentrated on any given subject. In these admirable discourses, the selection of illustration is as judicious and copious as the warmest enthusiast in the cause of science could deserve..[58]

While Walter Wade loved philosophy and history, Samuel Litton (Fig. 38) concentrated on pure botany, the anatomy of plants, the structure and function of various organs and only at the end did he discuss plant classification. In the practical classes at Glasnevin, Litton abandoned the Sexual System of Linnaeus, and taught his students about the Natural Orders of plants.[61]

Samuel Litton's appointment marked the end of John Foster's control of the Botanic Gardens. After 1826 Foster was not consulted, nor does it appear that he offered any advice. He may have maintained contact with John White, but Litton is not known to have had any correspondence with Foster. Eighteen twenty-six also saw the eclipse of Foster's political power with the near-defeat of his son in the June election. Foster had misread the state of the electorate, still believing that Catholics were politically naïve. He was by this time an old man of eighty-six and died two years later in September 1828.[62]

John Underwood and John White were now also old men who had worked in Glasnevin for over twenty-five years. Their capacities were declining and they neglected to maintain the Gardens in a reasonable condition. The glasshouses and lodges were in urgent need of painting and some of the greenhouses also needed repair. In November 1829, the Committee of Botany recommended that the wooden-framed glasshouses should be gradually replaced by houses with metal glazing-bars. Eighty pounds was required immediately to make the professor's house comfortable, but there was a shortage of funds and nothing was done except some painting and minor repairs to the lodges.

In May 1830 Dr Litton was asked to present a detailed report on the state of the Botanic Gardens, and of the arrangement of plants based on the prospectus published in 1800. At the same time there was closer supervision of repairs that were being carried out. It was soon discovered that the greenhouses were so badly decayed and 'of such defective original structure' that they needed complete rebuilding. Once more glasshouses constructed of iron were recommended, and plans and estimates were submitted by Messrs Mallett and Ryder, but there were no funds for building new glasshouses.

Litton's report was presented to the committee and was published as an appendix to the *Proceedings of the Society*.[63] He noted that the arrangement of the collections was essentially as it had been in 1800 although a section displaying plants arranged in Natural Orders ('Jussieuian Arrangement') had been added. The trees and shrubs were in good condition. The Gardens contained some extremely rare plants and Dr Litton was able to note that other important botanic gardens had been supplied with species from the Glasnevin collections. Litton said that the Cattle Garden did not deserve the time and labour spent on it, for although it was well stocked, many of the plants were growing elsewhere in the Gardens. The Hay Garden, which contained agricultural grasses, was composed of plots which were too small for experimental studies and the seeds tended to get mixed. Again, many of the grasses were planted out elsewhere in the systematic beds. The Dyer's and Medicinal Gardens were in good order and the plants were of considerable interest. Litton was critical of the rock mound which was supposed to contain alpine plants. He said it was quite useless and no longer ornamental and suggested that it should be dismantled. The collections of climbing plants and variegated plants were also neglected and Litton advised that they should be scrapped.

The report ended with a series of recommendations about the glasshouses and the arrangement of plants. Dr Litton pointed out that insufficient notice was taken of the public's desire for showy flowers, and too little attention paid to florist's plants. He suggested that more plants of purely ornamental value should be grown. The Arboretum was overcrowded and Litton strongly advised an extension to it. As for the maintenance of the Botanic Gardens, Litton observed that a third gardener was needed because the vacancy created when John Mackie had died had not been filled. Much time and labour would be saved by the installation of an hydraulic ram to raise water from the River Tolka to the highest point in the Gardens.

Some of these recommendations were accepted and work began immediately on implementing them. In June 1831 one hundred pounds was set aside for repairing the glasshouses. By March 1833

38. Samuel Litton, an engraving by H. Griffin from the *Dublin University Magazine* (February 1828)

the hydraulic ram had been installed and was in working order; it was supplied by James Sheridan and cost £42 5s. 2d. It pumped water from the river to a reservoir near the Long Range of glasshouses. This ram was the first of its kind to be installed in Ireland and continued to serve the Gardens for many years. A plan was proposed for an ornamental fountain in the reservoir and this was built.

As far as the recommendations for the plant collections were concerned, little action was taken. John Underwood was so infirm that he could not perform his duties satisfactorily; indeed, the Committee of Botany even considered giving him easier work on a reduced salary until a replacement name *Fuchsia macrostemma* var. *recurvata* (Fig. 40).[5] It was illustrated in the *Irish Farmer's and Gardener's Magazine* and also in the *Botanical Magazine*. It may have been Niven who produced a hybrid between two species of the bird-of-paradise flower, *Strelitzia × nivenii*.[6]

While working at the Chief Secretary's Lodge, Niven wrote the first article for *Irish Farmer's and Gardener's Magazine*, which was launched in November 1833. His contribution took the form of a letter on 'The villa plantations in the neighbourhood of Dublin'.[7] Niven began by wishing the new journal every success and remarked briefly on the rapid advances which gardening had made in Ireland during the past few years. Then he enlarged on a subject to which he attached great importance — the cultivation of gardens by the poor. 'How desirable then', he wrote, 'to see the twining woodbine, the neat trimmed hedges, and the blushing rose, with all the other useful appendages of industry and care, combining to make the poor man's cottage a happy home, and more than that, an index of his mind, take the place of the dunghill and the miserable mud wall, with all their accompaniments of indolence and wretchedness'. In the main body of the letter he criticised the solid screen of trees which surrounded so many of the houses on the outskirts of Dublin, '*belted and clumped* as ever a fortification or prison was surrounded by a wall'. This type of planting restricted the view to the foreground and the sky, and the middle distance was lost. To remedy this, he suggested cutting out trees which obscured an attractive view but removing them in such a way that the remaining trees appeared grouped in a natural manner.

Ninian Niven left his post in Phoenix Park and took up his new job at Glasnevin in late March 1834. He received the same salary as Underwood, one hundred Irish pounds a year. Niven was allotted one of the gate lodges and received a grant of forty pounds with which to furnish it and the Society promised to build a small kitchen at the rear of the lodge. The title of Niven's post was changed from head gardener to curator. The position of foreman was apparently established shortly after Niven's appointment, but it is probable that this was only a change of title, 'foreman' replacing 'undergardener'.

In the Minutes of the Council of the Royal Dublin Society for 26 June 1834 and 10 July 1834, it is recorded that a foreman (his name is not given) was to be moved into the gate-lodge vacated by John White, the original undergardener. In November 1834, the Committee of Botany established a new scale of wages and at the same time provided for two foremen to be engaged, at twelve shillings per week each. However, in the Minutes of the Committee for March 1835 there is a note about one foreman having to do the duties of both the indoor and the outdoor foremen. This person may have been Patrick McArdle, who joined the staff in 1831.

The earliest named foremen are Patrick McArdle and John Robertson, but again entries in the Minutes and on wages sheets are confusing and contradictory. McArdle is said to have been appointed *outdoor* foreman in 1841, while Robertson is said to have served almost seven years as foreman by May 1844. But, a wages sheet dated June 1838 only records Patrick McArdle as receiving the foreman's wage of twelve shillings. Robertson was then getting the pay of a labourer, eight shillings per week. A further complication is that two men, James Clinch and George Cornut, are noted on this wages record as receiving eleven shillings per week each.

Though Ninian Niven's appointment was welcomed generally, not everyone was delighted by the changes which he soon proposed. Edmund Murphy, editor of the *Irish Farmer's and Gardener's Magazine*, applauded Niven as a person with talent and energy who would promote 'a knowledge of scientific Agriculture and Horticulture in this country.'[8] He hoped that the practice of using large parts of the Gardens for the production of potatoes and vegetables 'for the use of those connected with the establishment' would cease and that the whole of the Gardens would be devoted to the purpose originally intended 'namely to that of an Experimental Garden'. Murphy also asked that the 'useless arrangement of plants which certain animals approve or reject' be removed as it was based 'for the most part on pure conjecture'. Niven responded to this editorial and Murphy's limited concept of the functions of a botanical garden by explaining the plans he had for the Gardens. He said that he wished to exchange plants with similar institutions so that Glasnevin 'could keep pace with the almost daily advancement in the value of collections'. Ninian proposed to eliminate the Linnaean arrangement of plants and to arrange all species according to the system of classification proposed

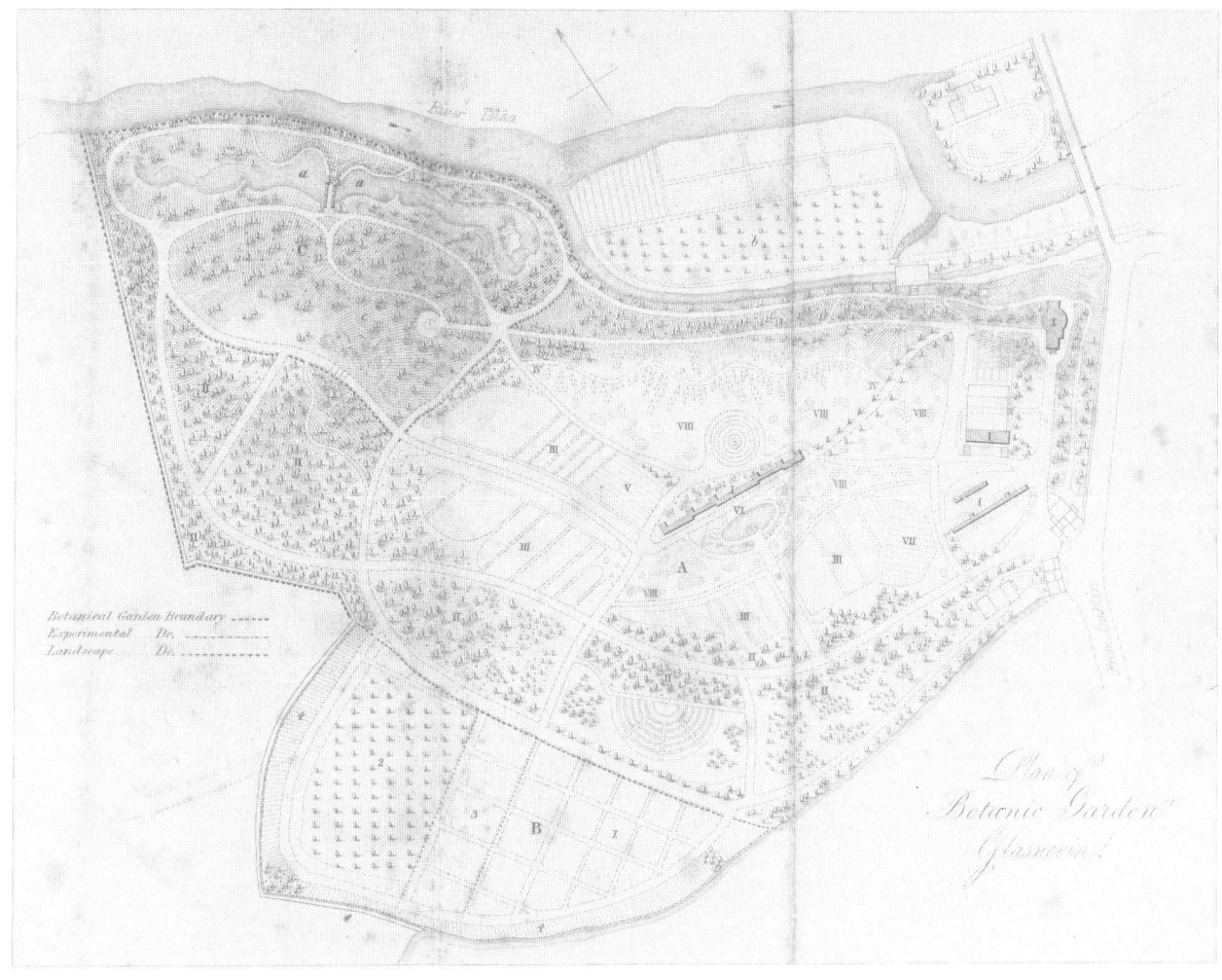

41. The Botanic Gardens in 1838; reproduced fom N. Niven's *The Visitor's Companion to the Botanic Garden*. (E. C. Nelson)

by Antoine de Jussieu. He also intended to form experimental plots to display the best mixtures of grasses for pastures and meadows, to remove the Irish and Cattle Gardens and to use some of the space in the glasshouses for forcing fruit and some of the walls to exhibit trained fruit trees. Beside the river he envisaged a few model cottages, let to the Gardens' labourers at a fair rent, where they could cultivate small plots. Finally, he hoped to improve the education of young gardeners.[9]

In a letter to the *Irish Farmer's and Gardener's Magazine*, Joseph Hamilton also welcomed the appointment of Niven, and his plans.[10] But Niven's proposal to build model cottages provoked at least one angry letter to the magazine. Was the Royal Dublin Society, asked the writer (J. C.), going to turn its Gardens into a 'kale-yard or turnip-field...to deform this fair spot with a number of poor labourers' cabins, surrounded as they ought to be with their respective plantations of leeks and cabbages...[and to let] the fair features of our magnificent garden, and science and beauty [be] defaced by pig-stys and potato-pots?'. Ninian Niven was likened to John Claudius Loudon who 'would throw down Westminster Abbey if it stood in the way of his wheelbarrow'![11] However, the most intemperate criticism came from Professor Samuel Litton who fumed against the ornamental beds near the Long Range. Dr Litton had suggested earlier that more purely ornamental flowers should be grown in the Gardens, yet he now said that while such frivolities were all right in the Chief Secretary's garden, they were unsuited to a botanical garden![12]

Once established at Glasnevin as curator, Niven began to pull the Gardens into shape. His improvements were mainly directed at refurbishing the depleted collections and increasing the general efficiency with which the institution was run. Comparison of the maps of the Gardens for 1818

42. The Chain Tent in 1838; wood-cut from Ninian Niven's *The Visitor's Companion* (see Fig. 43) (E. C. Nelson) slightly enlarged

43. The Chain Tent photographed by David Davison in November 1981. The weeping ash shown in Fig. 42 has been replaced by a central steel pole, and the whole tent is draped with *Wisteria*

(Plate 2) and 1838 (Fig. 41) show that he did not make sweeping changes in the general layout. The Irish and Cattle Gardens were removed and the Grass Garden was converted into an experimental plot for economic crops such as turnips, peas and potatoes. Ornamental plants were massed at the back of the Long Range. On the south side of the pond, which Niven restored, a chain tent was erected (Figs. 42, 43) with rustic pillars covered by fuchsias and climbing plants and centred by a weeping ash; this ash tree survived until 1870 when it was replaced by an iron pole. A semi-circular Rose Garden surrounded by a spruce hedge was laid out in front of the Long Range, and the South Field was converted into a general nursery and a fruit garden. Niven explained at length in *The Visitor's Companion to the Botanic Garden* (1838) the reasons for devoting a special part of the Gardens to the culture of fruit trees. Visitors could see all the varieties worth growing, and apprentices could be properly instructed in pruning, grafting, manuring and general management of orchards. Experiments could be carried on into methods of cultivation and training, and gardeners supplied with scions for the renewal of old orchards. The best fruits could be displayed in the headquarters of the Royal Dublin Society for the guidance of those who proposed to form new orchards.[13] When Niven resigned from the curatorship four years later, in 1838, there were over a hundred varieties of apples, sixty-three varieties of pears, thirty-nine different plums, nineteen varieties of cherries and eight different gooseberries in the fruit garden at Glasnevin.

The Mill Field was developed as a willow garden during the 1830s to provide sets for anyone who cared to ask for them. At this period, the growing of willow for canes was carried on extensively along the banks of rivers, and the Royal Dublin Society yearly offered premiums to encourage the industry.

By 1834 it was obvious that the original plantings in the Arboretum had been too close. Many of the trees were deformed for lack of room and light. Niven put into practice precepts he had expounded during the previous year and had some of the trees removed. This left more space for those remaining and at the same time opened up vistas to the Sugar Loaf Mountain, the spire of Saint George's church and the nearby mansion of the Bishop of Kildare. Niven was essentially a landscape gardener and he may have considered a vista to be more important than an individual tree. But, this did not accord with the ideas of others and he was not given a free hand. A month after his appointment Niven toured the Gardens with members of the Committee of Botany and pointed out those trees he wanted to remove. A few days later he was ordered to suspend felling.

Despite the overcrowded planting, there was still ample room in the Botanic Gardens for large gatherings, and the highlight of Ninian Niven's curatorship was the splendid *déjeuner* provided on Friday, 14 August 1835 for those attending the Dublin meeting of the British Association for the Advancement of Science, the fourth since the Association's formation in 1832. Although the three miles of road from Dublin to Glasnevin was 'excessively dusty, for our Dublin friends have not yet arrived at the luxury, the conveniency, of watering the roads to any distance from their capital', the pleasant gardens, and 'the still more admirable display of Irish beauty they contained', were ample compensation.[14] Twelve hundred people were accommodated in eighteen marquees which were pitched in a crescent. Two of the tents were reserved for officers of the Association, so that they could have their breakfast and return to Dublin before the rest. Some of the tables were not well-attended by the 'servitors', and the guests whose 'bad luck conducted them to these forsaken places had to do as well as they could *sans* coffee, *sans* tea, *sans* milk, *sans* everything'.[14] Elsewhere members of the Royal Dublin Society acted as stewards, and the bands of the 18th Regiment and the 7th Dragoons played 'some of the most admired pieces of music'. While the enthusiastic botanists, including Professor Graham from Edinburgh and Allan Cunningham of Australian fame, went to Howth after the feast to study the coastal flora, many other visitors stayed at Glasnevin until four o'clock, promenading in the grounds.[14]

Niven's short tenure of the curatorship was bedevilled by friction between himself and Professor Samuel Litton. Niven began his general reorganisation of affairs with the eviction of the McCoys from the basement of Litton's house. Dr Litton may have regarded this move as an intrusion into his domestic arrangements, but he does not seem to have protested about it. The ground floor and the first storey of the professor's house was occupied by Litton, his married sister, Mrs Margaret Cuthbert, her three daughters and a son. The McCoys had lived in the basement since about 1815; Mary McCoy was the housemaid responsible for looking after the apprentices' apartments, and her husband, Timothy, was a garden labourer. Apart from records of her yearly wages of £11 7s. 6d in the Society's accounts, there is no mention of Mary McCoy until 1822, when a member asked was it a fact that for years past she had kept a public laundry, washing the clothes in the Botanic Gardens and drying them in the hothouses? And, was it a fact that her husband was in charge of the coal for fuelling the furnaces in the hothouses? Six years later she was warned by the Committee of Botany that if she did not keep the apprentices' quarters in a more cleanly state she would be dismissed. In 1831, Mrs McCoy was told to restrict herself and her family to one room in the basement and to carry out whatever orders Litton might give her. Shortly after his arrival Niven complained to the Committee of Botany about the McCoys' behaviour, and Mrs McCoy was told to leave the Gardens together with her husband and two daughters. But the Society relented and, in view of her long service and the fact that her banishment was in part due to the misdemeanours of her family, gave her a pension of five pounds a year and re-employed her husband. Mrs McCoy died in 1841.

It is, of course, speculation that Dr Litton was displeased with Niven's action in removing the McCoys from the basement, and their replacement by apprentices, on Niven's suggestion. But, Litton may have preferred the McCoys, for the records of the Society show that the young men were an unruly lot. There is, however, nothing speculative in the saga of the differences of opinion on the correct conduct of Gardens affairs between the academic, Samuel Litton, and the gardener, Ninian Niven.

On the available evidence it would seem that Dr Litton had a good relationship with the former staff of the Gardens. In 1830, when, according to outside reports, the Gardens was in a decrepit state, Litton had much good to say of it, though he observed that lack of money precluded the purchase of new plants and the failure to replace John Mackie, who died in May 1821, had resulted in understaffing. Nor did Litton voice any criticism of Underwood. When John White was retired in 1834 on a pension, Professor Litton insisted that White's presence as a general aide at his lectures was indispensible and White continued as attendant until his death in 1837.

Before the end of his first year at the Botanic Gardens Niven asked the Society to state precisely what were his reponsibilities and what were Litton's, relative to the giving and receiving of plants, the teaching of pupils and apprentices, the coal allocation and the appropriation of ground for the use of the curator and the professor. But before a ruling could be given a major dispute erupted between Niven and Litton, over the botanical arrangement of plants in the Gardens.

When Ninian took over as curator, most of the plants were arranged according to the Linnaean system. In a letter to Professor William Hooker, Niven said that he had found the Glasnevin Gardens

> totally destitute of any Natural Arrangement...I conceived that the best method of occupying the comparatively limited space of ground [available] so as to include as much of the above principles as possible, would be, to exhibit a lineal arrangement of plants according to their natural orders, so as to include...complete British arrangement, distinguishing those plants peculiar to Scotland, England and Ireland, by appropriate labels, whilst on the opposite side of the curvilinear walk...I propose exhibiting, merely a few of the most prominent types of each natural order in the system...'[15]

Niven used the occasion of the first Irish meeting of the British Association for the Advancement of Science in August 1835 to explain his scheme for rearranging all the plants according to 'Natural Orders' (Fig. 44); he read a short paper to Section D (Zoology and Botany).[16] The new arrangement was strenuously condemned by Samuel Litton, in a three thousand word document which he presented to the Society.[17] It was folly, Litton contended, to rearrange the Gardens according to the new system when that system could well be superseded in ten years time by another classification. If that happened then instead of the motto being 'Toujours en avant' it would be 'Toujours en derrière', and if Niven's arrangements were acceptable to the Society then he hoped that the Society would discharge

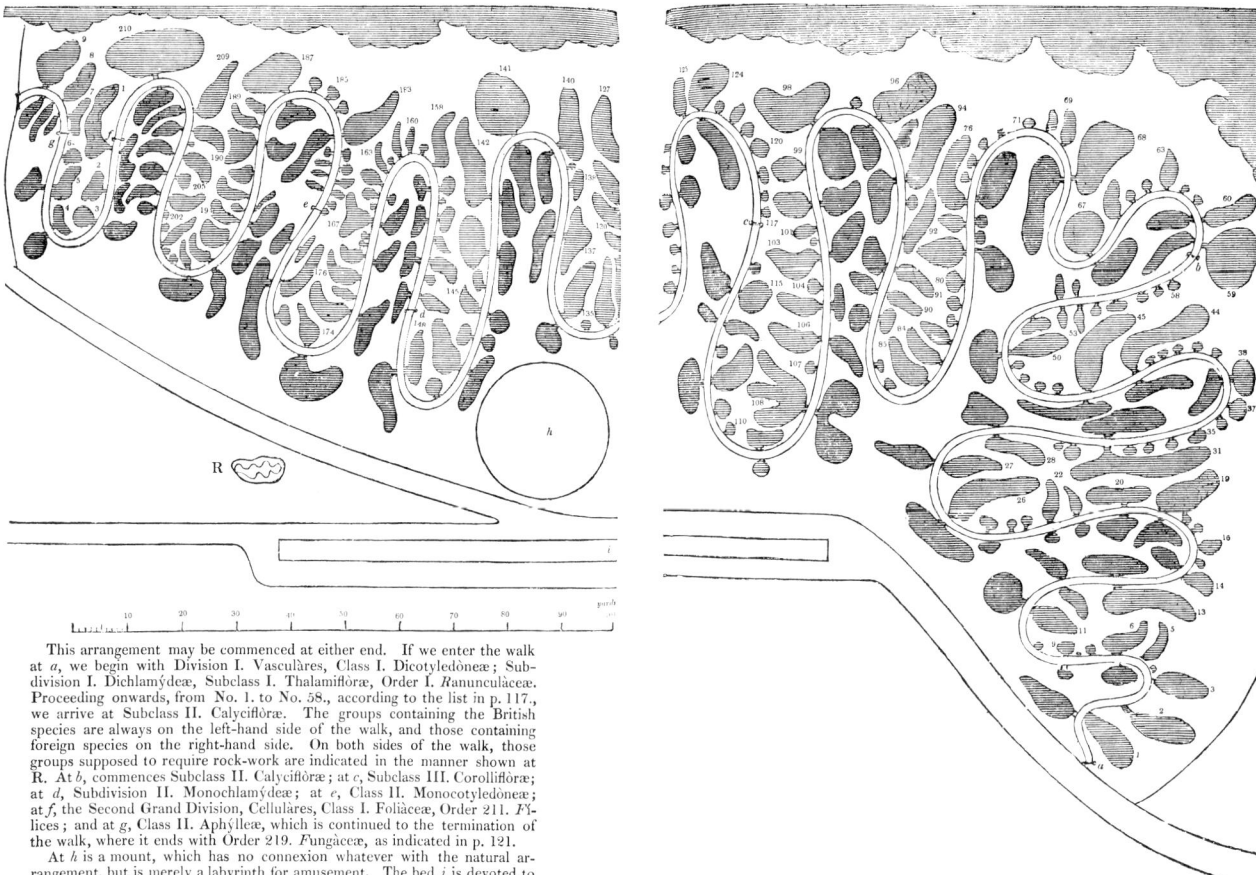

This arrangement may be commenced at either end. If we enter the walk at *a*, we begin with Division I. Vasculàres, Class I. Dicotyledòneæ; Subdivision I. Dichlamýdeæ, Subclass I. Thalamiflòræ, Order I. *R*anunculàceæ. Proceeding onwards, from No. 1. to No. 58., according to the list in p. 117., we arrive at Subclass II. Calyciflòræ. The groups containing the British species are always on the left-hand side of the walk, and those containing foreign species on the right-hand side. On both sides of the walk, those groups supposed to require rock-work are indicated in the manner shown at R. At *b*, commences Subclass II. Calyciflòræ; at *c*, Subclass III. Corolliflòræ; at *d*, Subdivision II. Monochlamýdeæ; at *e*, Class II. Monocotyledòneæ; at *f*, the Second Grand Division, Cellulàres, Class I. Foliàceæ, Order 211. *F*ýlices; and at *g*, Class II. Aphýlleæ, which is continued to the termination of the walk, where it ends with Order 219. *F*ungàceæ, as indicated in p. 121.

At *h* is a mount, which has no connexion whatever with the natural arrangement, but is merely a labyrinth for amusement. The bed *i* is devoted to ornamental shrubs.

44. 'Plan for the exhibition of a Natural Arrangement of Plants, in the Glasnevin Botanic Garden' by Ninian Niven, as presented to the meeting of the British Association for the Advancement of Science in Dublin 1835. (Reproduced from J. C. Loudon's *Gardener's Magazine* 12 February 1836) (National Botanic Gardens, Dublin) The plan was not executed at Glasnevin

him of all responsibility and protect him from censure, if censure should arise. Litton took the opportunity to observe that the Botanic Gardens was essentially an apparatus for instruction in scientific botany and 'not a promenade for the mere refined amateur such as the garden of the Chief Secretary in the decoration of which Mr Niven has acquired such deserved celebrity'.[17] The Professor needed more space for specimens for his lectures and Niven was filling the Gardens with flowers.

No action was taken until June 1837, when a subcommittee was formed to report on the duties of the curator and professor which were then delineated clearly. Niven was to have complete control over the management of the hothouses, plantations and ground, the staff, carts, horses, propagation of trees and plants and the saving of seed. He was to keep accounts, be in charge of coal, and given power to dismiss labourers and suspend undergardeners. He was to attend on strangers seeking admission and to see that no plants were pulled or injured. The Society's Committee of Botany was to send him his orders in writing and he was to file these orders. The curator was to be answerable only to the Society. The professor was to give public lectures and to prepare an essay every November on discoveries or improvements in botany which the Society would print for public benefit. It was also the professor's responsibility to keep the meteorological records at the Gardens and to make a monthly return of them to the Society. He could suggest improvements and experiments. Finally, the professor was in charge of naming the plants in the Gardens and the production of catalogues. The plants for the professor's lectures had always been supplied from the Gardens and the Committee of Botany made a ruling on this matter also. The professor was to give three days' notice of any plants he needed and the curator had discretionary power to refuse if he had only one plant or if the cutting of a plant would injure its growth. If a dispute arose between the two men, it was to be referred to the Committee of Botany but it was hoped that 'goodwill will prevail all round and the Garden always be put first'.

When Niven became curator in March 1834 the plant collections were depleted. The only information that survives today about the collections is a count which Niven made of plants in the glasshouses in April 1834. He found that there were 2,549 plants in the Upper (Long) Range, 1,593 in the Lower Range, 605 in the Pelargonium House and 253 in the Acacia House; a total, including duplicates, of 4,990 potted plants.[18]

To expedite the restocking of the Gardens, the Society decided to allot Niven seventy-five pounds to buy plants. He went to England for six weeks in September and October 1834 and visited estates in Scotland and England, including those of the Duke of Northumberland, of Lady Stanley at Hoole near Chester, and of John Acton in Kew, and the botanical gardens in Edinburgh, Glasgow, Liverpool, Manchester and Birmingham. Niven also visited Dickson's nursery in Chester, and those of Whalley, Skirving and Cunningham in Liverpool. He returned from this tour with a thousand plants as gifts and over two hundred and sixty which he had purchased. Drummonds of Stirling gave the Gardens a collection of all known varieties of wheat, oats and barley, as well as foreign seeds used in agriculture in their respective countries.

The Earl of Mountnorris was so pleased with improvements effected in Glasnevin that he suggested in January 1835 that Niven should go to the Mountnorris family seat, Arley Hall in Staffordshire, and select more plants for the Gardens. Niven went to England in the following month and visited not only Arley Hall, where he obtained six hundred plants, but also Wentworth, Chatsworth and the botanical gardens in Birmingham, Sheffield, Manchester and Liverpool, from which places he obtained another five hundred plants.

Niven, like Underwood, was a horticulturist not a botanist, so he did relatively little work on the native flora of Ireland. In an attempt to improve the collection of native species, he visited County Mayo, and Connemara and the Joyce Country in west Galway between 19 August and 2 September 1836, travelling on foot for over two hundred miles in search of plants.[19] On the summit of Mweelrea in County Mayo, Niven collected the dwarf willow, *Salix herbacea*. In Connemara he recorded new habitats for the mountain avens (*Dryas octopetala*) and the spring gentian (*Gentiana verna*). At Roundstone he met the young local amateur botanist, William McCalla, who had discovered a new species of heather almost two years previously. McCalla showed Niven this plant (*Erica mackaiana*) recently named after James Mackay of Trinity College Botanic Garden. Niven obtained plants of the Irish heath (*Daboecia cantabrica*). These were all brought back to Glasnevin for cultivation. Niven also collected, pressed and dried herbarium specimens of ferns and mosses.[20]

A few months before his resignation, Niven made his last trip to Great Britain in his capacity as curator of the Botanic Gardens. He went in the late autumn of 1837, and returned with about two hundred new plants from various botanical and private gardens in Scotland and from the West Highlands. Edinburgh Botanic Garden donated a banana plant which was satisfactorily established in the Octagon House.

One of Ninian Niven's innovations was the publication of a circular giving instructions on the collecting of plants and seed in foreign countries.[21] He sought to encourage people to donate interesting plants to the Gardens so that the collections would increase without too much cost. He was successful in this and the records he kept indicate that plants and seeds began to arrive in quantity at the Gardens from members of the Society and from people living abroad. A considerable collection of South American plants was donated over many years by the Earl of Arran and by John Tweedie. These private donations and the exchange arrangements made with other botanical gardens meant that the collections of plants at Glasnevin grew steadily; in all, in the four years 1834 to 1837, Glasnevin gave 1,810 plants to other gardens and in return received over 3,600. A number of interesting new species were established in the Gardens; for example, in 1838, Glasnevin had a single plant of Douglas fir (*Pseudotsuga menziesii*). In view of the present importance of this tree in commercial forestry, Niven's comment on it is worth quoting:

> Of this, as yet comparatively rare pine, we have only one small specimen. From the way in which we have seen this species luxuriating in England, we have no doubt it will be found a tree well adapted for cultivation in many parts of this country.[22]

45. Title page of Ninian Niven's guide to the Botanic Gardens published in 1838. The 'Banana House' depicted in the vignette is the Octagon House erected in 1819 (see also Fig. 36) (E. C. Nelson)

46. (opposite) 'View from Front of Hothouses' in the Botanic Gardens, Glasnevin, drawn and engraved by J. Kirkwood, from Ninian Niven's *The Visitor's Companion*. This is a fanciful view perhaps meant to represent a vista of the Long Range and pond in the centre of the Gardens (see Fig. 41) (E. C. Nelson)

Those lines are taken from the major work that Ninian Niven published during his curatorship, *The Visitor's Companion to the Botanic Gardens*, which was produced in 1838 (Fig. 45). This little book contains a number of somewhat fanciful engravings of the Gardens (Figs. 42, 46) and in about fifty thousand words not only describes the Gardens but also provides a great amount of information about the plants to be seen there. It is sprinkled with biblical quotations. The review of this book in the *Irish Farmer's and Gardener's Magazine* was favourable and advised every visitor to Glasnevin to buy a copy. The reviewer gave

> unqualified praise [for] the style in which the work has been got up. It reflects much credit on the publishers Messrs Curry and Co., the plates by Mr Kirkwood and the letter press by Mr John S. Folds are fully equal to the best books brought out in London...[We give] our unqualified approbation, not only of the arrangement of the subjects in this elegant little work, but of the intrinsic usefulness and interesting nature of these subjects...[The] descriptions are conveyed in an elementary and pleasing language and will be found suited to the capacity of every reader.[23]

Ninian Niven also published several articles in horticultural journals while he was in charge of the Botanic Gardens. One, on the physiology of plants, reported his experiments on elm saplings and other plants; he found that in the stems of the plants he had studied there were two currents of 'nutritive

VIEW FROM FRONT OF HOTHOUSES.
PUBLISHED BY WILLIAM CURRY JUN & Cº

juices', ascending and descending.[24] In another article he proposed that a botanical garden should include a section with different types of rocks and soil and their appropriate flora, properly labelled, growing on them.[25]

In 1835, the Royal Dublin Society awarded Niven a silver medal for an essay on the failure of the potato crop. This also gained him a prize of twenty guineas from the Agricultural Society of Ireland.[26] The failure was due to 'dry rot' disease, now known to be caused by *Fusarium*. Niven rejected the idea that 'a minute Fungus or parasitical plant' was the cause, for he believed the fungus was 'the consequence' of the disease. He proposed that the crop failure arose 'from an atmospheric influence, acting upon the cut, or bruised Potato' and then suggested that hill farmers should be encouraged to produce seed potatoes so that damage 'unavoidably sustained by imported potatoes should be obviated'. Niven also suggested growing more potato plants directly from seeds. He followed this article in 1837 with another on experiments on autumn and spring planting of potatoes, and in 1846 with a second pamphlet on a disease of the potato.[27]

Niven resigned from the curatorship in August 1838. It was his accomplishment during the four and a half years he held the curatorship that he retrieved the Gardens from the poor state in which he found it. An index of the improvements he made, as well as of the growing popularity of the Gardens, is the increase in the number of visitors. In 1834, the year of his appointment, they numbered 7,000; by 1838 the figure had risen to 10,000.

When Ninian Niven left the Gardens he set up as a nurseryman and landscape gardener at The Garden Farm, Drumcondra, where he remained for the rest of his long life. In 1864, Garden Farm was described as being a combination of 'farm, a good half of market-garden, a fourth of nursery garden, a slice of alpine ground, and two or three of experimental horticultural buildings with a slight dash of the botanic garden, and a flavouring of the school for young gardeners—the whole spiced with fruit growing...'. Vines continued to be one of Niven's major interests, and he had several elaborate vineries in his garden (Fig. 47), from which he sent excellent grapes and strawberries to the Dublin markets.[28]

One of his first commissions as a landscape gardener was the planning of public gardens at Monkstown. His plans were included as an appendix in a *Prospectus of the proposed public gardens at Monkstown*

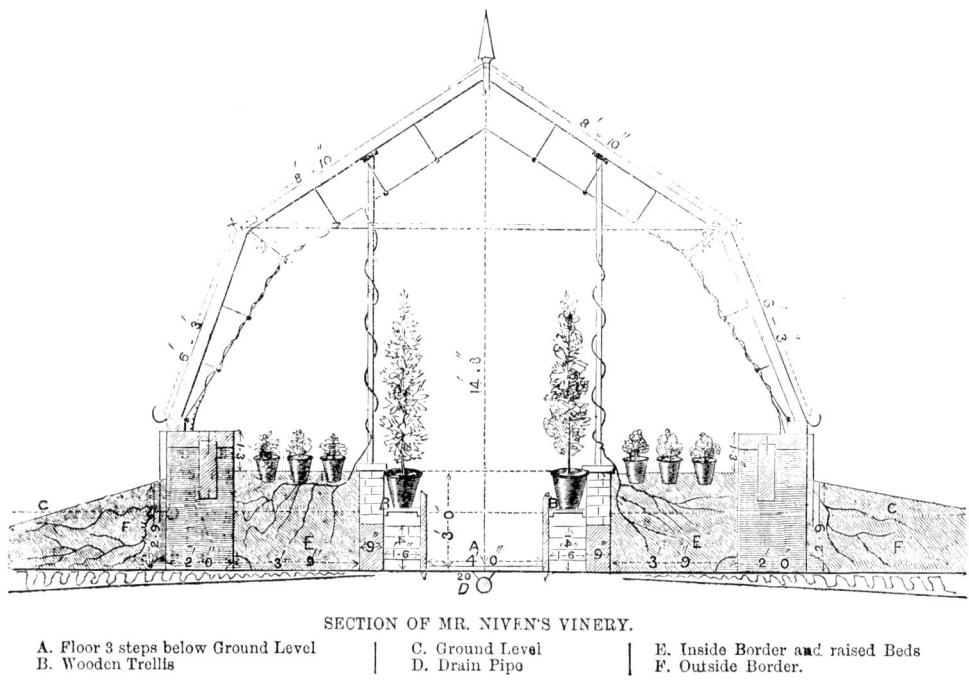

SECTION OF MR. NIVEN'S VINERY.

A. Floor 3 steps below Ground Level
B. Wooden Trellis
C. Ground Level
D. Drain Pipe
E. Inside Border and raised Beds
F. Outside Border.

47. Vinery at Garden Farm, Drumcondra; from the *Gardeners' Chronicle*, 24 December 1864 (slightly enlarged) (National Botanic Gardens, Dublin)

Castle.[29] This scheme was not carried out but his completed commissions include the gardens at Nutley, Santry Court, the Royal Marine Hotel at Kingstown—'the Brighton of the Irish Metropolis' (Dun Laoghaire)—Athgarvan in County Kildare and Baronscourt in County Tyrone. He also laid out the parterre in front of the Vice-Regal Lodge in Phoenix Park, Dublin (now Áras an Uachtaráin) and the garden of the Chief Secretary's Lodge (now the residence of the ambassador of the United States of America to the Republic of Ireland).[30] There is a cryptic note in the *Irish Farmers' Gazette* of October 1853[31] to the effect that it was a pity that Niven's design for the floral arrangements in the Winter Garden for the Great Exhibition had not been adopted. But he did plan and supervise the gardens of the International Exhibition of 1863. Many of his designs were based on the formal style of gardening popular in France, which he studied during a visit to Paris at the end of the 1830s.

For many years Ninian Niven was closely connected with the Royal Horticultural Society of Ireland. At a dinner during the 1838 Show, replying to a toast by James Mackay to 'The introduction of rare plants and to the health of Mr Niven', Niven said 'from a feeling of old acquaintance and many other circumstances connected with the Society, I think I have some little right to begin to feel myself becoming indigenised'.[32] He served as honorary secretary of the Royal Horticultural Society from November 1847 until the beginning of 1853. It was not an easy time, especially during 1848 when the Society was attacked severely in the horticultural press for its 'state of decadence' and for apparent irregularities and unfairness in the award of prizes.[33]

Niven was given the task of trying to improve the Society's image after the controversies had died down. Unruly conduct was all too common at shows, and the behaviour of the crowd at the Autumn Show in 1849 came close to a riot. A 'noted character' was observed stuffing fruit and vegetables into his pockets and 'his unmentionables' and departing with a melon wedged into his top-hat. Niven had to report to the Society a few days later about the affair.[34] This incident has to be viewed against the background of poverty which afflicted Ireland, especially in the aftermath of the famine caused by the failure of the potato crop in the years since 1845.[33]

The potato crop had been destroyed by a disease which had not been seen in Europe before. In 1846 Niven published an open letter addressed to the Duke of Leinster titled *The Potato Epidemic and its probable consequences*.[35] It was one of the 'saner and more scientific' commentaries on the potato blight, written in Niven's inimitable style, sprinkled with remarks deferential to the Irish Establishment and with the occasional biblical quotation. He described the scene on the quays of the Liffey, as the destitute waited for the ships to take them away from the famine and death to a new life in North America. Niven's horticultural experience allowed him to differentiate between 'dry rot' disease, 'black leg' disease and the blight, but he was unable to identify the cause of the problem. He was among the majority who believed that the blight had an 'atmospheric origin', a physical rather than a biological cause.[35]

> **Agricultural and Horticultural Instruction.**
>
> N. NIVEN, formerly Curator of the Royal Dublin Society's Botanical Garden, Professor of Landscape Gardening, Secretary to the Royal Horticultural Society, &c., &c., begs leave respectfully to inform his friends and the public that he is making arrangements on his Premises here, to afford Instruction, both Theoretical and Practical, on Agricultural and Horticultural Science, to a class of young Gentlemen, visiting Pupils from the city or neighbourhood, and with this view he purposes instituting a Course of Lectures to be delivered by him on the Tuesday and Friday of each week throughout the Season, between the hours of Eleven and One o'clock, commencing in the first or second week of May, and ending in November. Terms Eight Guineas (including all charges) for the course—payable in advance. Besides the above, Mr. NIVEN will have no objection to impart Instruction on these most useful and interesting subjects, on the Premises, to any portion of the Pupils of public or private Establishments in the city, as may be agreed upon. Open to Professional Engagements, during the winter and spring months, as heretofore.
>
> N.B.—Early notice of intention to attend will oblige, addressed by letter, Garden Farm, Drumcondra.

48. Ninian Niven's advertisement from the *Irish Farmers' Gazette*, 1849. (National Botanic Gardens, Dublin)

In the 1840s, Niven established himself as a 'Professor of Landscape Gardening'. He advertised in local newspapers (Fig. 48) offering a course of instruction to young men in practical and theoretical horticulture and agriculture. Thus Niven continued his interest in education and the training of young gardeners. His philosophy of education had been stated earlier:

> It shall be my duty to inculcate the greatest assiduity to their business; and whilst I insist on a certain advancement in a knowledge of plants and of botany, shall not permit them to cultivate this branch of their business to the exclusion of the not less important one of general gardening.[36]

In his later years, Ninian Niven took to writing poetry and in 1869 he published a collection of poems under the title *Redemption Thoughts*.[37] The verses are essentially religious in outlook, low in poetic merit and heavily weighted with allusions to trees and plants.

> Of vegetable life—equally to man
> Inexplicably in his finite state
> Each individual therein hath its
> Microscopic spores, its germ, its seed.
> 'Mongst trees, what monarchs rear their lofty heads!
> The oak of Mamre ancient, Cedrus grand
> Of Lebanon (a remnant thereon still),
> The tow'ring conifers of distant zone
> The "Wellingtonias" so marvellous!
> Its altitude three hundred feet or more,
> Spiring heavenward!....

Niven died on 18 February 1879 following a heart-attack, having spent the previous day marking trees in his nursery. His obituary notice in the *Gardeners' Chronicle* described him as a landscape gardener with very few equals.[38]

An intensely religious man, Ninian Niven undoubtedly saw the Botanic Gardens as the pale image of Eden,[39] in the same way that the Reverend Robert Walsh had perceived the place twenty years earlier.

> What scene of beauty, happiness, and peace
> Must ancient Eden—Paradise have been!
> Laid out, arranged, and group'd by master-hand,
> The glorious Architect of universe!
> Whose works, amidst Creation's ruin, still
> So lovely are, and marv'llously sublime;—
> What must it then have been ere Adam fell!
> Who, in the garden exquisite was placed
> To dress, to keep, enhance (if possible)
> By rich artistic skill; order perfect
> Ev'ry part complete, unparallel'd...

Niven offered his resignation to the Committee of Botany in the middle of August 1838. On 20 August the Committee submitted his letter to the Council of the Society noting the resignation 'with very great regret'. The reasons for his resignation were not disclosed—one can only surmise that, despite an increase in salary from one hundred Irish pounds to one hundred and fifty English pounds, he found it intolerable working with Professor Litton, and perhaps he resented the petty restrictions imposed by the Committee of Botany and wished to be his own man.

The Society realised that Niven's departure was a serious loss and paid tribute to the 'zeal, energy and talent' which he had displayed in restoring the Gardens 'to its present state of perfection'. In the four and a half years he was in charge Niven invigorated the collections and introduced many new plants. On his departure there were more than 8,500 potted plants in the glasshouses, representing almost 2,500 species; this indicates a doubling of the collections during his curatorship. There were nearly 5,000 species and varieties of herbaceous plants (including annuals) out-of-doors, eighty-six named varieties of potato and 336 different *Dahlia* cultivars. The Arboretum and shrubberies contained over 1,000 different plants.[40]

However, the state of the Gardens on Niven's departure was far from pleasing to Professor Litton who continued to fight on. A year after Niven relinquished the curatorship, Dr Litton wrote to the Council of the Royal Dublin Society, not to the subordinate Committee of Botany:

> In the original beauty of its laying out by the late Mr Underwood and in the neatness, order and tasteful decoration introduced by Mr Niven, it is, I believe not surpassed by any similar establishment. The arboretum may be considered as a good collection of trees, many of which are very rare; the herbaceous departments and conservatories have many plants of great value and beauty; and the esculent and agricultural garden, as well as the orchard, through the zeal of the late curator, are at present both useful and instructive. But it has many defects, and as these must chiefly be noticed by a scientific botanist, it is on this account more particularly incumbent on me to direct your attention to them.
>   The systematic department of herbaceous plants is very defective. When the Gardens were originally formed, it was, I believe, in this respect, one of the most complete in Europe; but the species have gradually been disappearing, and in some of the classes are now reduced to a very small number...the vacancies thus produced ought to be filled up. Many plants, indeed, have been introduced by our late active curator, but they have been principally plants recently brought into Great Britain and consequently, though valuable in themselves, have not supplied the defect pointed out.[41]

Litton suggested that the collection of hardy and other plants was incomplete and inadequate for various purposes. The hothouse collections lacked species of economic value, and some of the more valuable plants had suffered during work on the glasshouses. Additions were also needed to the collections of orchids, palms, tropical ferns, Cape heaths and various other groups. He estimated that it would take about £150 to fill these gaps. But Professor Litton was alone in his criticism of Ninian Niven's curatorship. Niven had found the Gardens neglected and the collections depleted, but he had left it in a 'creditable and comparatively perfect state'.

By 23 October 1838, the Royal Dublin Society had received three applications for the curatorship; they were from Daniel Ryan of Templemoyle, Andrew Murray of Liverpool, and David Moore of the Ordnance Survey. Moore had been an unsuccessful candidate in 1834, but on 29 October 1838 he was appointed Niven's successor and took up the post on 24 November that year. The situation facing the new curator was a little less intimidating than that which had confronted Ninian Niven.

8
David Moore
The first decade, 1838-1848

LIKE the previous curators of the Botanic Gardens, David Moore came from Scotland. His forebears had worked on the land for many generations and had held the same farm in Abernyte parish in the Carse of Gowrie near Dundee since at least 1600. Traditionally the family name had been spelt Moir, but about 1830 the whole family adopted the spelling Moore, and it was said that David, who had sometimes used the form Muir, welcomed the change as he believed it would help to camouflage his Scottish origins.[1]

David, born on 23 April 1808[2] in Dundee, named for his grandfather, was the eldest son of Charles Moir (born 1782), a gardener, and his wife Helen Rattray (1784-1832). Nine more children were born into the family, but two died in infancy. Of the six surviving boys two had distinguished careers in botany and horticulture, culminating in directorships of important botanical gardens — David at Glasnevin and Charles (1820-1905) at Sydney in Australia. The latter was the second boy in the family called Charles, his parents following the then prevalent custom of naming a later child after one who had died.

David Moore was brought up near Dundee and was fortunate to receive instruction in botany from Douglas Gardiner, Conservator of the museum attached to the city's Rational Institution. Gardiner was a widely-read man and had a 'wonderful' botanical garden. He was described as a 'true genius and a true gentleman', but unfortunately was 'very hard up for the means of living and bringing up his family'.[3] He influenced at least one generation of Dundonians, stimulating their interest in natural history. His nephew, William Gardiner (1808-1852), an umbrella-maker and a contemporary of David Moore, wrote several botanical books including *The flora of Forfarshire* published four years before his early death.

When he was old enough David Moore was apprenticed to Mr Howe, head gardener at the estate of the Earl of Camperdown near Dundee. At that period Camperdown House possessed one of the finest gardens in Scotland, with well-stocked conservatories, shrubberies and borders.[4] Here Moore learnt the basic skills of gardening, and after two years as an apprentice was promoted to foreman. He left Camperdown two years later, moving to various other 'good places for instruction' including James Cunningham's 'extensive botanic nursery' at Comely Bank near Edinburgh. For the last six months of his domicile in Scotland David lived at Borthwick,[5] south-east of Edinburgh, and then in November 1828 he migrated to Dublin to take up the post of foreman in Trinity College Botanic Garden.

Following the law-suit between Professor Edward Hill and the University in 1803, Trinity College had lost possession of the embryonic botanical garden at Harold's Cross. However, on 5 July 1806, three acres were leased at Ballsbridge for the formation of a new College Botanic Garden and James Townsend Mackay was appointed curator. Mackay was also a Scot, a native of Kirkcaldy, who had come to Dublin in March 1804 as College gardener and botanical assistant. He did not have any connection with the Harold's Cross Botany Garden nor with Professor Edward Hill. Mackay took up his post on 25 March, and looked after various gardens on the campus, including the derelict Physic Garden behind the Anatomy Theatre. He also assisted at the lectures of the new Professor of Botany, Dr Robert Scott. During these years Mackay travelled to various parts of Ireland collecting and studying native plants, thus laying the foundations for his outstanding work on the native flora. He quickly established a fine collection in the new College Garden, which soon gained a considerable reputation in Ireland and abroad. By the mid-1820s the College Botanic Garden at Ballsbridge was regarded as a better garden than that at Glasnevin—John Claudius Loudon noted in 1834 that the College Garden 'though small...contains the richest and most varied collection in Ireland', whereas Glasnevin 'like most of the other botanic gardens...is very imperfectly kept up'.[6]

In his position as foreman at the College Botanic Garden David Moore took charge during Mackay's absences. He began to write notes and articles for horticultural journals, including the *Irish Farmer's and Gardener's Magazine*. In February 1834, Moore published his first substantial article, on a group of plants that was, in later years, to exercise more and more fascination for him, 'An account of a new *Catasetum* from Brazil, with observations on the importation and treatment of "Orchideoeus Plants"'.[7]

In September 1832, David Moore's mother had died. She had written to him a month previously, and, as far as one can judge from the wording of this distressing letter, David had not written home since his departure from Scotland four years earlier.[8] Perhaps as a result of the inevitable family dislocations following his wife's death, Charles Moore sent his young son Charles, aged twelve and a half, to join David at Trinity College Botanic Garden as an apprentice.

In February 1834, David Moore applied for the curatorship of Glasnevin Botanic Gardens. James Mackay, believing that Ninian Niven would be appointed to Glasnevin, canvassed for Moore to fill the probable vacancy at the Chief Secretary's Lodge—it is not known if David Moore wanted that job, or even if he applied for it—but he was not appointed to either post. In June 1834, he joined the Ordnance Survey as botanist,[9] replacing Dr Thomas Hopkirk (1785-1841), another Scot, who had resigned two weeks after his own appointment due to ill health. Dr Hopkirk remained in Belfast and became a friend and adviser to David Moore.[10]

David Moore was based in Belfast and worked under the direction of Captain Joseph Ellison Portlock (1794-1864) who was in charge of the geological work of the Survey. Moore's duty was to carry out surveys of the floras of counties throughout Ireland and to collect specimens of plants, 'together with whatever other objects of Natural History come in his way'.[11] During his time with the Ordnance Survey Moore worked in counties Antrim and Londonderry, and made copious notes on the flora of the region. Moore did not confine his work to flowering plants; he also studied the mosses, liverworts, lichens and seaweeds. Some of the plants he was able to identify himself, but many had to be sent for naming to more experienced botanists, including James Mackay and Dr William Hooker. At the beginning of his work for the Survey, Moore's time was fully occupied and he continued to be a poor correspondent. James Mackay wrote to him on 13 November 1834, scolding that he had heard nothing for over one month, adding that 'Charley is now going on very well'.[12]

An interesting assessment of David Moore, at this time of his life, was given by William Harvey, who, thirteen years later was to become the Royal Dublin Society's Professor of Botany. In April 1835, Harvey told Dr William Hooker that Moore was the

> young man employed by the Ordnance Board...to collect the vegetable productions of Ireland...He is engaged to go through the whole of Ireland and fifteen years from the present time are allocated for his tours, so that by the time his engagement ceases he will have a very full knowledge of Irish botany. He is both zealous and acute and quite fond of Cryptogamia—and I look to his discovering many new species. Already he has given our flora *Asplenium lanceolatum*[13] which is a good beginning. Write him a lecture on drying plants, he will take it in good part I have no doubt. He is but a young collector, though a fair botanist having been for many years as assistant at the College Gardens to our friend Mackay.[11]

Hooker wrote, as requested, beginning a correspondence with David Moore that continued for thirty years.

During the four and a half years David Moore worked in the north of Ireland he discovered several plants which had not previously been recorded in Ireland. A sedge, *Carex buxbaumii* (Plate 5), which he gathered on the shores of Lough Neagh is now extinct,[14] and a grass, *Calamagrostis stricta*, also from Lough Neagh's shores, is extremely rare. Both these plants were illustrated in the Ordnance Survey's memoir on the parish of Templemore published in 1837,[15] to which Moore contributed botanical information. The illustrations were prepared by a young, Dublin-born artist, George du

Noyer (1817-1869), best known for his geological work, who was employed by the Ordnance Survey as a draughtsman. David Moore observed that du Noyer had no experience of botanical drawing, yet his work done under Moore's supervision was of an exceptionally high standard.[16] As well as the watercolours of the grass and sedge from Lough Neagh, and two supposed new seaweeds from the north coast, du Noyer drew and painted roses, brambles (*Rubus* ssp.), apples and fungi.[17] Moore hoped that the roses would be included in the Templemore memoir as they could 'form an interesting feature of the botanical section...We have given portions of the shoots along with each species where such formed a very essential part of their character'.[18] Du Noyer's original watercolours of plants are now housed in the National Botanic Gardens, as are the herbarium specimens collected by David Moore (Fig. 49). Moore's manuscript, listing the plants he had found in counties Londonderry and Antrim, also survives in the Gardens. Some of Moore's plant records were published by James Mackay in *Flora Hibernica*, the first comprehensive flora of the island, in 1836.

In 1836, William McCalla, from Roundstone in Connemara, was appointed to assist Moore but seven months later he was dismissed for passing some of Moore's unpublished information to other botanists.[19] Shortly after this Charles Moore, then aged sixteen, joined his brother as an assistant botanist, having completed his training as an apprentice and journeyman in the Trinity College Botanic Garden.[20]

While employed by the Ordnance Survey, David Moore married Hannah Bridgford, the daughter of Thomas and Sarah Bridgford of the Spafield Nursery, Ballsbridge. The wedding took place at the Scots Church in Saint Mary's Abbey, Dublin, on 7 April 1836 — James Mackay and the bride's father witnessed the marriage certificate.[21] Their life together did not last long, for, tragically, Hannah Moore died at Glasnevin of typhus in December 1840 leaving a nine months old baby and the two years old Isabella; no further trace of these two children has been found in the Moore family records, so they may have been brought up by the Bridgfords.[22] David Moore retained his connections with the Bridgfords: of the £500 spent in Irish nurseries during his curatorship just under half went to the Bridgford's Spafield Nursery. This firm was already in existence in 1832, specializing in tulips, and in the 1850s held a respectable collection of conifers. By 1840, Thomas Bridgford and Sons had opened a shop at 52 Lower Sackville Street, but by 1892 the business had transferred to Saint Stephen's Green West.

By 1838 Moore was becoming restless in the Ordnance Survey. The travelling and the long absences from home did not make his family life easy, and he was incurring much extra expense. In January he asked Sir William Hooker about his chances of joining a plant collecting expedition to America that he had heard was being organized, but this turned out to be only rumour.[23] Then in May he showed interest in the curatorship of Sydney Botanic Garden following the resignation of Allan Cunningham in disgust and frustration at being 'a mere cultivator of official cabbages and turnips'.[24] Dr John Lindley, Professor of Botany in the University of London, advised that the post would not be 'a garden of roses to him who gets it'.[25] However, Cunningham was asked to withdraw his resignation by the new governor of New South Wales and David Moore was spared the necessity of making any decision on the matter. Moore also showed interest in a proposal that a botanical garden should be made in Bermuda and wrote to Hooker saying that as long as a post abroad was permanent and 'settled' he would accept it; the climate did not matter, but religious instruction had to be available.[26] A month later Ninian Niven's resignation from Glasnevin Botanic Gardens was announced, and Moore turned his sights once again to Dublin. This time, when he applied for the curatorship, Moore was more experienced and able to present an impressive list of sponsors: Sir William Hooker, Regius Professor of Botany at the University of Glasgow, Professor Robert Graham who held the chair of botany at the University of Edinburgh, James Lawson Drummond, Professor of Anatomy at the Royal Belfast Academical Institution, the curators of three botanical gardens — James Mackay of Trinity College Botanic Garden, Stewart Murray of Glasgow and William McNab of the Royal Botanic Garden, Edinburgh — and Captain Portlock of the Ordnance Survey. Sir William Hooker wrote of Moore: '...in the whole circle of my acquaintance, there is no one whom I could recommend with more confidence and satisfaction...his experience as a Cultivator, his Botanical and

49. Pages in David Moore's Ordnance Survey herbarium; on the left is a specimen of *Lathyrus palustris*, on the right *Rosa* × *hibernica*; Charles Nelson 1985 (National Botanic Gardens, Dublin)

General Scientific Knowledge, his devotedness to his profession, his excellent character...and his acuteness and variable knowledge are well known to me...' Mackay also gave 'the fullest testimony' of David Moore's fitness and qualifications. Murray declared that he had known Moore for fourteen years, and that the extensive correspondence which Moore carried on with many leading botanists in Europe could be of great benefit to Glasnevin 'because he carried the benefit of their contributions along with him'. In this belief Murray was right, for in a few years Moore was to be a welcome visitor in many European gardens and nurseries. McNab thought that there were 'very few men so well qualified for the situation', and Professor Graham stated that Moore was eminently suitable in every way for the curatorship.[27]

In late October 1838 David Moore was elected curator at Glasnevin with an initial salary of one hundred Irish pounds and a house with free fuel and light. He received twenty-two pounds for removal expenses. Moore took up his appointment on 24 November 1838 and remained at Glasnevin until his death forty-one years later. His brother Charles succeeded him as botanist to the Ordnance Survey.

Before his departure, Niven was instructed by the Royal Dublin Society to make all necessary arrangements for the change of curator. He prepared inventories of garden tools and exchanged letters with David Moore about the apprentices.[28]

One of the first tasks which David Moore undertook in his new position was the listing of the plants in the greenhouses and frames. His manuscript catalogue survives in the National Botanic Gardens, and provides a record of the state of the indoor collections at the end of Niven's curatorship. It indicates that there were many southern hemisphere plants in Glasnevin, the genera *Banksia*, *Acacia* and *Erica* (Fig. 50) being particularly well represented. Also, as a result of Niven's contacts with South America, there were seventeen different *Verbena* species. The largest genus in the glasshouse collections was *Mesembryanthemum* with ninety-eight distinct species. There were numerous un-named cacti. The collections contained nearly 2,500 species, and many duplicates. The most surprising aspect is the large number of mosses which were cultivated, sixty species from eleven genera being represented. In contrast, there were only eighteen species of ferns, although these did include the native maidenhair fern (*Adiantum capillus-veneris*) and the Killarney fern (*Trichomanes speciosum*).

Apart from the normal adjustments that always need to be made when one official takes over from another, at Glasnevin there was also the matter of apprentices to be considered. Once they had completed the course, apprentices from Glasnevin were much sought after by owners of private gardens. But while they were being trained in the Garden, the apprentices gave a great deal of trouble, cutting lectures, absenting themselves from the Gardens, drinking, insulting Society members and being 'idle and inattentive to their studies in botany'. By the end of his first year at Glasnevin Niven had asked the Royal Dublin Society for permission to admit pupils of a better social class than those usually taken, boys who would come to be taught and who would not receive any wage. He thought that apprentices should live inside the Gardens under 'the eye and control' of the curator, and suggested that they should be lodged in the basement of the professor's house in place of the McCoys. Support for Niven's ideas on improving the standard of the apprentices came from the House of Commons Select Committee in June 1836.[29] This Committee advised that the Gardens should be made 'as much as possible a school for young gardeners seeking instruction in horticulture', that a better type than the ignorant boys formerly taken should be recruited and that their time of service should be increased to four years. They should pay a total fee of twenty pounds and receive wages increasing from three shillings a week in their first year to seven shillings in their fourth year if they merited it. During their stay at the Gardens they should be put through each department as quickly as they were able to absorb the instruction.

As the apprentices paid their fees to the curator, when Ninian Niven left, in November 1838, he returned to each pupil whatever proportion of his fee remained unused. David Moore agreed to take Niven's pupils, all eight of whom signed the memorandum which transferred them to Moore's charge. On 1 December, the new curator issued a memorandum to the pupils informing them that he would 'devote part of an evening in each week during the winter for botanical conversation'[30] when he would teach them the rudiments of botany. This was intended to prepare the pupils for the lectures given by Professor Samuel Litton, in the spring and summer. Moore demanded from each pupil 'on the last Saturday of every month, a written list of all the plants he had learned...which I shall test by bringing him specimens of some of the plants noticed in his list'. The first meeting was held on Monday 10 December and the first lists were submitted on Saturday 29 December. Moore invited any pupil who had a reasonable objection to these instructions to hand him the objections in writing. Finally he declared that the more anxious the pupils were to learn 'the more will I interest myself on their behalf'.[30]

In 1845 the terms of the apprenticeships were altered slightly. The first year's wages were increased to four shillings a week, and the final year's to eight shillings. Pupils still had to supply their own bed-clothes and linen as well as whatever utensils they required. The letter of contract sent to one prospective pupil gardener, Thomas Scott, stipulated that he would be entitled to lodgings and fuel, and would have the 'share of a bed with another person'.[31]

Moore had only held office for three months when he was faced with two problems. One was the storm damage caused by the Big Wind of 7 and 8 January 1839, and the other was his involvement in a feud with Professor Litton.

The Big Wind of 1839 has passed into the realms of legend and tradition in rural Ireland. It was the custom of old people applying for a pension at the beginning of this century and unsure of the date of their birth to say, 'I could hold a potato in me hand on the night of the Big Wind'. The storm, 'one of the most violent which had blown from the face of Heaven within the memory of the oldest inhabitant', began during the day and by ten o'clock in the evening had become a severe

50. Pages from David Moore's manuscript catalogue of the plant collections in 1838 (this opening includes *Erica*, and indicates that there were 68 different species from southern Africa) (National Botanic Gardens, Dublin)

westerly gale. It wreaked havoc all over Ireland, leaving many villages and towns without a roofed house. In Dublin the trees along the Grand Canal were pulled up by the roots and hurled across the water to the opposite bank; the splendid row of elms at the Royal Hospital in Kilmainham was completely destroyed; thousands of trees were uprooted in Phoenix Park and furniture, flung out of the upstairs windows of burning houses, was carried away like leaves in the wind. The *Aurora Borealis* glowed brightly in the northern sky for the greater part of the night. Towards four o'clock in the morning the wind had eased a little, and by the following afternoon it had completely died away and snow fell.[32]

The wind did less damage at the Botanic Gardens than might have been expected. The wall along the Glasnevin Road was partly destroyed, injuring a policeman sheltering there, but inside the Gardens the most expensive item due to storm damage was sixty pounds for 400 tons of gravel to resurface the paths which were cut up when the fallen timber was carted away.

Moore considered that the storm actually improved the picturesque appearance of the grounds as several of the toppled trees had obscured the view on the north side. In many cases these trees were of little scientific interest, and he proposed to replace them with rare species. However, Moore observed, 'it was a pity that so many trees fell on the side of Glasnevin village as the village was now exposed to view from the Gardens and the Gardens to cold north-east winds'.

He replanted several important trees that had been blown down, chiefly poplars, oaks, ash and pines and nearly all survived the experience, including a specimen of *Pinus pungens* which had been planted in 1805. The Big Wind brought down so many trees all over the country that the price of timber was depressed and less was received for the Gardens' wind-thrown trees than would have been obtained under more normal conditions.

Moore showed considerable interest in the development of the Arboretum. In 1842 he declared that future operations in the Gardens should be directed at improving the Arboretum and extending it by reducing the area occupied by hardy herbaceous plants. He continued the work Niven had begun in rearranging plants according to the Natural Orders, but as he pointed out, it was not practical to move fully established trees and so he planted new specimens in positions best suited to their needs rather than according to botanical order. The western end of the Gardens was developed as the main part of the Arboretum, but as the soil was shallow, large pits had to be dug and filled with good loam, especially on the slope above the pond, to ensure that the new trees would flourish.

The original plantings in Glasnevin were too crowded. Ninian Niven removed a few trees, and Moore made a further clearing in 1844 taking out duplicates and unsightly, sickly specimens. As Moore obtained more specimens he found that, in spite of the thinnings, he was running short of space and repeatedly appealed to the Society to acquire the ten acres that lay beyond the western boundary wall. The Arboretum was eventually extended into this area, but not while the Royal Dublin Society controlled the Gardens.

The second problem that David Moore faced, immediately after his arrival in Glasnevin, was a protracted series of squabbles with Professor Litton. On Niven's departure, Dr Litton's household appropriated the cowshed that had been allotted to the curator, and Moore complained to the Royal Dublin Society. The Society instructed Litton to return possession to Moore and at the same time rebuked the professor for allowing his horse to roam loose in the grounds 'to the detriment of the Gardens'. A bequest of books to the Gardens in 1839 provided Samuel Litton with another opportunity to discomfit Moore.

In view of the Society's desire to make Glasnevin an educational institution it is not surprising that a library was formed at the Gardens a short time after foundation.[33] The first purchase, in 1799, was of twenty-four pounds' worth of second-hand books, and William Sole's *Menthae Britannicae*, published in Bath in 1789. A few years later, Thomas Pleasants gave one hundred pounds' worth of books and by the end of the 1830s the Society had spent a further sixty pounds on the library. In 1839, John Robertson, an eminent horticulturist from Kilkenny and a member of the Society, left his library of 400 books for the use of the Glasnevin staff. These volumes were put with the others in the lecture room, but the roof leaked and they were damaged by water. Dr Litton refused a request from the Society to take the books into his house and they remained in the leaking lecture room until 1846, when a reading room, furnished with tables and seats, was provided at the rear of the newly-built Curvilinear Range.

At the time of the Robertson bequest, Professor Litton was made librarian of the Botanic Gardens, on his personal security of five hundred pounds. David Moore was allowed full use of the books and could take out not more than six volumes at a time, giving Litton a receipt for them. On Moore's order the foremen could borrow one book at a time, with the exception of the finely illustrated volumes. Anyone injuring or mutilating a book had to replace it. Six months later the regulations were relaxed: both Moore and Litton were to have keys to the bookcases, and the apprentices were given access to the library. But Professor Litton refused to let either David Moore or the apprentices use the library and eventually, after a year of disputes over the matter, Litton relinquished all responsibility for the books and Moore was placed in charge of the library.

The library was well-used by the gardeners and apprentices after this relaxation of the rules. A register of loans, beginning in 1852, survives in the Gardens' archives,[34] and shows that a wide range of books was available. Thomas McDonagh, for example, read Caius Pliny's *Natural History* as well as works on engineering, plant physiology and zoology. However, several of the borrowers eschewed such learned texts. John Hegarty preferred much lighter reading, and so he availed himself of the novels and story-books. He read Thomas Crofton Croker's *Irish Fairy Tales, Rob Roy*, and a book on pirates! William McArdle read *Robinson Crusoe*, a volume on self-instruction for gardeners, and one titled *Evenings at Home*.

David Moore's early years as curator were not easy, yet the obstructive actions of Professor Litton were only minor irritations when compared with the crisis that loomed in 1841. The Lord Lieutenant refused to propose to Parliament that the grant of £5,300 for the Royal Dublin Society should be renewed. One possible result of this, envisaged by various concerned people, was the dismantling of the Botanic Gardens at Glasnevin. David Moore confided in Sir William Hooker that

> ...I feel myself most uncomfortably circumstanced, not knowing whither this establishment will be broken up (as it must be, if the grant be really withheld) or whither it may be put under the management of some other board, which I can't wish to see the case, and I think it not unlikely, as it must surely be too monstrous to anihilate [sic] such a generally acknowledged useful concern as this, because other branches of the Society are not fulfilling what government considers the desired object.[35]

From 1830 onward the Royal Dublin Society became increasingly subjected to government interference. Tensions also developed within the Society between those members who favoured, and those who opposed, the liberal policies being pursued by the Whig Government. A Select Committee of the House of Commons was appointed in 1836 to investigate the management and activities of the Society. While recognizing the importance of the Society as a national institution, the Committee recommended various changes, which included a suggestion that the Botanic Gardens should be made more freely available to people who were not members of the Society.[36] It also suggested that newspapers and political periodicals should not be provided in the Society's reading room. In general, the Select Committee approved of the way in which the Society fulfilled its role, but the division between the Government and the Royal Dublin Society was deepened by its report. The Society accepted the report and attempted to implement the recommendations, but not to the complete satisfaction of the Irish Administration. After the return of the Tories to power in 1841, the Lord Lieutenant was advised not to renew the annual grant until a commission set up later in the year reported on how it could best be used by the Society, or alternatively if it was desirable to form an entirely new institution.

Public uneasiness over the consequences of 'misunderstanding' between the Administration and the Society found expression in the *Irish Farmer's and Gardener's Magazine* in the spring of 1841. The editor asked how could the Gardens survive in its present form if the difference between the government and the Royal Dublin Society resulted in the cancelling of their grants?[37] The Glasnevin Botanic Gardens, according to J. C. Loudon, was 'not only the most extensive in Europe, but the most comprehensive in its plan'. Would its 'great natural landscape beauty', and the fine specimen plants in the conservatories 'make way for geraniums, primroses, and any other production which would command a ready sale', and would the Gardens be made the 'invidious rival' of Dublin's well-managed and respectable nurseries?[37]

Concern was also expressed in London and the crisis provoked an editorial in the newly established weekly journal, *The Gardeners' Chronicle*. On 17 April 1841 Dr John Lindley observed that 'a great outcry has been raised in Ireland at the supposed intention on the part of the Government of suppressing the Botanical Garden of Glasnevin, one of the finest establishments of its kind in Great Britain'.[38] If this was true, then he would have to comment in no uncertain terms on such a discreditable act.

Dr Lindley made extensive inquiries about the matter, and concluded that such an outcome was extremely improbable. He was confident that the commission members recognized the importance of the Gardens even if 'some of the members of the Dublin Society talk of letting it for market-gardens or for a cemetery, in support, we imagine, of their newsroom...where idlers, congregating to read the [political newspapers] elect each other into committees without the slightest reference to the fitness of the parties for the position in which they are placed'.[38]

David Moore was able to inform Sir William Hooker in June that the commission had reported most favourably. But the problem was not yet resolved and by October Moore was again very despondent. Supplies for the Gardens had all been stopped due to lack of money. Moore felt that it was a national disgrace that the Botanic Gardens should be allowed to suffer neglect 'for the mere gratification of a political party...How it may end I know not, but such has been the annoyance to me, that I would gladly quit if I can't get anybody to answer me and let them fight as they think best'.[39] In March 1842 the situation had changed for the better—the Lord Lieutenant was satisfied with alterations made by the Society to its management and the grant was restored. David Moore was then hopeful that many repairs could be carried out and especially that the dilapidated glasshouses could be replaced.

The difficulties that the Glasnevin Gardens faced between 1836 and 1842 were paralleled at Kew Gardens which also went through a period of uncertainty. In January 1838 a committee of three—Dr John Lindley, Joseph Paxton, gardener to the Duke of Devonshire, and John Wilson, gardener to the Earl of Surrey, Lord Steward of the Royal Household—was appointed to inquire into the management of the Royal Gardens at Kew. The 'existence of Kew as a scientific establishment came perilously near to an end' and it was not until the middle of 1840 that its continuance as a botanical garden was certain. It had by then 'successfully weathered the severest crisis in its history', and under the guidance of Sir William Hooker, who was appointed director in March 1841, it developed into an institution of incomparable international importance.[40]

After the Royal Dublin Society reached its accommodation with the Administration and the grant was secured, the Society's Committee of Botany proceeded with plans for the replacement of the decaying glasshouses.

It was a felicitous time for the building of new hothouses, as the first half of the nineteenth century saw revolutionary changes in the manner of their construction. The first conservatories at Glasnevin, such as the original range, were of wood with a solid roof. Greater understanding of the light requirements of plants led to the substitution of glass for wood in the roof, as in the Long Range. But the construction of lofty, curvilinear conservatories was subsequent to two technical advances in the manufacture of iron and glass.[41]

Changes in the technique of iron production, including the substitution of coal for wood charcoal in the smelting process, and the elimination of impurities, produced a better quality iron. In the 1810s John Claudius Loudon developed and patented a wrought iron glazing bar for glasshouses, which was the single most important technical innovation leading to the construction of curvilinear conservatories. These new glazing bars could be produced much more easily than cast iron ones, as they were extruded through a die.

Until the mid-1830s only two kinds of glass were available. Broad glass was made in sheets of four square feet, but was uneven in texture, thickness and colour. Crown glass could be produced in slightly larger sheets, if the bulbion bulge in the centre was not cut out, but this glass was green and the sheets were slightly curved. Changes in the methods of glass manufacture resulted in the production of good quality sheet glass that was flat and had even texture and colour. However, glass was an expensive item until 5 April 1845, when the tax on it was reduced—the cost of sheet glass then fell from one shilling and two pence to only two pence for one square foot, and the cost of crown glass was reduced by four-fifths.

The Royal Dublin Society's case for a new range of glasshouses at Glasnevin was strongly enhanced early in 1842 when Jacob Owen, the architect to the Board of Works, reported that it was a matter of great urgency to replace the derelict glasshouses. The Epiphyte Range, near the entrance, was in a state of total decay and could not be repaired, and the Long Range was so dilapidated that it would be a waste of money to try to repair it. Only the Octagon House was still sound.

In February 1842, the Society resolved to erect new greenhouses and the Committee of Botany was instructed to look into the matter. It estimated that £2,500 was required for a range of houses to accommodate the whole collection, but as the Epiphyte House was in a state of imminent collapse, the Committee suggested that a glasshouse, costing about £1,000, should be erected as soon as possible for the epiphytes. The Lord Lieutenant, Earl de Grey, was approached and he showed interest in the proposal but the Society was refused a government grant. In April 1842, a subscription list was opened in order to raise part of the cost of the new glasshouses before an application was made to Parliament for a special grant.

At the same time, David Moore was instructed to travel to England, not only to obtain plants, but also to obtain plans for suitable buildings. Among the gardens he visited was Syon House, near London, the property of the Duke of Northumberland.

On 20 June, David Moore presented a plan for a new epiphyte house to the Committee of Botany, and was then asked to consult Duncan Ferguson, master of architectural drawing to the Society, about the preparation of detailed plans and estimates. This work took six months to complete and in the meanwhile the tender plants had to be cared for in crumbling glasshouses—it became necessary to prop up the buildings despite the previous decision not to spend more money on them. In October, Moore wrote to Sir William Hooker:

> I was in hopes something would have been done for us this season at Glasnevin, but, there now remains no hope. I have been struggling against the stream to keep matters ongoing.[42]

Duncan Ferguson, Dr Litton and David Moore were summoned to a meeting of the Committee of Botany in December, after which the architect was instructed to prepare final plans for an epiphyte house, one hundred feet long, twenty feet wide and thirteen feet high, constructed partly of wood and partly of iron, and glazed with sheet glass. This would eventually form part of a range extending for four hundred feet, with a central house and two side wings, one of which was the proposed epiphyte house. When Ferguson had finished his plans they were exhibited in the Society's Conversation Room and Moore was instructed to begin clearing the site for the building. An invitation was sent to Lady de Grey, asking her to lay the foundation stone in May 1843 when a public promenade would be held.[43]

However, in March 1843, Duncan Ferguson was asked to revise his plans as the estimates were thought to be too high. Once the revised plan had been considered and accepted, the Society advertised for tenders—a notice appeared in *Saunder's Newsletter* on 25 April. By 8 May, sixteen contractors had submitted tenders, and on 13 May the Committee of Botany received a submission from a Dublin iron-master, Richard Turner, and a special meeting was called for 17 May. Turner had proposed a conservatory constructed entirely of iron, that is, without any wood. The Committee accepted this idea. Richard Turner presented plans and specifications to another meeting of the Committee one week later, and following this the Council of the Royal Dublin Society agreed to abandon the original scheme. Duncan Ferguson was instructed to prepare a new set of plans for a conservatory of wrought iron, one hundred feet long, eighteen feet wide and eighteen feet high, glazed with patent glass. It is very likely that the new plans were based closely on a design submitted by Richard Turner.[44]

51. East wing of the Curvilinear Range completed by William Clancy. The door lintel bears the inscription 'Royal Dublin Society 1843' and a plaque with William Clancy's name on it is also on the front of this greenhouse; Charles Nelson 1985

On 9 June an advertisement appeared in *Saunder's Newsletter* inviting new tenders, but only four were received. Richard Turner's tender of £890 was undercut by one of £810 from William Clancy and Clancy was awarded the contract. Meanwhile the Society had raised £680 by subscriptions, largely through the efforts of Henry Kemmis, a barrister and vice-president of the Royal Dublin Society, and on its completion the new glasshouse was named after him.

Building began in July 1843. The contract stipulated that the work was to be completed by 1 November, but Clancy was dilatory and his experience of this kind of work seems to have been limited. The house (Fig. 51) took one year to complete and by that time Clancy was almost bankrupt. The final cost was £924 12s. 0d including £180 for glazing, which was subcontracted to Messrs Dawson and Mitchell.

The completed house was, in essence, a lean-to structure with a curvilinear roof and bowed ends, backed on the north side by a brick wall behind which were apartments for the gardeners and apprentices, an office and a store room. On 5 August the Committee of Botany met at Glasnevin to inspect the newly finished glasshouse and to consider the heating system. It had been proposed to instal a hypocaust system—hot air circulated through the house and back wall by means of flues— devised by Dr Anthony Meyler of Dublin.[45] An estimate of sixty pounds for installation was approved and, by December, David Moore was able to report that the heating was functioning satisfactorily.

In the meantime, on 16 April 1844 the Society had applied to the Treasury for a grant to complete the whole range of glasshouses, and received a positive response. On 6 May, Frederick Darley, who designed the library of King's Inns and Trinity College's Magnetic Observatory in Dublin, was asked by the Council of the Royal Dublin Society to prepare plans of the second phase, taking into consideration the almost-completed Kemmis House. Duncan Ferguson was annoyed and complained about the switch in architects, to be told by the Society that he had only been engaged to supervise the building of the one glasshouse. Ferguson, in his complaint, referred to the 'perfect execution of

52. The entrance to Richard Turner's Hammersmith Ironworks at Ballsbridge, Dublin, decorated for the visit of Queen Victoria (from *Illustrated London News*, August 1849, by courtesy of the Director, National Library of Ireland, Dublin)

work' at the Botanic Gardens, but the Society noted that 'contrary opinions had been conveyed' to it. On 12 June Frederick Darley submitted his completed plans and these were approved. Estimates were prepared and submitted to the Treasury with a request for a grant of £2,000 for the final stage of the work.

The grant was approved on 12 August 1844, and working drawings were prepared by Darley before tenders were sought. By 13 January 1845, six firms had submitted bids including £1,650 from Courtney and Stephens of Dublin and £1,945 from Richard Turner. However, when the cost of heating systems was included, all the tenders exceeded the £2,000 approved by the Treasury, so Darley was asked to amend his plans. Richard Turner provided a new tender of £1,695 for a west wing, identical to the Kemmis House, and two straight corridors that would eventually link the wings and the central house. Turner's new price was accepted. On 3 February 1845 the contract was prepared, and work began in April.

The Turner family had been involved in the iron trade in Dublin since the mid-eighteenth century when one Timothy Turner (died 1765) was an ironmonger. His son, also called Timothy (died 1785), supplied various items for Trinity College, including the ironwork for the stairs in the Provost's House. About 1813, Richard Turner, the uncle of Richard the glasshouse builder, had opened an ironmonger's shop on Saint Stephen's Green, which Richard junior eventually inherited and then expanded through profits gained in building speculation. Richard junior was born about 1798, and had accumulated sufficient custom and capital by 1834 to build the Hammersmith Iron Works (Fig. 52), beside Trinity College Botanic Garden at Ballsbridge. This foundry operated until 1876, when all the firm's business moved to a second factory that had been built at Oxmantown in 1859. Richard Turner (Fig. 53) died in 1881 but the iron works continued to operate until forced to close following the death of his son, William, in 1889.

The earliest glasshouses by Richard Turner were built for private gardens, as at Colebrook (c. 1833-1837) and Belview (c. 1835-1837) in County Fermanagh, and Marlfield (c. 1835-1840) at Clonmel in County Tipperary.[46] Richard Turner's first public commission for curvilinear conservatories was the two wings of the Palm House in Belfast Botanic Garden. These glasshouses, each sixty-five feet long and twenty feet in width and height, were begun in June 1839 and completed the following year at a cost of £1,400.[47] Why the fine central dome, erected in 1852, was not built by Turner but by Messrs Young of Edinburgh is not known.

Little is recorded about Richard Turner himself, and for this reason a letter from William McNab, curator of the Royal Botanic Garden, Edinburgh, to David Moore written in May 1846, is of interest. 'I suppose', wrote McNab, 'I may expect a sharp blowing-up from Turner the first time he visits this part. He is certainly a most extraordinary man, but I can scarce think I should like to have much to do with him in the construction and building of that house. However I wish him all manner of good speed for all that'.[48] John Charles Lyons of Ladiston, Mullingar, knew Turner well—they shared a common interest in greenhouse construction—and he commented to Sir William Hooker that 'Dick Turner...is an ingenious, tasty, clever fellow, without a depth of science'.[49]

Richard Turner was already engaged in the construction of the Great Palm House for the Royal Botanic Gardens, Kew, when he tendered for the second stage of the range at Glasnevin. He had gone to London in January 1844 to exhibit a model of his design for the Winter Garden Conservatory in the garden of the Royal Botanic Society at Regent's Park, and while in London learned about a proposal for a palm house at Kew. Turner spent several days discussing ideas with Sir William Hooker, Director of Kew Gardens, who was most impressed. Hooker told Lord Lincoln, who was First Commissioner of Woods and Forests, that Richard Turner knew 'more about Hothouses and Greenhouses and the best principles of heating them than any man I ever met'.[50] Turner then prepared some preliminary plans for Hooker, and on 3 February he was introduced to the architect Decimus Burton at a meeting of the Board of the Commissioners of Woods and Forests, which controlled Kew. During the next few weeks Turner and Burton did a lot of work together, modifying and refining the plans. Eventually they agreed on a design and on 23 May 1844 Richard Turner was awarded the contract to erect the Great Palm House. Turner assured Hooker that the new glasshouse

53. Richard Turner, a hand-coloured photograph taken about 1860, now in the National Botanic Gardens, Glasnevin, having been presented by Christopher Dobson in 1979

would be 'perfection for the purpose for which it will be erected...the best and choicest house yet in Europe'.[51]

At Glasnevin in the summer of 1845 work on the west wing and corridors did not go ahead as rapidly as was expected. The completion date stipulated in Turner's contract was 30 September 1845, but on 13 October Frederick Darley, the architect overseeing the work, reported that the building was far from being finished. Sir William Hooker had been asked to advise about heating the new wing, and he had recommended a hot-water system. Richard Turner tendered for the installation of the pipes and boilers, and his bid of £190 was accepted. Construction continued throughout the winter and was finally completed in March 1846, at a total cost of £1,985 0s. 8d.[49]

By this time almost £3,000 had been spent on the glasshouses, of which £2,000 came from the government grant of 1844. In May 1846 the Royal Dublin Society stated that it could not raise any more money for the completion of the range and asked for another government grant to enable the central dome to be built. In the original plan Turner had envisaged a circular pavilion, sixty feet in diameter, surmounted by a dome fifty feet high.[52] Plans and specifications for this were not approved until October and tenders were received in December but all of them exceeded the sum available. In March 1847 the Society asked Frederick Darley to redesign the central pavilion, and he seems to have consulted Turner, as many features of the completed house appear in other glasshouses that Turner built. They proposed a central palm house with a rectangular floor-plan, surmounted by a lantern giving a height of about forty feet. Turner placed a tender of £1,900 for this final phase and this was accepted on 17 May 1847. Work began in June and the pavilion was completed, except for the heating system and flagging, in January 1848. As the Society had again run out of money for this project, the heating system was not installed until June 1850, after the government had provided a further £500.

The government's generosity in providing funds to complete the project was prompted by renewed pressure from the Society, which itself was prompted by such comments as that in *The Illustrated*

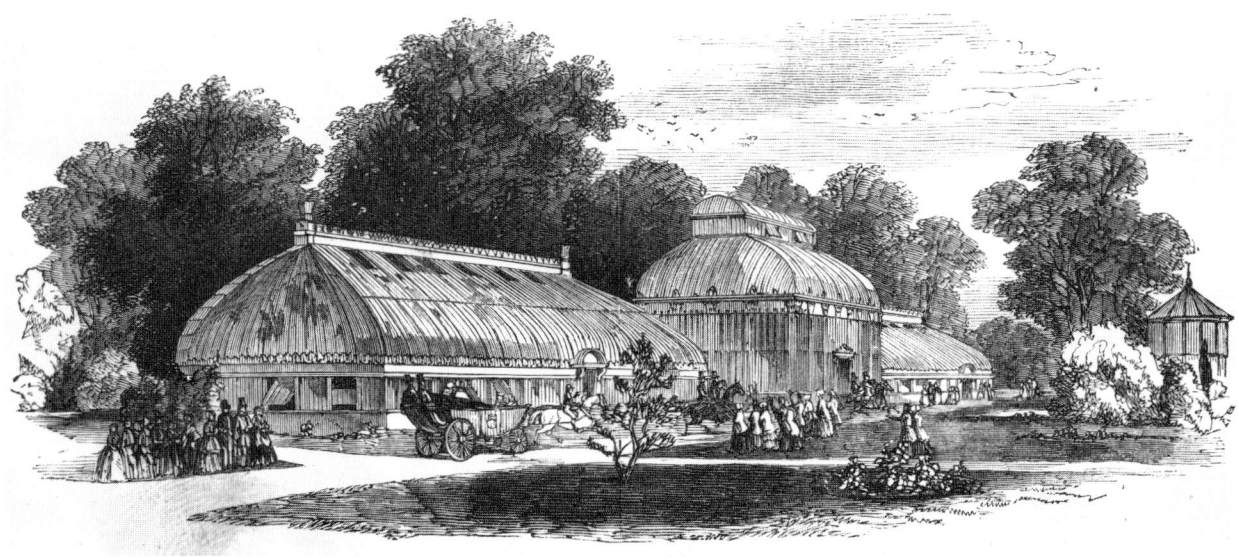

54. Engraved view of the completed Curvilinear Range, published in the *Illustrated London News* of July 1849, following the visit by Queen Victoria and the Prince Consort. This engraving was modified (the coach and people were deleted) and included in David Moore's guide to the Botanic Gardens first published in 1850. (The Octagon House is shown on the far right) (Reproduced by courtesy of the National Library of Ireland, Dublin)

London News of 11 August 1849, following the visit by Queen Victoria and Prince Albert to the Botanic Gardens at Glasnevin: '...we trust the leading members of the Royal Dublin Society will mark their esteem of the honour thus paid [by the Queen], by getting the noble house lately built properly heated, and filled with the plants intended for it.'

The Royal visit (Fig. 52) was reported in the same journal; it took place on the Queen's first day in Dublin.

> Her Majesty the Queen visited the Royal Dublin Society's Botanic Garden on Monday afternoon [6 August] accompanied by His Excellency and the Countess of Clarendon. The Prince and His Excellency rode on horse-back and led the way after the outriders. The Countess of Clarendon was in the carriage along with Her Majesty [Fig. 54] and the Ladies in Waiting. The carriage with the Royal children and two ladies immediately followed.
>
> The intention of Her Majesty to honour the Botanic Gardens with a visit was so little known, that very little preparation was made to receive her. Notwithstanding the unexpected honour...a considerable number of members of the Royal Dublin Society were in attendance to wait on Her Majesty.
>
> Mr Moore, the Curator, was introduced by the Duke of Leinster when he accompanied Her Majesty and His Royal Highness Prince Albert round these beautiful gardens, pointing out the objects most worthy of attention. Her Majesty and the Prince appeared much delighted; and the curator remarked that Prince Albert appeared to possess accurate knowledge of the principal trees and shrubs, as well as the more scientific departments. His Grace the Duke of Leinster pointed out the more attractive objects to the Queen as she passed through the gardens. The ladies and gentlemen present received the Royal party with the enthusiastic welcome characteristic of the country, which they appeared thoroughly to understand and appreciate. The first visit of Her Majesty to one of the principle scientific institutions, which is calculated to do so much good...augurs well indeed.

The entire Curvilinear Range (Fig. 54), 350 feet long, provided a series of separate greenhouses, each having a different environment, so that plants from several distinct habitats could be suitably housed. For many years the west wing was used for cacti, succulents, Cape heaths and a general collection of sub-tropical plants. Palms, bananas and other tropical species were accommodated in the central pavilion. The east wing—the Kemmis House—contained the epiphytes removed from the old Epiphyte House.

The Epiphyte House, then more than thirty years old, had been demolished shortly after William Clancy completed the east wing in 1844, and building material salvaged from it was used to make two propagating pits, each fifty feet long. A small stove-house that stood a few yards to the north of, and parallel to, the old Epiphyte House was also removed—in Ninian Niven's day it contained a cocoa tree, some cycads and date, sago and betel nut palms. Some unattractive shrubs growing

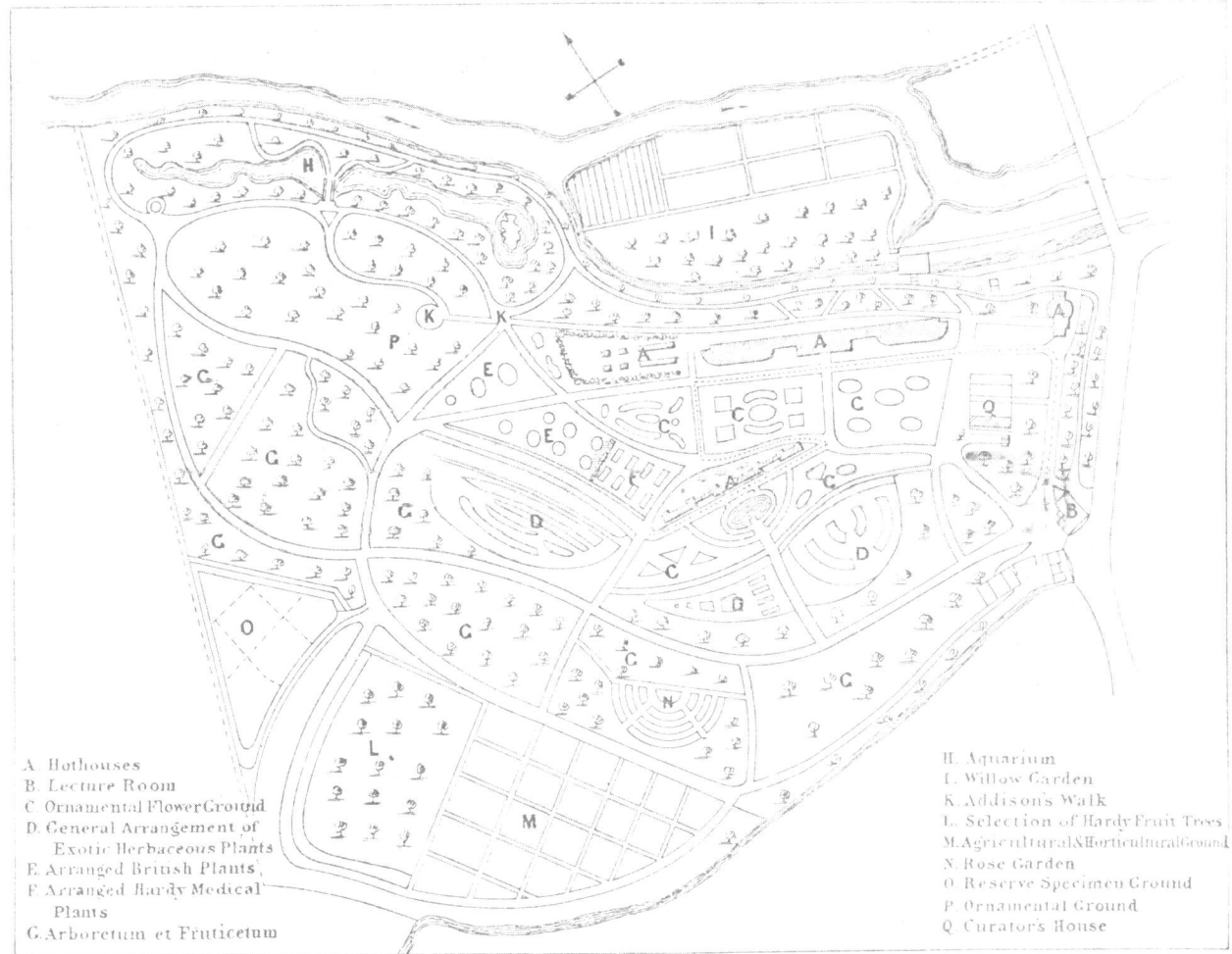

55. Plan of the Botanic Gardens in 1850, showing the new Curvilinear Range and reorganised paths; reproduced from David Moore's *Handbook for the Botanic Gardens...Glasnevin*
(Reproduced by courtesy of the National Library of Ireland, Dublin)

near the Garden's entrance were dug out, and a new path—the Main Walk—was laid out from the main gates, past the professor's house to the Octagon House, and thence along the front of the new Curvilinear Range (Fig. 55).

Since his days as foreman in the College Botanic Garden, David Moore had been interested in the cultivation of orchids. The Glasnevin collection of tropical epiphytic species expanded rapidly in the 1840s as new orchids were imported in glazed cases, especially from South America. The mania for orchids that burgeoned in Great Britain in the middle of the nineteenth century did not spread to Ireland, and in 1845 there were only three important orchid collections here. The largest and most comprehensive one was maintained by John Lyons of Ladiston, near Mullingar, who wrote, printed and published the first manual on the cultivation of tropical orchids.[53] Lyons and Moore knew each other well and Lyons exchanged orchids with the Botanic Gardens.

Not only was the first manual on orchid growing published in Ireland, but Ireland can claim another first: between 1844 and 1849 four species of orchids were cultivated from seed to flowering stage for the first time, under David Moore's direction in Glasnevin. Orchid seeds, which are minute, had been germinated in the Chiswick Garden of the Royal Horticultural Society in 1832, but the seedlings had not been allowed to mature and were pickled in alcohol. The Very Reverend William Herbert, Dean of Manchester, raised orchid seedlings before 1846, but it is not known if the plants survived to bloom.[54]

56. David Moore's article on germination of orchid seeds from *The Gardeners' Chronicle*, 1 September 1849 (National Botanic Gardens, Dublin)

ON GROWING ORCHIDS FROM SEEDS.

At the present time there are few subjects connected with plant growing on which there is less recorded information than that of growing Orchids from seeds, which appears the more remarkable when the great interest our ablest cultivators have taken in growing this singular tribe is considered, along with their tardiness of increase by division of the plant, and their intrinsic value. I am not aware that there is any case on record of hybridisation having been effected among Orchids, though there seems no doubt that such could be accomplished by careful manipulation, an inference I draw from reasoning analogically on experiments made here to get seed.

Observers on this subject will have perceived that many of our indigenous Orchids appear to seed freely, whilst comparatively few exotic species among our cultivated collections produce seed, circumstances suggestive of the idea that the latter require artificial assistance, which can be readily afforded, by carefully applying the pollen masses to the viscid face of the column and rostellum. But whether the seeds of hardy Orchids he generally imperfect, or the necessary circumstances requisite for vegetation and the subsequent growth of the young plants wanting, we certainly do not find crops of young Orchids growing spontaneously in various stages of growth, as occurs with most other endogens, though experience has proved to me that when Orchid seed does vegetate under favourable circumstances, a very large number of the myriads of extremely minute seeds contained in the ovaries are perfect, whether artificially impregnated or not.

Within the last five years, seedlings of the following species have been raised in the Orchid-house at Glasnevin, namely, Epidendrum elongatum and crassifolium, Cattleya Forbesii, and Phaius albus, the seeds of which all vegetate freely.

The manner of sowing the seeds, and treating the young seedlings, has been to allow the fine dust-like seed to fall from the ovaries as soon as they show symptoms of ripeness, which is readily known by the ovaries bursting open on one side. When this takes place, they are either taken from the plant and shaken gently over the surfaces of the other Orchid-pots, on the loose material used for growing them in, or on pots prepared for the purpose, after which, constant shade, a steady high temperature, with abundance of moisture, are all requisites which are absolutely necessary to insure success. In the course of eight or nine days after sowing, the seeds, which at first had the appearance of a fine white powder, begin to assume a darker colour to the naked eye, and if looked at with a Coddington, or even a simple lens, evident signs of approaching vegetation may be perceived, which increase until the protrusion of the young radicle and cotyledon takes place, which varies from a fortnight to three weeks. From this period of their growth the young plants grow rapidly, and the rootlets lay hold of whatever material is supplied to them. If the seeds happen either accidentally or intentionally to be made to vegetate on bare wood, as in some instances has been the case here, the young roots extend themselves in different directions, adhering closely to the bark, and make great progress compared with the growth of the stems, thus affording beautiful examples of the manner in which epiphytical plants fix themselves so firmly on the highest boughs of lofty trees in tropical forests, as well as accounting for the isolated positions they frequently occupy in their natural state.

The principal difficulty to contend with in rearing the young seedlings has been found to consist in their treatment during the first year, particularly the winter months, when they are very liable to perish, if anything approaching to extremes of moisture, drought, cold, or even heat be permitted; though a steady medium of all these requisites is necessary. The second year's growth has been one during which the plants made much progress, and the only two kinds which have been brought to a flowering state have bloomed the third season. These are Epidendrum crassifolium and Phaius albus, the latter being now in flower, exactly three years from the sowing of the seeds. *D. Moore, Glasnevin, Dublin.*

At Glasnevin seeds of four orchid species were 'allowed to fall or were shaken on to the loose material used for growing orchids on, and on to the surface of other orchid pots'. Germination took place within two or three weeks. The seedlings grew rapidly and eventually flowered—both *Phaius albus* and *Epidendrum crassifolium* produced flowers three years after germination.

David Moore reported the successful outcome of this experiment in an article published in the *Gardeners' Chronicle* on 1 September 1849. He realised the significance of his work, commenting that 'at the present time there are few subjects...on which there is less recorded information than that of growing orchids from seeds...I am not aware that there is any case on record of hybridisation having been effected among orchids, through there seems no doubt that such could be accomplished by careful manipulation, an inference I draw analogically on experiments made here to set seed'. He had noticed that cultivated orchids rarely set seed, and suggested that they required artificial pollination—this could be done 'by carefully applying the pollen masses' to the column and rostellum. At that time it was not understood that orchids have exacting pollinator requirements and in cultivation the specific pollinators are almost invariably absent. Neither was it known that orchid seeds will germinate only in the presence of a symbiotic fungus.

David Moore's paper stimulated numerous people to write to the *Chronicle* about the cultivation of orchids. The Glasnevin experiment prompted other gardeners to try to germinate orchid seeds, and eventually to cross-pollinate species and raise the hundreds of thousands of artificial hybrids that exist today.

It is noteworthy that the development of the glasshouses at Glasnevin and the experiment on the propagation of orchids were conducted during the Great Famine. Small-scale sporadic and local failures of the potato crop had occurred in the previous decade, but in 1845 an unknown disease, called potato murrain in contemporary accounts, appeared in Europe.[55] It was reported from Belgium in late June, and, by the middle of August, it had appeared in England. Alerted by reports of the murrain published in horticultural journals, David Moore watched for the signs of the disease in Glasnevin. He noticed it on potatoes in the Botanic Gardens on 20 August. A few days later the murrain was reported from

Castlebellingham, in County Louth, and by the end of the month the disease had been detected in County Fermanagh. During late August and early September Moore received many samples of diseased tubers from people in the Dublin area. By mid-October the whole of Ireland was affected and the potato crop was almost completely destroyed.

Moore prepared a preliminary paper on the disease, and it was published on 10 September 1845 in the *Irish Farmer's Journal*. He reported that he had found microscopic fungi on decaying potato tubers, and agreed with other botanists and agriculturists who believed that the disease was caused by the unusually damp weather and that the fungi was merely exacerbating the decay of tubers. Moore carried out some experiments at Glasnevin on the digging of tubers and their storage. He sent a report to the Council of the Royal Dublin Society on 29 October and suggested that all potatoes still left in the ground should be dug out immediately and stored in dry, airy places. The Society asked him to continue working on the problem, and to test whether diseased potatoes could contaminate sound potatoes if brought into contact with them. On the following day, David Moore appeared before the commission which had been set up by the Irish Administration to inquire into the disease.

Moore was not the only person carrying out experiments on potatoes. The Dublin Natural History Society, of which he was a member, heard papers from William Andrews and Dr O'Bryen Bellingham. Dr Bellingham concluded that the disease was caused by the 'highly charged electrical state' of the atmosphere. Andrews had found fungi on diseased tubers, and, like David Moore, he concluded that these had invaded the tuber after it had been weakened by another factor. He thought that the murrain was accelerated by sudden atmospheric changes and the presence or absence of sunlight and warmth.

In January 1846, Miles Berkeley, an English clergyman and amateur botanist, published a paper in which he stated that a minute parasitic fungus (*Phytophthora infestans*) was attacking potato plants and causing decay. His opinion was criticised in the horticultural press and was rejected by botanists. However, Berkeley adhered to his theory.

By April 1846, Moore's experiments on overwintering potatoes were ended, and on 14 April he reported to the Royal Dublin Society that diseased tubers would decay irrespective of the conditions under which they were stored. Plants raised from diseased tubers had quickly 'become infected with the fungus' and he warned that the success of the crop in 1846 would depend on the selection of uninfected seed potatoes. David Moore concluded by noting that the murrain had already appeared that year, but he still accepted the idea that the fungus was only an 'accompaniment to the disease'.

A few days later he wrote to Berkeley sending specimens of diseased leaves—'The inclosed will show you that we are to have a visit from your friend [*Phytophthora*] this season again'. Berkeley's reply prompted David Moore to reconsider his ideas about the origin of the epidemic. Moore had observed that potato plants grown in a glasshouse had been attacked by the disease in January 1846, and this convinced him that the disease could not have been caused by weather. In late July he wrote to the *Irish Farmer's Journal* expressing his conviction of the correctness of Miles Berkeley's theory. He also wrote to Berkeley saying that he could no 'longer deny myself the pleasure of congratulating you about the justness of your views relative to the [*Phytophthora infestans*] being the *cause* of the potato disease and *not* the effect as I had supposed. Nothing could be more satisfactory [sic] proved to my mind...' Moore added that he had observed 'the pest' on specimens of potato plants recently sent from different parts of Ireland—'What may be the result of this is too awful to contemplate...'

Moore and Berkeley were lone voices. Most horticulturists were not convinced by the 'fungal theory' and the supporters of the 'atmospheric theory' remained voluble. However, Dr John Lindley was prompted to print an editorial in the *Gardeners' Chronicle* on 8 August 1846 about David Moore's acceptance of the 'fungal theory'. Lindley commented that

> Mr Moore's conversion cannot fail to give new weight to the high authorities with whom the [fungal theory] commenced and we look with much interest to his further explanations. We admit that his experience and intelligence are such as to demand the most respectful attention. Nevertheless, we are as much unconvinced as ever...'

Moore reasoned from his observations that any treatment which was effective against other fungi, for example those which attacked leaves of peach trees or peas, might perhaps halt the potato blight.

He also suggested that the fungus spread from plant to plant by spores which 'may be imbibed by the rootlets or through the [leaves] of the seedling potatoes'. His ideas were partly correct. At this time botanists were unsure about the method of attack by the fungus and they had not found a cure for the disease.

In August 1846, an Australian plant (*Anthocercis ilicifolia*), growing in one of the greenhouses at Glasnevin, died; it belongs to the same family as the potato (Solanaceae). Moore examined its leaves and found that they were infected with the blight fungus. He reported his observation to Berkeley on 24 August and added this dramatic postscript to his letter—'Our potato crop is lost without exception I believe throughout Ireland'. A few weeks later David Moore found *Phytophthora* attacking tomato plants in the garden of the Vice-Regal Lodge in Phoenix Park.

The Irish potato crop was utterly destroyed in 1846. The prospects for the 1847 crop were regarded with great anxiety by Moore and Berkeley. David Moore had carried out experiments on steeping potato tubers in solutions of various salts before planting them, in the hope that one of these treatments might halt the blight, but his experiments were unsuccessful.

Moore was instructed by the Committee of Botany to carry out experiments designed to test the theory that the murrain was caused by excessive amounts of electricity in the atmosphere. The potato plots at Glasnevin were surrounded by 'tractors of electricity at distances capable of attracting it and conveying it into the earth'. The 'tractors' consisted of copper wires cut into lengths varying from twelve to eighteen feet, and raised on poles placed about twelve yards apart. This electro-culture experiment was regarded by Moore as a farcical waste of time and space, and he reported bluntly that the 'tractors' had not the 'slightest effect one way or another'.

In his annual report for 1847, David Moore noted his regret that insufficient funds had been made available for work on the disease. He pleaded for more money so that the Glasnevin Botanic Gardens could be 'rendered more useful to the country than it has hitherto been'. The financial support did little more than keep the institute 'in a state such as to prevent public censure'. Moore complained that he had too few assistants and thus could not devote as much time as he wished to research on the murrain. It is clear that he was hampered by the unwillingness of members of the Committee of Botany of the Royal Dublin Society to give the 'fungal theory' serious consideration—they persisted in adhering to the 'atmospheric theory' and instigated the farcical 'electro-culture' experiments.

David Moore came closer than most other botanists to discovering a treatment for potato blight, because he correctly understood its cause. He tried steeping seed potatoes in bluestone (copper sulphate) solution, but this treatment was ineffective as it did not kill the spores resting in the tuber tissues.[55] Many years later it was discovered that spraying the foliage with Bordeaux mixture (which contains copper sulphate) was the only effective treatment.

Although the destruction of the potato crop itself did not seriously afflict the middle-class people in Dublin, the 'famine fevers' that accompanied the famine did affect the whole community. The primitive sanitation in Dublin encouraged epidemics of typhus and dysentery, which reached their peaks in June 1847 and continued unabated until February 1848. It is probable that the death of David Moore's second wife, Isabella,[56] on 7 November 1847 was caused by one of these diseases, for David had complained about the insanitary condition of the gate-lodge in which the family lived—the floor was below ground level and the walls were running with damp. For a second time Moore was faced with bringing up a young family by himself.

Meanwhile, Dr Samuel Litton had continued to perform some of his duties as Professor of Botany to the Royal Dublin Society, but in the early 1840s he fell foul of the Society. He neglected to take the meteorological observations, and ignored repeated requests from the Society for a report on a grafted beech tree that had been toppled on the night of the Big Wind in 1839 at Collon on the Foster estate. In 1846 the Society considered the possibility of demolishing the Professor's house as it was in poor condition with a leaking roof, and even discussed whether or not Litton was entitled to live there. When Dr Litton wished to use a room at the rear of the new Curvilinear Range for his lectures instead of the lecture-room in his house, the Society refused him permission.

Samuel Litton died on 4 June 1847 following a heart attack while in his carriage. He left no record of original botanical research, and his only publications are insubstantial précis of several lecture series. His reputation as a good lecturer survived him—*Erinensis* recorded that Professor Litton's lectures were 'master-pieces of popular composition [and]...possessed all the freshness and integrity of original composition, carrying the audience along...without those narrational interruptions of undisguised plagiarism which so eminently distinguished the clumsy compilations of some of his contemporaries'.[57] Dr Litton's herbarium was purchased from his sister, Mrs Cuthbert, for twenty pounds and is now incorporated into the collections in the Glasnevin Botanic Gardens, but he is not known to have contributed to our knowledge of the native flora. Despite his ambivalent, querulous nature, Samuel Litton was respected by fellow botanists, and Sir William Hooker dedicated *Littonia*, a small genus of climbing lilies from South Africa, as

> A tribute to the memory of...a deeply learned and amiable man, and a popular lecturer. The modest appearance of this plant [*Littonia modesta*] may further serve to indicate his unassuming and retiring disposition which, as has been recorded by the Council of the Royal Dublin Society, "prevented his taking that rank in general society to which his acquirements entitled him".[58]

9
'An era in gardening'
David Moore's last three decades
1848-1879

WITHIN a week of Samuel Litton's death, a 'host of candidates'[1] declared their interest in the vacant professorship and there was a great amount of canvassing in the Royal Dublin Society and the botanical fraternity in Dublin. Among those who immediately declared their interest were Dr Thomas Taylor, Professor of Botany at the Royal Cork Institution and an authority on mosses and liverworts; William Andrews, a founding member of the Dublin Natural History Society, who had studied botany under Professor George Allman at Trinity College in 1844; and Dr John Aldridge, one-time pupil of Sir William Hooker and a keen student of plant physiology and chemistry, who at first had the support of William Henry Harvey. On 10 June 1847 Dr William Harvey, who was curator of the herbarium in Trinity College, Dublin, wrote to Hooker on Aldridge's behalf asking for a testimonial and noting that his friend was 'in every respect the best fitted' for the professorship. On the following day, in a letter to Joseph Hooker, William Henry remarked that

> I have had a narrow miss in doubling my income, and lost it all by my own goosishness. Dr Litton is dead and his chair to be filled. I had a notion of standing, but while I thought on the matter a friend called and asked for my vote for Dr Aldridge. Time will tell. Scarcely was the early friend gone, after catching his worm—when my friends began to pour in, with such a report of the votes as made it appear that I had only to stand, to walk over the course. But alas! the poor horse was Doctored—and could not move. So all I could do was to speak as wanly as possible for poor A[ldridge]—and recommend them all to vote for him. Many will do so, but the interest is divided.[2]

On 13 June, Harvey repeated that he could not be a candidate but by the following evening he had changed his mind. 'You will be much surprised to learn,' he told Sir William Hooker, 'after what I have already written that I am a Candidate for the Professorship in Dublin Society—duly declared and ready to take the field against all opposers'. By this time Harvey realised that Aldridge had little chance, especially as Dr George Allman, Professor of Botany in Trinity College, had announced his candidature. William Harvey sought testimonials from some of the most eminent European botanists[3]—Adrien de Jussieu, Adolphe Brongniart, Joseph Decaisne, Alexander von Humboldt, Augustin-Pyramus de Candolle and Stephen Endlicher—and, like the other candidates, canvassed members of the Royal Dublin Society.

The Committee of Botany decided that applications for the professorship would close on 1 November 1847, that each candidate would be required to give a probationary lecture on botany in the Society's Kildare Street lecture-theatre, and that the new professor would be elected in January following. Later, the Committee also decided that it would be advisable to limit the duration of the appointment to seven years and that the professor should no longer live in the house at Glasnevin.[4] David Moore and his family had removed to the house after Professor Litton's sister had vacated it, and the house was later assigned to the curator as his official residence.

Nine candidates applied for the professorship, including Dr Allman, Dr Aldridge and Dr Harvey.[5] The series of lectures commenced on 3 December. William Harvey delivered his on 10 December and described the event in a letter to Sir William Hooker:

> I got on much better than I had expected—spoke out big—and had very little flutter—but am told I read some parts too quickly. The fact was I had prepared too much—and was fearful of not getting through within the time...It is now over—and I believe the Professorship is over too—for I am told, from the Best informed quarters that my connexion with the College offers an insuperable bar to my election. They will have nothing from TCD—an Angel would be bid fly away. The Secretary [L. E. Foot] says

he does not want science — they merely want popular lectures to amuse the young. <u>Mine</u> was quite in that style — and he said he approved it, but I verily believe such books as Phycologia and Nereis have seriously <u>damaged</u> me. Had I written "The Botanist's Penny Whistle" impression would have been better. But no matter we can live without them.[6]

Harvey's dismissal of his own prospects was altered as rapidly as his earlier decision about his candidature had been. On 13 December, he wrote to Hooker saying that he now understood his lecture had left a favourable impression. The Royal Dublin Society's honorary secretary, Lundy Edward Foot, had told Dr Harvey's friends — who promptly told Harvey — that if Dr Allman could be persuaded to withdraw Harvey would win the election. Equally, if Harvey withdrew Allman would be elected for 'which ever of us goes to the Poll will have the support of the "House Party" — a powerful one. Up to the moment of the lecture Mr Foot (and the House) were for Aldridge. If both Allman and I stand we both go to the wall'.[7]

However, Dr Allman ruined his own cause when he gave his probationary lecture on 20 December. 'He has a dreadful manner of <u>mouthing</u>', Harvey told Hooker, 'and ultra-high-bombasticizing — which make even his good things sound ridiculous'.[8] By Christmas Day, when William Harvey yet again wrote to Sir William Hooker, the Council of the Royal Dublin Society had privately declared in Harvey's favour. But the Council had no direct control over the election — it was the members of the Society as a whole who elected the professor. Dr Allman would not withdraw from the contest and competition remained fierce, with canvassers for all the candidates active throughout early January. William Harvey was delighted with the support he received, even from strangers, and remarked that the conduct of his canvassers was 'highly creditable to them, (showing that there are some anti-jobbers in Ireland), and to me such support is, of course, the most gratifying, as it feeds the creature's vanity'.[9] But, on 6 January 1848, Dr Allman wrote to the Council of the Society retiring from the contest, and stating, in flattering terms, his support for Dr Harvey. Voting took place on 27 January, and later that day William Harvey wrote to Sir William Hooker:

My Dear Friend
    Give me joy — as I am sure you will — at the result of today's poll
        Harvey 202
        Steele 71
        Bellingham 61
        Aldridge 26
the rest nowhere. So I have been elected and triumphantly after all. It was a great scene of excitement — about 500 persons present, but only 385 voted — the largest number remembered at any former election.
    I have no more news — but am your
        always affectionate
        W. H. Glasnevin (elect.)
        I hope you like my new title
I have a dozen other letters to write. Your letter to Simon Foot and one from Dr John Lindley to the same worked wonders in my favour.[10]

William Henry Harvey was just a few days short of his thirty-seventh birthday on his appointment and, like Walter Wade and Samuel Litton, he was to remain the Society's Professor of Botany until his death. He was of Quaker stock — although later in life he joined the Anglican church — born at Summerville on the banks of the River Shannon near Limerick, on 5 February 1811, the youngest son of Joseph Massey Harvey, a prosperous businessman. He was educated at the Quaker School in Newton, near Waterford, and when he reached the age of thirteen was moved to another Quaker school at Ballitore, County Kildare, where the pupils were happy in an institution run on the Quaker principles of love and non-violence. William Harvey became a keen naturalist and he attributed his love of botany to the companionship when he was very young of an old gentlewoman, a friend of his mother, who would carry him into the garden and teach him the names of flowers.[11] The head teacher of Ballitore was James White[12] who had a comprehensive knowledge of the classics and natural sciences. He encouraged the young Harvey's botanical interest, although William himself wrote that while '...I am very fond of botany...I have not much opportunity of learning anything, because I have only to show a plant to James White, who tells me all about it, which I forget the next minute'. On the other hand, one of his cousins noted that 'William cares for nothing but weeds'.[11]

57. William Henry Harvey, a pencil portrait by F. W. Burton (Reproduced by courtesy of the National Gallery of Ireland, Dublin)

58. Octagon House with Victoria House (in foreground) an engraving from *The Gardeners' Chronicle*, 1884
(National Botanic Gardens, Dublin)

When Harvey left Ballitore School he went into his father's business, but maintained his interest in natural history. It was during this part of his life that he laid the basis of some of his later research and publications by studying the seaweed and shells found on the coast, especially about Miltown Malbay. On his father's death, William Harvey was appointed to the post of Colonial Treasurer at the Cape of Good Hope, thanks to the patronage of Lord Monteagle, and he arrived there in 1836. At once he began to study the flora of the Cape. He made innumerable field trips—in March 1837 he wrote to a friend that 'I have indeed taken so many excursions lately that I almost fear I shall earn the soubriquet of Her Majesty's *pleasurer-general*, instead of the more dignified one of *treasurer*.'[13] During his years at the Cape, Harvey prepared hundreds of drawings of native plants, some of which he subsequently lithographed. His book, *Genera of South African Plants*, which formed the foundation for *Flora Capensis*, was ready for publication two years after his arrival. William Harvey resigned as Colonial Treasurer in 1842 and returned home to Ireland, but he kept a deep interest in the botany of the Cape Colony for the rest of his life. On 1 May 1844, Dr Harvey succeeded Dr Thomas Coulter as curator of the herbarium in Trinity College,[14] where one of his first tasks was the sorting and distribution of the many duplicate specimens of Mexican and Californian plants that had been left untouched by the ailing Dr Coulter. Harvey retained the curatorship after his election as the Royal Dublin Society's Professor of Botany.

Professor Harvey (Fig. 57) paid his first official visit to Glasnevin on 5 February 1848, when he identified and named three plants with the aid of books from the library and specimens in the Gardens' herbarium. He began to plan a lecture series and its starting date was advanced to April to facilitate his other work in the herbarium of Trinity College. However, in early April 1848 the Royal Dublin Society's headquarters, Leinster House, was taken over as a military barracks in anticipation of serious trouble from the Young Irelanders, so the botany lectures were postponed. In a letter to Hooker, William Harvey remarked 'I am happy to say the Salary goes on—and I have already pocketed the first quarter!'[15]

By early May the situation had become calm, and lectures commenced. For a few years Harvey gave lectures in both the Botanic Gardens and Leinster House. In 1849, some of his lectures were given in the newly completed Central Pavilion of the Curvilinear Range, but, after the heating system had been installed and plants had been brought in, this venue ceased to be available. By the mid-1850s the lecture-room in Glasnevin had become so dilapidated that all Harvey's lectures were transferred from the Botanic Gardens to the Society's headquarters. During the years of William Harvey's professorship attendance at the botany lectures declined and by 1860 only about fifty people enrolled. After Dr Harvey's death in 1866 the Royal Dublin Society ceased to sponsor lectures in botany for the general public.

As well as giving lectures in Dublin, Professor Harvey was obliged to speak in other Irish towns and cities, and, while he carried out these duties conscientiously, he regarded them as a nuisance—they interfered with his taxonomic research.[16] He was a hard-working botanist. In his formative years he had acquired the habit of rising at five o'clock in the morning, working until breakfast, and then returning to his research, which continued until about five o'clock in the evening. In later years, William Harvey had his own enormous collections to study as well as the large number of specimens that came to Trinity College as gifts from fellow botanists. In addition to examining herbarium specimens, and sometimes labelling them and mounting them for storage, Harvey drew about nine hundred illustrations for his various publications, including one hundred plates for *Thesaurus Capensis*, which complemented the early volumes of the great *Flora Capensis* that he wrote in collaboration with Otto Sonder. While Harvey took little part in the botanical and horticultural research work of Glasnevin Botanic Gardens, he did evince keen interest in the fabric of the Gardens, at least in the early years of his professorship.

The fine Octagon House, erected in 1819, was, in its day, the tallest structure of its kind in Europe. A small wooden porch, twenty feet long and fifteen feet high was added to it in 1834, and at about the same time four metal pillars were inserted inside the house to support a circular gallery. The heating system was improved, too. For a wooden conservatory the Octagon House had a remarkably long life. At 7.30 in the evening of May Day 1843, sparks from its furnace chimney ignited the roof timbers, but the blaze was tackled so efficiently by the Gardens' labourers that the building was saved. The Society later gave five pounds to the men in recognition of their efforts. As a result of this fire, the Royal Dublin Society insured all of the Gardens' buildings.

After the fire, general repairs were carried out and a Polmaise system of hot-air heating was installed, the first use of this type of heating in Ireland. But by 1850, when a hail storm destroyed its roof panes, the Octagon House was deteriorating rapidly. The plants, which by that time included fan palms, a dragon tree (*Dracaena draco*) and some papyrus planted in a magnificent Chinese vase—probably one of the pair bequeathed by Thomas Pleasants—were moved to the Central Pavilion of the Curvilinear Range.

David Moore advised the Society to demolish the Octagon House (Fig. 58), but Professor Harvey eloquently put forward the case for retaining it, and pointed out that there were alternative courses open to the Society. He suggested that the house could be reduced in height by ten feet, and used to accommodate a pond for the giant Amazonian waterlily; but he withdrew this plan, as the Dublin Society's funds would be inadequate for heating the water in the pool. Harvey reminded the Society that the Octagon House had been built originally to house a Norfolk Island pine, and he noted that the specimen which the Gardens then possessed would soon have to be beheaded and spoiled if it remained in its present quarters. He proposed that the Octagon House be repaired by re-glazing the roof with new hail-resistant sheet glass, and that the Norfolk Island pine be moved into it. Heating could be discontinued, thus effecting an economy, and Australian plants and some of the newly-acquired Himalayan rhododendrons could be grown in it. The porch could be filled with South African heathers, *Pelargonium* species and Epacridaceae. So strongly did Harvey feel about the Octagon House that he was prepared to head a subscription list to save the glasshouse with a donation of ten pounds.

A subscription list was opened, repairs were carried out, and in January 1852 the Octagon House was ready to receive those semi-tender plants that did not require heat, and the Norfolk Island pine. Ten years later heat was re-introduced, and the Octagon House was then used for tree-ferns. It was demolished in 1886 to make way for the first purpose-built fern house, having graced Glasnevin for sixty-seven years.

Shortly after he returned from a visit to the United States of America in the spring of 1850, Professor Harvey was asked to hold a 'Grand Botanical Conversazione' at Glasnevin. As the botanical sensation of 1850 was the giant waterlily from South America, Harvey decided to obtain a leaf for display at this event. It was not growing in Glasnevin in May 1850, so Harvey wrote to Sir William Hooker asking for a leaf from the plant at Kew. Hooker despatched one of the massive leaves—they can grow up to six feet in diameter—to Dublin, and it was displayed at the Gardens on 30 May, floating upside down in a large sponge-bath, to show the remarkable structure of the undersurface. The leaf caused great excitement—everyone who attended the Conversazione stopped to marvel![17]

This astounding waterlily was discovered in 1801 by Thaddeus Haenke and his missionary colleague, La Ceuva, growing in the River Marmore, a tributary of the Amazon. Although the two men reported their discovery to the Spanish government, this vegetable wonder remained little more than a vague rumour to most botanists for the next thirty-five years. Aimé Bonpland found the waterlily near Corrientes, in the Argentine, in 1819, and another French botanical explorer, Alcide Dessalines de Orbigny, collected specimens of it in 1827, which he sent to Paris together with drawings. In 1832, the German botanist, Eduard Poeppig, confirmed that a gigantic waterlily grew in the Amazon region; he described it and named the plant *Euryale amazonica*. Three years later, the Royal Geographical Society of London sent Robert Schomburgk to British Guiana, and he also found the waterlily—he sent drawings and specimens to London and these were exhibited at a meeting of the Botanical Society of London on 7 September 1837. There followed a protracted argument about the plant's name—Schomburgk requested that it should be named after Queen Victoria. Reflecting the controversy, Ninian Niven remarked[18]—quite pointlessly—at the dinner of the Royal Horticultural Society of Ireland in the spring of 1838, that if Ireland was lucky enough to get a living plant of this giant waterlily before England, 'we should have much gratification in naming it *Victoria regina*!' But the task of getting seeds of the Amazon waterlily to Europe proved difficult. In June 1846, Thomas Bridges brought some to London, and while two germinated, the plants died during the winter before they could flower. In the following year more seeds and some young plants were despatched from South America, but they did not survive the journey. In 1846, Dr Rodie of Demerara put some seeds in bottles of water, and these reached London in excellent condition; numerous seedlings were raised at Kew, and then were distributed to botanical gardens throughout the world. Two seedlings were brought to Dublin by a special courier on 23 August 1849; one went to Trinity College Botanic Garden, and the other came to Glasnevin. David Moore received the 'rare and much talked of *Victoria regia*'[19] on 24 August, and it was planted in a water-tank that had been specially constructed in the Long Range. Although the waterlily grew well it did not flower, and during the winter, it died.

Over the next few years David Moore tried, without success, to grow and flower this waterlily. In June 1850, he obtained a young plant from Joseph Paxton, gardener to the Duke of Devonshire, in whose garden at Chatsworth the waterlily flowered for the first time outside its native habitat early in November 1849. Again, the waterlily did not bloom at Glasnevin. Moore got seeds from the British Ambassador in Monte Video, Henry Gore, in 1851, and also young plants from Syon House, residence of the Duke of Northumberland, and from Chatsworth, but the waterlily still would not flower at Glasnevin. In 1852, the Royal Horticultural Society of London donated a seedling from its garden in Chiswick, after one sent from Syon House had died *en route*. As there is no record of a gift in 1853, it may have been decided not to attempt growing the waterlily again until a special greenhouse could be built. David Moore had emphasized this need in his annual report, dated 25 November 1852:

> I would further beg to notice the want felt at present of an Aquarium House [to accommodate] the tropical aquatic plants...more especially *Victoria regia*, for which a house is set apart in most public gardens...With your present appliances it must always be suffered to perish on the approach of winter; the treatment it requires being so different from that of the other plants with which the house is occupied, they cannot be cultivated together. An ornamental small building suited for the purpose could be erected at a very moderate expense.

Professor Harvey supported David Moore's request and on 24 June 1853 a horticultural fête was held in the Botanic Gardens to raise money towards the cost of a stove aquarium. Four thousand people attended, and the road from Dublin to Glasnevin looked like Epsom on Derby Day, so great was the number of carriages. Two marquees were fitted up as Turkish tents, and there were numerous pavilions where refreshments were served. The Royal ensign flew on the right-hand side of the entrance gates to greet the arrival of the Lord Lieutenant. Six regimental bands attended and played different programmes; their unison rendering of the Peterhoff Waltz and *La Fête des Lilacs* at 4.30 in the afternoon was especially well received, while the playing of the National Anthem was 'one of the finest things heard for a long time'. Everybody had an enjoyable afternoon at 'one of the most brilliant assemblies of rank and fashion...seen for many years'.[20] The Lord Lieutenant left at half past five, having admired the orchids and the giant waterlily. The fête cost the Society sixty-five pounds, but, by charging admission, it was able to raise £260, about half the cost of the 'stove aquarium', the balance being met by the Government.

59. Victoria Pool — the pool has been drained and the central well is filled with well-rotted compost ready for the planting of a young waterlily. The heating pipes are visible. Photograph taken c. 1900 (National Botanic Gardens, Dublin)

The 'Victoria Regia House', or stove aquarium, was designed by Duncan Ferguson, who had worked on the east wing of the Curvilinear Range ten years earlier. The construction of the new glasshouse began in 1854. When completed it was forty-four feet long by forty feet wide, and contained a circular pool for the waterlily (Fig. 59). Originally, iron uprights supported a wooden roof, but this was replaced by an iron ridge-and-furrow roof nine years later. The Victoria House, with its elaborately decorated facade (Fig. 58) was ready to receive plants by the end of May 1854. That year, seedlings of *Victoria* were raised at Glasnevin from seed donated by the University of Oxford Botanic Garden, and by the Duke of Northumberland, but although the waterlily grew well, it still did not blossom. The citizens of Dublin had to wait until 1855 before they could see the giant waterlily bloom in Glasnevin.

The prestige of being the first Irish garden to flower this plant went to the botanical garden in Belfast. Daniel Ferguson, the curator, returned from a visit to London in the autumn of 1851 with seeds of *Victoria amazonica* from both Syon House and the Royal Botanic Society's garden in Regent's Park. He erected a special glasshouse in Belfast at his own expense, although the company which ran the Royal Botanic Garden later refunded him the cost, seventy-five pounds. *Victoria* grew well in 1852 but did not flower. Ferguson obtained another plant in 1853, and in July the giant waterlily did at last bloom, in Belfast.[21]

At Glasnevin, two other important buildings were erected during the 1850s — a museum and the first palm house.

60. East Front of Leinster House, headquarters of the Royal Dublin Society, 1852; in this engraving the Iron House is shown in right foreground. The illustration originally appeared in *Proceedings of the Royal Dublin Society* 87 (Reproduced by courtesy of the Director, National Library of Ireland, Dublin)

In his report for 1851, David Moore argued that a Museum of Economic Botany should be established at Glasnevin, on a similar plan to the museums already existing in the botanical gardens at Kew and Edinburgh. Indeed, Samuel Litton had suggested that such a museum would be a desirable asset but nothing was done during his professorship. In November 1852, Moore repeated his suggestion, and the following year the Royal Dublin Society acquired a prefabricated iron house which had been erected for the 1850 Exhibition of Manufactures on Leinster Lawn (Fig. 60), in front of the Society's headquarters.[22] The Iron House had to be removed to make way for the Great Industrial Exhibition in 1853, so it was dismantled, moved to Glasnevin, and reassembled just inside the entrance near the left-hand gate-lodge. Moore realized that the building was 'not the most appropriate structure' for a botanical museum, but it sufficed (Figs. 61, 62). It was furnished with sixteen upright and fourteen horizontal display cases, and heating stoves, at a cost of £250. Glasnevin already possessed a considerable collection of artifacts and vegetable materials, and this was moved into the display cases. The Society granted the Committee of Botany fifty pounds to purchase a collection of fibres, barks, gums, oils and sections of wood that had been exhibited by the British Guiana Department in the Great Industrial Exhibition. In 1856, a collection of cut and polished wood samples, fruits and seeds was acquired from Jamaica, and all the raw fibre products and medicinal plants displayed by Mauritius at the London Exhibition of 1865 were bought for Glasnevin. Sir William Hooker donated a specimen of the remarkable primitive cone-bearing plant, *Welwitschia mirabilis*, which grows only in the Namib Desert in south-western Africa. By the 1880s the Economic Museum had been filled with all sorts of bric-a-brac made from plant materials, Indian canoes, Chinese papers, 'a beautiful chemisette, made from the fibre of Pine Apple Plant', spices, preserved fruits, pine cones and drugs, including a 'fine ball of opium presented...by the Honourable the East India Company'.[23]

The Royal Dublin Society also decided, in 1852, to replace the unsatisfactory Long Range, which was then just over thirty years old, and commissioned the architect, Jacob Owen, to design a new range comprising a tall central house for large palms, and two low, flanking wings. The first wing of the Palm House Range was completed in 1857, and the central house and second wing were ready three years later, the total cost being about £4,000. This new Palm House had a rectangular floor plan, sixty feet long and eighty feet wide, and was sixty feet high, twenty feet higher than the Central Pavilion — the Old Palm House — of the Curvilinear Range. The foundations of the new house were twelve feet deep. A rockery was constructed in the centre of the new Palm House, using stones sent by Professor William King of the Department of Geology in Queen's College, Galway. The east

61. (above) The Museum Building at Glasnevin as it appeared in 1884, and 62. (below) as it is in 1985 (photograph by David Davison). The engraving is taken from *The Gardeners' Chronicle*, October 1884. (National Botanic Gardens, Dublin)

63. Palm House built in 1860; from *The Gardeners' Chronicle* October 1884. (National Botanic Gardens, Dublin)

wing accommodated the orchids, and a collection of *Camellia* was housed in the west wing. The palms, some of which were over twenty feet tall, were moved without accident from the Curvilinear Range into the central house, and David Moore later reported that they had been successfully established in their new home. Cycads, agaves and a bread-fruit tree were also planted in the Palm House, and a mango and a cocoa tree (*Theobroma cacao*) both fruited in it a few years later.

This poorly-designed glasshouse (Fig. 63) was singled out for criticism by William Robinson, following a visit to Glasnevin in 1864. He wrote a two-part article for the *Gardeners' Chronicle* on the Gardens, and in the first part stated that the Octagon House was 'almost classical compared to the astonishingly ugly new Palm House' which was 'externally a hundred degrees too ugly'. The Curvilinear Range and the Palm House were to one another 'as Dove to Dodo, as white-winged Yankee clipper to a swinish Monitor, as Gibson's Venus to a hideous gutta-percha faced Amazon of Dahomey'.[24] Robinson returned to the attack in the second part of his article; the ugliness of the house seemed to fascinate him. He said he had met 'a son of the soil' on his way to Glasnevin, who spoke of 'the beautiful plants and quare flowers' and of the Palm House 'like an ould gable end'. This confirmation of his own opinion further loosened Robinson's pen.

> There is an intensity of ugliness about this building, seen from the north side, which quite appals one. It is as if the demon of bad taste had been let loose, fully accredited, to attack the enemy in his strongest camp, and accordingly set to work and built this temple in the central scene of a beautiful garden, in the midst of the elegant leaf-builders, and over a host of graceful Palms'.[24]

But William Robinson[23] was full of praise for the plants inside the Palm House and the manner in which they were managed and cared for.

The design of the central house was so faulty that it swayed in high winds, and stays had to be inserted five years after it had been completed to brace the building. During 1871, major repairs were made; heavy iron girders and supports were put in to replace the original teak uprights and beams, which had decayed. More props had to be inserted in 1873, and these saved the building from serious damage during the winter gales. Early the following year, the woodwork of the front portion of the Palm House was found to be rotten, and further repairs costing almost £1,000 were carried out. Patched and propped, the Palm House lasted until the autumn of 1883, when it was so extensively damaged by storms that it had to be demolished and replaced.

When Thomas Tickell acquired the property at Glasnevin in 1735, there was already a house on the site. During the 1850s and 1860s, this house, then the curator's residence, was also substantially repaired. It still stands today (Fig. 64), the oldest building in the Botanic Gardens, apparently in its original form with a basement and two storeys. The lecture room and Pelargonium House that had been built against the north and south sides respectively in 1800 were removed during the 1850s. The Society wanted to demolish the house and build a new residence for the curator, but the government refused to meet the cost of a new building — £700 — and repairs to the existing building were started in the summer of 1867. For some reason the contractor removed the roof, and left the unroofed house standing for six months, while rain poured in and the crumbling walls deteriorated. Eventually work was resumed, and the house was turned 'back to front'. Originally, the front door faced east and was reached by a flight of steps, the back of the ground floor was taken up by two rooms, and an inside staircase led from the basement up to the first storey. In the renovations the front door was moved to the west side, beside the bow window, and the original hall split into two rooms. The original front entrance was bricked up, and an outside staircase added, which led down to the basement at the rear of the house. The red-brick facing of the west side was removed, and the key-stone above the centre window was inscribed with the date 1868. Upstairs, the doorways were made higher, as they had been too low for a tall man to pass through without stooping.

The primary function of a botanical garden is to grow as wide a range of plants as possible within the limitations of the site and facilities. Since its formation in 1795, the collections at Glasnevin had increased steadily, and this trend continued under the direction of David Moore and William Harvey. Professor Harvey made two foreign journeys, during which he carried out his own scientific research, and collected large numbers of herbarium specimens. Harvey also gathered some seeds for Glasnevin, and for Trinity College Botanic Garden. In the autumn of 1849 Dr Harvey visited the United States of America, where he had been invited to deliver a course of lectures at the Lowell Institute in Cambridge, Massachusetts. He took the opportunity to travel south, through the Carolinas to Florida, and spent much of his time studying marine algae. He sent back seeds and insectivorous plants, including the Venus fly-trap (*Dionaea muscipula*), to Glasnevin. In 1853, Professor Harvey was granted three years leave by Trinity College, and by the Royal Dublin Society, and he undertook a more ambitious journey to Ceylon, Australia and the South Sea Islands. During his absence, Dr George Allman, who still held the chair of botany in Trinity College, deputized as Professor of Botany to the Royal Dublin Society. Harvey did not collect as much seed on this journey — Glasnevin received only sixty packets of seed from him during his absence.[25]

David Moore (Fig. 65) travelled abroad too, but not as far afield as Harvey, and he also used his journeys to forward his scientific studies. Between 1839 and 1876, Moore made at least twenty trips overseas. The expense to the Royal Dublin Society of these excursions to Britain, France, Germany, Italy, Spain, Portugal, the Low Countries, Iceland, Scandinavia and Russia totalled about £800. As the Society paid for the expeditions, there was an onus on Moore to return with plants, dried specimens and seeds for Glasnevin, and there were other tangible benefits. The personal contacts that the Glasnevin curator made with gardeners and botanists in other countries ensured the continuing exchange of plants and seeds. During the four decades of David Moore's curatorship, the bulk of Glasnevin's new plants was obtained as gifts from botanical collectors and gardeners, and by exchange with private individuals and European gardens and nurseries; only about fifty pounds had to be spent each year on buying plants. The accession records for 1844 to 1876 show that approximately 12,000 plants and over 15,000 packets of seeds were donated to Glasnevin, and in return about 18,000 plants and 4,000 packets of seeds were despatched to other gardens.

It is extremely important that a botanical garden should be able to exchange plants and seeds, and personal contacts between curators help in establishing and maintaining this interchange. The situation in the Royal Botanic Garden, Belfast, was starkly different. The Garden was established on its present site at Stranmillis in 1829 by the Belfast Botanic and Horticultural Society, and until 1895 was funded mainly by levying an entrance charge. Successive governments refused to give grants on the grounds that only botanical gardens in cities received state aid, and Belfast did not become a city until 1888. So the Belfast Botanic Garden struggled along, and the Society tried to boost entrance receipts by staging exhibitions completely unsuited to a botanical garden, such as balloon ascents, dog shows, dancing Zulus and simulated explosions. Belfast Botanic Garden did receive gifts of plants and seeds from other gardens, including Glasnevin and Kew, and also from private individuals. For

64. The Director's residence, National Botanic Gardens; photograph by David Davison, November 1981

example, in the early 1830s, Thomas Drummond, who had been curator there in 1829, sent seeds from Texas. But his successors did not have the opportunities to travel overseas to establish the personal contacts that had helped David Moore enlarge the Glasnevin collections. However, Belfast had more money than Glasnevin for the purchase of plants — about one hundred pounds per year — yet its plant collection contained little of interest to offer other gardens in exchange. In 1895, Belfast Corporation took over the Garden and since then it has functioned as a public park.[26]

The Glasnevin Botanic Gardens was fortunate in that it did not depend on income from visitors, and thus there was no need to stage events to boost attendance. However, at various times diversions, such as fêtes, promenades and flower shows, were held in the Gardens. While the Great Industrial Exhibition of 1853 was in progress, a horticultural fête was held at Glasnevin to raise money to build the Victoria Regia House. The following summer another fête was arranged, which was attended by nearly 3,000 people.

In 1857, the British Association for the Advancement of Science met in Dublin for the second time — the first meeting there had taken place in 1835. On this second occasion, as on the first, delegates were entertained at the Botanic Gardens. The Lord Lieutenant and the Lord Mayor of Dublin attended. Military bands played. Thirteen marquees provided shelter and refreshment and the general public were admitted. Altogether over 7,000 people were present that afternoon.

Promenades were introduced in 1860. The Lord Lieutenant attended one in May 1863, but the weather was bad and only a hundred non-members and 113 children turned up. The Lord Lieutenant stayed for an hour. Subsequent promenades were also poorly attended, and these diversions were discontinued in 1893.

The number of people visiting Glasnevin annually remained almost constant between 1840 and 1860. There had been a six-fold increase in numbers during Ninian Niven's curatorship, but no further

65. David Moore
(c. 1865)

increase during David Moore's first two decades. For its first two years (1800-1801) the Botanic Gardens was not open to the public unless a pass was obtained from the Professor of Botany; this was valid for one visit only, which had to be on a Tuesday or a Friday. In 1802 the conservatories were also opened to the public, but visitors had to be accompanied by one of the Gardens' staff. The regulations were gradually altered and relaxed, and by 1852 the public was admitted—no passes were required—on all days except Saturday and Sunday. As an experiment, entrance fees of sixpence for an adult and threepence for a child were charged on Tuesdays and Thursdays in 1854 and 1855, but these were dropped in 1856. In 1857 the Gardens was opened to the public on Saturdays.

During the 1850s, moves were made to have the Botanic Gardens opened on Sunday. One of the instigators was James Haughton, a Quaker cornmerchant and temperance advocate, who, in 1851, proposed Sunday opening to the Royal Dublin Society. It was widely believed at the time that if the working classes were provided with a park or a botanical garden to visit, men would go to them with their families instead of drinking in public houses. In Belfast, the Botanic Garden was opened to the public on Sunday for this reason, but subsequently it was closed during the hours of evening church services because of 'scandalous goings-on among the bushes'.[26]

A public committee was formed in Dublin to press the case for Sunday opening, and this committee published a pamphlet asking 'Why exclusiveness and Sabbatarianism should be the rule at Glasnevin Botanic Garden [sic]whilst under Her Majesty, at Kew, free admission is the rule on all days of the week, *AND ON SUNDAY*'.[27] There was a spate of letters to the Dublin newspapers on the issue and both sides produced petitions. Six thousand people signed a petition against Sunday opening. But counterbalancing this, Sir William Gregory, at the time Liberal-Conservative Member of Parliament for Galway, was able to present to parliament a petition bearing sixteen thousand names supporting Sunday opening.[28]

The government took the view that, as the Royal gardens at Kew and Hampton Court were open on Sundays, the people of Ireland were entitled to a similar amenity. So, in March 1861, the Society was instructed to open its Botanic Gardens on Sundays. The Society replied quickly that Kew and Hampton Court were Crown property and completely state-supported, and therefore different from Glasnevin, which was neither Crown property nor entirely state-supported. Furthermore, the Society noted, the British Museum was not open on Sundays!

The government countered that it cost the state only twenty pounds a year to open Hampton Court on Sundays, and £156 to open Kew. The amount of money annually granted to the Royal Dublin Society for all purposes was £6,000, and subscriptions from members only came to £1,336. In addition, £2,371 had been given to the Society in 1860 for conservatories, and the Society was now asking for a further £2,382. The government trusted that the Society would 'afford to the working class' access to its Botanic Gardens on Sunday.

The Royal Dublin Society reacted by saying that although the government talked about the great liberality of parliament in the education of the people, and the improvement in the country, the Society had not hitherto been given any great share in the increased liberality, and, in fact, received less from the state than it had sixty years previously. During the past six years it had received £35,415 from the government, and by its own efforts had raised £21,811. In the same period the British Museum, which closed on Sundays, had been given £660,370. With reference to the charge against the Royal Dublin Society of religious bias, the Society pointed out that it was the only chartered body 'embracing within its ranks the educated classes of Ireland of any denomination'. There were 'reasons of *local* and special application in connection with the *peculiar situation* of the Botanic Gardens of Glasnevin which present cogent reasons' why the Society viewed Sunday opening 'with grave apprehension'.

The government remained unmoved. In June 1861, the Royal Dublin Society explained that its opposition to Sunday opening on religious grounds was not the sole reason for keeping the Gardens closed on Sundays. The Botanic Gardens shared a common boundary with Glasnevin Cemetery, and Sundays were the usual days for burials, which were attended by great gatherings. The Society was afraid that these crowds would spill over into the Gardens and cause damage.

At a meeting the following month, the Society's members were told that if the Royal Dublin Society continued to oppose Sunday opening, the government would withhold its grant. A vote was taken and 125 members out of the 236 present voted in favour of Sunday opening. Thus after ten years of agitation, the Society opened its Gardens on Sundays.

Opening hours were fixed from 2.30 in the afternoon—to allow the staff to attend morning church—until 7 o'clock in the evening in summer, and until sunset in winter. On 18 August 1861, the Botanic Gardens was opened for the first time on a Sunday. During the next three months 78,132 visitors came on Sundays alone. October saw the greatest number present on a single Sunday, 15,100, which was about as many as came to Liverpool Botanic Garden on a Sunday. The police magistrate for the Glasnevin district reported to Sir William Gregory that drunkenness had declined since the Botanic Gardens opened, and that instead of congregating in the public houses 'the Dublin folk...went to the Gardens and enjoyed themselves there together with their families and returned home in peace and quiet'.[28]

To give the staff a day free from interference to carry out their normal duties the Gardens was closed on Mondays, and it was not until 1869 that it was opened to the public seven days a week. To compensate for any expense Sunday opening involved, the government allowed the Society an extra thirty-five pounds a year.

The fears of the Royal Dublin Society that its Gardens might suffer damage turned out to be unfounded—the government provided a police sergeant and six constables every Sunday free of charge. David Moore reported that the visitors had behaved 'in a most orderly and decorous manner'. The most he could complain of was that young boys leapt over the flower beds and ran through clumps of plants. One problem arising from the huge increase in visitors on Sundays was the crush of people in the wings of the Curvilinear Range. The passages were too narrow to accommodate the crowds and some plants were accidentally knocked off the shelves. On 9 November 1861, Moore wrote to the Committee of Botany complaining about the narrowness of the pathways in the Range, and, on 13 December, members of the Committee went to the Gardens to assess the situation. Moore was instructed to consult the architect of the Board of Works, Frederick Villiers Clarendon, to discuss the possibility of extending the wings. Moore later recorded in his annual report, dated 13 January 1862, that Clarendon thought the wings could be 'readily widened by adding a corresponding side to the north of that which now faced south'. No immediate action was taken. In 1864, in his article for the *Gardeners' Chronicle*, William Robinson reported that it was intended 'to improve the handsome Curvilinear Range of lean-to houses by doubling its width and converting it into a roomy span-roofed range'. But it was not until early 1868 that any action was taken.

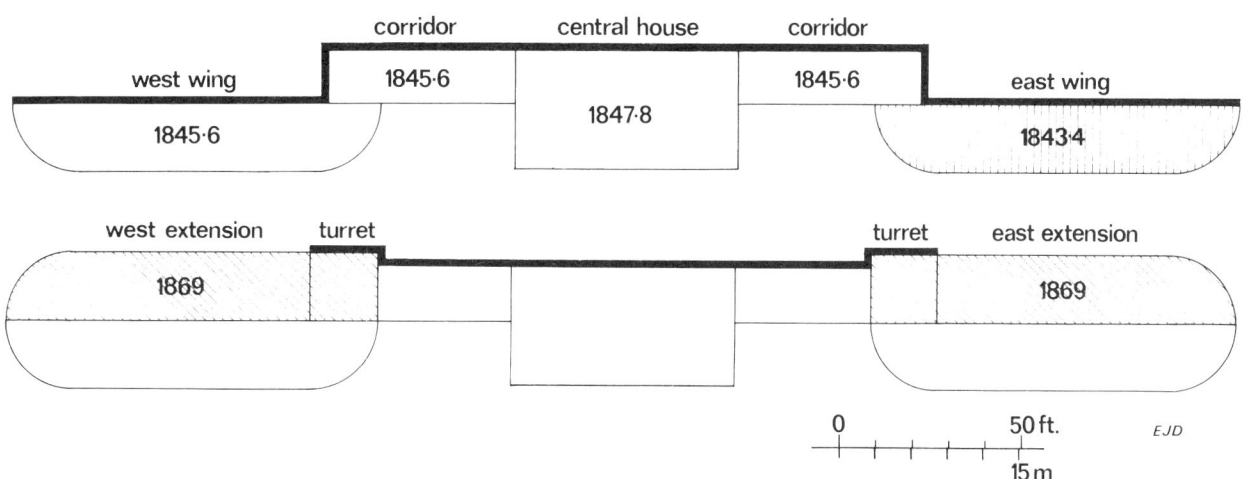

66. Ground plans of the Curvilinear Range; upper figure shows the progress of the Range as built between 1843 and 1848 (see chapter 8 for details, pp. 109-111); lower figure shows the plan of the completed Range with additions made in 1869—this is the plan of the Range today. (Diagrams by E. J. Diestelkamp)

In January of that year, plans were submitted to the Royal Dublin Society by William Turner, son of Richard Turner, indicating how the wings of the Curvilinear Range could be doubled in width. William had succeeded his father as head of the Hammersmith Iron Works, but Richard was still active. The plans for the widening of the Glasnevin Curvilinear Range were indeed produced by Richard Turner (Figs. 67, 68).[29] On 24 June, the Committee of Botany requested the Council of the Society to submit the Turners' plans to the Board of Works. The Commissioners of the Board of Works approved the project early in September, and the Committee decided to proceed with the extensions, but stipulated that the rear walls of the wings should not be demolished until the new structures were in place—this was necessary to protect the plants. The contract was signed on 21 October 1868.

The alterations to the Range involved the erection of a curvilinear structure on the north side of the wall, identical in detail to that already there on the south side. Once this new ironwork was

67 and 68. The original plans by the Turners for the extension to the Curvilinear Range, dated August 1868. (Reproduced by courtesy of the Commissioners of Public Works)

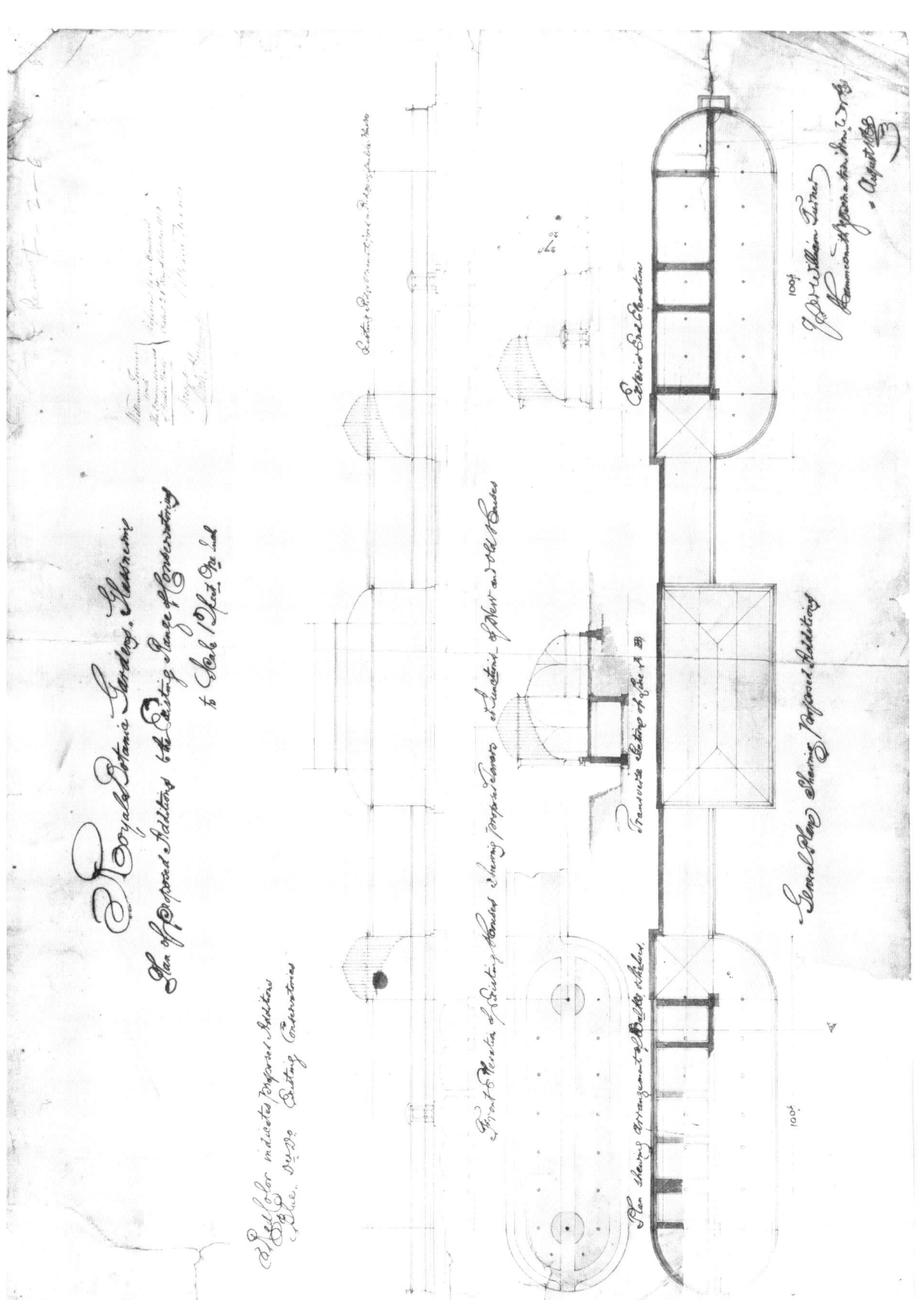

69. The Curvilinear Range today (photograph by David Davison, November 1981)
Note the turrets added in 1869 (see Figs. 67 and 68)

in place, the wall against which the original part leaned could be demolished. This plan was audacious, but highly successful (Fig. 66). After the rear wall was removed, the span roof was supported by a series of butterfly trusses, 'an ingenious and intuitive approach'[29] which Richard Turner had previously suggested in his unsuccessful design for the 1851 Great Exhibition buildings in London. New doors were placed at the east and west ends of the Range, two elegant turrets joined the wings and corridors, and new, wider pathways were laid out in each wing.

The extension work was finished by October 1869. In his annual report that year, David Moore recorded that the heating was working satisfactorily, and that the extensions had greatly improved the appearance of the Curvilinear Range, Richard Turner's greatest glasshouse in Ireland.[30]

As it was primarily fear of damage that lay behind the Society's reluctance to open the Botanic Gardens on Sunday, this is a suitable point at which to look back and note how the public had behaved during the seven decades since the Gardens' foundation. As in all botanical gardens there are two principal sources of possible damage — vandalism and petty thieving.

The Glasnevin Gardens was protected by a boundary wall running along the public road, by the River Tolka on the north side, and on the south side a small stream and hedge separated it from Glasnevin Cemetery. A nightwatchman and such of the staff as lived in the Gardens gave some additional protection at night. Ninian Niven kept a guard dog when he was curator, and during the 1840s, one of the gardeners was provided with a single-barrelled gun to increase security. These arrangements worked well. The only major theft during this period was of lead from a cistern. The thieves were caught, and the two policemen who arrested them were awarded two guineas by the grateful Society.

The main cause of loss was the petty theft of plants, and the taking of cuttings. In the early days, when attendances were small and all visitors were escorted around the Gardens by one of the staff, there had been little pilfering. Ninian Niven said that very occasionally someone would try to bribe one of the men to give them cuttings. However, by David Moore's day the number of visitors had increased, and so had the pilfering. To contain this, William Pope was appointed Gardens' constable at a small annual payment. David Moore recalled an incident involving two ladies. One was detected with 'a choice pelargonium nearly two feet high' under her coat, and the other had filled her pockets with cuttings. The local magistrate fined them ten shillings, and the fine was remitted to the Royal Dublin Society. During the 1850s there were complaints about people bringing in baskets of provisions for picnics in the grounds. The thousands of visitors who came on Sundays after August 1861 were, on the whole, well-behaved. 'I am not aware', commented David Moore, 'of a single flower having been plucked by them, nor a plant broken; neither have the police been obliged to expel any person for bad conduct'. The worst that happened was that, during dry weather, the grass verges and corners became trodden down. To cope with this problem forty cast-iron notices reading *Keep off the grass* were ordered from the Saracen Foundry in Glasgow, at a cost of three shillings each.

William Robinson supplied an interesting description of the Sunday crowds in his account in the *Gardeners' Chronicle*:

> They come from the odoriferous purlieus of The Coombe; from the stews of Church Street; from the many fair streets and suburbs of Dublin and from the many foul alleys on the banks of the Liffey—for to do Dublin justice, there is no air so savoury in the neighbourhood of Petticoat Lane East as that which may be inhaled within a hundred yards of its grandest squares from
> "...rows called Paradise
> Which Eve might quit without much sacrifice"—
> from all these they come to feast on the beauties of tree, and shrub, and herb, and enjoy the fragrance...Yes, there on Sunday afternoons the "intelligent workman"—a type of rarity in many places, but not by any means so in Dublin—may admire [the beauties of the Gardens].[24]

While the quality of a botanical garden is, in great measure, dependent on the ability of the curator, the successful implementation of his ideas will depend not only on having a sufficient number of staff, but also on their ability. About thirty men were employed during the latter part of Moore's curatorship, including a gate-keeper, a garden-messenger, a clerk, and a fireman to stoke the furnaces. The gate-keeper was expected to exclude idlers and stragglers from the Gardens, to see that the resident employees signed a pass-book and were in before 10.30 in the evening, and to ensure that people did not sneak out with purloined plants or cuttings. The garden-messenger until about 1860, Paul Boshill, came to Glasnevin in 1804 aged nearly thirty, and was singled out for mention by William Robinson in 1864. Robinson noted that Boshill had hid in a chimney when rebels approached in the 1798 Rising. 'He could', Robinson declared, 'mow before my grandfather first opened his eyes, and is still able to cut round any man in the Gardens...'[24] although, in 1864, Boshill was almost ninety years of age.

Foremen were first appointed when John White, the original assistant gardener, retired. One foreman, called the indoor foreman, was given charge of the conservatories, and the second foreman had responsibilities for the outside departments and propagation. The first outdoor foreman was Patrick McArdle, who retired in 1868, and was replaced by two men, William Parnell as outdoor foreman, and Wilhelm Keit as propagator.

Julius Wilhelm (or William) Keit (Fig. 70), a German, had trained in Dresden and worked in Linden's nursery in Brussels during the early 1860s. He came to Ireland to manage the garden of the 1865 International Exhibition for the Dublin Exhibition Palace and Winter Garden Company. After the Exhibition concluded he went to England, subsequently returning to Ireland in 1868 to take up the position of propagator in Glasnevin.[31]

The appointment of Keit and Parnell brought the Royal Dublin Society into confrontation with the Commissioners of the Board of Works. The Commissioners decided that the two men should be examined by the Civil Service Commissioners to determine their fitness for the posts. The Society suggested that as Keit was German he should be examined in reading and writing in German and English and that Parnell should be examined in reading, writing and elements of botany. After discussion, it was agreed that the test should be confined to reading, writing and simple arithmetic, and that Keit be allowed to choose between reading a simple passage either in German or in English. The Commissioners also insisted that, instead of being deemed eligible for a pension, the two men should get three shillings a week extra on their wages.

David Moore was pleased not only with Keit's work, but also with the man himself, commenting to Dr Joseph Hooker that Keit was '...respectful but not cringing. I have never known a man whose moral conduct stood higher in every respect than his does, besides he is a shrewd man'.[32] When David Moore sent his two sons, Frederick and David, to school in Hanover in 1869, Keit accompanied them on the journey.[33] Wilhelm Keit only stayed at Glasnevin for five years, and, in 1872, he

70. Julius Wilhelm Keit (Reproduced by courtesy of Dr L. E. Codd, Botanical Institute, Pretoria, and Owen Keit, Lyttleton, Transvaal)

left to take up the curatorship of Natal Botanic Gardens, Durban. David Moore had recommended Keit to Hooker for this position, although he was aware that it would be difficult to obtain another man of Keit's calibre for Glasnevin.[34]

After he had settled in Durban, in spite of the immense amount of work needed to bring Natal Botanic Gardens to an acceptable standard, Wilhelm Keit found time to send plants to Glasnevin. In 1874 he despatched a large case containing 'splendid stems of tree ferns',[35] and four years later, a collection of bulbs. It is probable that this latter consignment included the dwarf, dark blue flowered *Agapanthus* which Wilhelm Keit had named *Agapanthus mooreanus* after his former director—it was sometimes named *Agapanthus umbellatus* var. *mooreanus* and its identity has remained a mystery, although it may have represented a subspecies of *Agapanthus campanulatus*.[36]

Wilhelm Keit's successor as propagator was John Jack, who was in turn succeeded by Thomas Buckley. Little information is available about these two men, but Buckley left Glasnevin in January 1878 to become head gardener somewhere in Ireland. According to a memorandum in the Gardens' archives, David Moore wrote to James Veitch, owner of the Royal Exotic Nursery, Chelsea, asking him to recommend a suitable replacement for Buckley. Alexander Moore was the person suggested, and he was given an allowance of £2 12s. 0d to move from Chelsea to Dublin. But nowhere in the surviving documents is there any mention of the fact that Alexander Moore was Dr Moore's nephew. While working as propagator in Glasnevin, Alexander Moore shared free accommodation in quarters which were said to be comfortable, but would have been more so '...if the old man with whom he shares the apartment were in other quarters—this may probably be ere long accomplished'.[37] Alexander Moore subsequently moved to the Royal Botanic Gardens, Kew, in May 1880, and from there to the New South Wales State Nursery at Campbelltown, where he was superintendent until his death, aged 29, on 14 March 1884.[37]

71. William Parnell's herbarium volumes; upper one contains grasses; lower one contains mosses.
Charles Nelson 1985 (National Botanic Gardens, Dublin)

At Glasnevin, the outdoor foreman with whom Keit worked was William Parnell, described by David Moore as a good botanist, and a man 'of the most upright and moral character'. Keit's weekly wage was twenty-four shillings, and Parnell received thirty shillings. But, as he was married with a young family, William Parnell found it difficult to make ends meet on such a wage, and, in 1873, he also wanted to move from Ireland to a garden overseas. David Moore wrote to Dr Joseph Hooker asking for help in obtaining a colonial appointment for Parnell '...in some temperate climate where his family could get out in the world and where he would...give much satisfaction'.[38]

William Parnell was born on 3 September 1833, and had served a four-year apprenticeship in the Glasnevin Botanic Gardens, attending lectures given by David Moore. Two volumes containing herbarium specimens which William Parnell prepared during his apprenticeship have survived (Fig. 71). The smaller and earlier volume, dated 1850, is inscribed

> The following notes & preserved specimens of British grasses have been collected by William Parnell, during his stay in the Botanic Gardens with a view to making himself master of that interesting part of the System of Botany.[39]

It contains almost one hundred and twenty specimens of grasses, with full descriptions, the correct scientific names, and localities of collection. There is a long, poetic introduction — it may not be entirely original — that concludes: 'Hoping the following pages may be the means of enticing his fellow pupils to greater efforts, the Author subscribes himself. William Parnell'.[39]

In his introduction, Parnell ventured the following moral proposition:

> Let any man who chooses admit the voluptuous beauty of an Italian landscape as he would the charms of a lovely woman with modesty, but for me I love those quiet rural scenes in the same manner that I would a modest female with whose virtues I am acquainted. There are women as their [sic] are landscapes that do not strike the eye or heart at a first glance but who upon a longer intimacy disclose virtue

after virtue and charm after charm, until before we are conscious of it we find them irrevocably fixed in our affections and wonder why we did not at first perceive their loveliness...Every grass that he passes affords him ample contemplation and wonder and leads him up from Nature unto Nature's God...Every new grass that we become acquainted with is a source of pleasure only known to him who has felt it to be able to explain to some dear friend the many different forms of grasses that meets the eye...I feel an unwillingness to conclude but words can ill express the pleasures of such a study equally suited for the high as the low.'[39]

The second volume is much larger, and contains hundreds of specimens of mosses, again correctly named and with descriptions. It bears a manuscript title page dated 1853, and the inscription 'Preserved specimens of British mosses with the generic and specific characters and places of growth interspersed with original remarks by William Parnell'.[40] It also contains an introduction which concludes with the verse

> They whisper hope, they comfort Man,
> Whene'er his faith is dim;
> For he that careth for the moss,
> Careth much more for Him.[40]

On the page immediately after the last set of specimens, someone — perhaps David Moore — has written in pencil 'Well done, good and faithful servant &c'. These two herbarium volumes show that Moore encouraged eager young apprentices to study botany, and that under his guidance such students could produce substantial volumes of work which exceed in quantity and competence the work of most similar lads today.

Parnell left Glasnevin in April 1852 on the completion of his apprenticeship, and went to the Royal Botanic Gardens, Kew, as an assistant gardener, but he was soon moved into the herbarium to help Dr Joseph Hooker. Hooker had gone to the Sikkim Himalayas in 1847 on a plant collecting expedition, and had returned to England with many dried specimens, including numerous rhododendrons. Parnell assisted Hooker to sort and mount these specimens. In November 1853, William Parnell was recommended as a temporary assistant to another Kew botanist, George Bentham, and spent six months helping him. During the summer and autumn of 1854 Parnell resumed working outdoors at Kew, but in November he returned to Ireland.

At first, Parnell was employed at the Vice-Regal Lodge in Phoenix Park, Dublin. During the next nine years he worked in several other large gardens near Dublin, and was mainly responsible for looking after their greenhouses. His employers included George Roe of Nutley, Henry Carroll of Ballynure and Nathaniel Hone of St Dolough's.[41] In February 1858, while working for Alexander Campbell in his nursery at Glasnevin, and laying out grounds for other patrons, Parnell applied unsuccessfully for the curatorship of Liverpool Botanic Garden. William Parnell resigned from his post at St Dolough's in November 1863, so that he could return to Glasnevin Botanic Gardens as a gardener. He was appointed outdoor foreman in November 1868, and remained at Glasnevin for the rest of his working life. In 1879, after the death of David Moore, he sought the curatorship of the Botanic Gardens, but was not successful.[42] Parnell had gone as far as he was to go in the world of gardens. David Moore had passed a sound Victorian comment on him: '...he possesses many of those qualifications which render a man both a good servant and a good member of society'.[43] William Parnell died on 28 November 1906.

The indoor foremen under David Moore were, successively, John Robertson, David Orr, John Lemon and William Pope. Robertson was appointed foreman in the second half of 1838, and received twelve shillings a week. Nothing can be discovered about his background, nor when he left Glasnevin, but by 1856 David Orr was the indoor foreman. Orr, who may have been of Scottish origin, lived and worked in Belfast from about 1840 until 14 May 1849, when he joined the staff at Glasnevin — the date of his subsequent appointment as foreman is not known. When John Lemon was brought from London in December 1866, to work as indoor foreman and plant propagator, David Orr reverted to the position of assistant foreman and assistant propagator, but he continued to live in a house near the Gardens which was rented for him by the Royal Dublin Society. In a memorandum dated

72. Dr David Moore, an engraved portrait published in *The Gardeners' Chronicle*
(National Botanic Gardens, Dublin)

June 1879, Orr was described as a very old man, but well versed in plants. He had a special interest in bryophytes, and assembled large collections of mosses and liverworts; his extant herbarium contains material from Scotland and Ireland collected between 1830 and 1880. David Orr retired from Glasnevin on 1 October 1879, and died in Dublin on 1 February 1890.[44]

Orr's successor as indoor foreman, John Lemon, was appointed on 3 December 1866, and received three pounds towards the cost of moving from London to Dublin. At Glasnevin he occupied a furnished room in one of the gate lodges, and his weekly wage was £1 6s. 6d. Lemon worked as foreman and propagator until March 1868, when William Keit was appointed propagator. Shortly after this Lemon left the Botanic Gardens, and on 1 January 1869, William Pope was appointed indoor foreman. Pope was nineteen when he came to Glasnevin in 1853 as a labourer; both his father and grandfather had worked in the Gardens, and his son was to follow in years to come.

David Moore, as the Curator, was responsible for the day-to-day running of the Royal Dublin Society's Botanic Gardens. In 1854, the Society's professors, including the Professor of Botany, William Harvey, were removed to the new Museum of Irish Industry, and the Botanic Gardens was effectively left without a figurehead. Although Dr Harvey had no role in the management of the Gardens, he did act as a scientific adviser, and was the nominal director of the Gardens. It is a measure of the diminution of his role in Glasnevin that Dr Harvey is almost invariably referred to by biographers and historians as 'of Trinity College'—Professor David Webb, one of his successors in the University's chair of botany, has noted that William Henry Harvey's '...devoted labours...made the name of Trinity College familiar in many scientific centres where it was otherwise unknown...There are many men in Capetown or Pretoria, in Liverpool or Lund, in Adelaide or Berkeley...who, when the name of T.C.D. is mentioned, will say: 'Ah, that is where the Harvey collections are'.[45]

It was left to David Moore to act as the figurehead in Glasnevin, and his status was therefore enhanced. In April 1865, Dr Moore asked the Committee of Botany to change his title from 'Curator' to 'Director'. The Committee approved the alteration, and the following month the Society's Council ratified this decision. But, the change was not accepted by the Department of Science and Art until 29 November 1869, when the Lords of the Committee of Council on Education gave approval. At the same time it was decided not to merge the two salaries Moore received, one as Director of the Gardens and one as Curator of the Botanical Museum. The two offices, with their respective salaries, would remain separate.[46]

David Moore's reputation as a botanist had grown steadily since his move to Glasnevin in 1838. With his work for the Ordnance Survey in counties Londonderry and Antrim behind him, Moore maintained a strong interest in native plants. He made numerous field excursions into the remoter parts of Ireland, and discovered several species not previously recorded. On the shores of Lough Derg, near Portumna, in 1843, he found the Irish fleabane (*Inula salicina*) (Plate 6)—this species has never been found elsewhere in Britain or Ireland. In 1854, Moore collected the rare pyramidal bugle (*Ajuga pyramidalis*) on Inishmore in the Aran Islands. Several plants that Moore discovered were named after him. In 1851, at Rockfield, County Wicklow, he collected a horsetail that Edward Newman named *Equisetum moorei*—it is now thought to be a hybrid. A dwarf willow that Moore found in 1868, among moss on the summit of Muckish Mountain in County Donegal, has eluded all the botanists that have since searched the mountain plateau for it. Fortunately, Moore had collected cuttings and the willow survives in cultivation under the name *Salix* × *grahamii* var. *moorei*.[47]

David Moore (Fig. 72) had a broad range of interests; as well as flowering plants he collected lichens, mosses, liverworts and seaweeds. His study of the lower plants—cryptogams—had its origins in the work he did for the Ordnance Survey. But Moore shrewdly recognized his own limitations. He conducted a substantial correspondence with botanists in Europe and North America, sending them specimens of Irish cryptogams for identification, or for confirmation of his own tentative identifications. In 1856, he published a short paper 'Observations on the mosses of Ireland' in the *Journal of the Royal Dublin Society*, and in succeeding years he wrote further papers on mosses and liverworts. His work on Irish cryptogams was recognized as being of considerable importance by other European botanists. One of his correspondents, Dr Oswold Heer, proposed David Moore for an honorary degree at the University of Zurich. In February 1863, David Moore travelled to Switzerland to receive the degree of doctor of philosophy, an honour bestowed on the man '*qui variis scriptis botanicis de scienta naturali optime meruit*' (Fig. 73). Dr Moore continued to work on cryptogams, and published catalogues of native mosses in 1873 and of Irish liverworts in 1877.[48] The information which he amassed on Irish flowering plants from 1834 onwards was used in *Contributions towards a Cybele Hibernica*, which he compiled with Alexander Goodman More and published in 1866.[49] This work was based on the personal field notes of the two men and the earlier publications of other botanists. There are no species descriptions in *Cybele Hibernica*, only data on the distribution of plants in Ireland.

David Moore also produced a guide-book for the Botanic Gardens, which was published in 1850. It included an engraving of the newly completed Curvilinear Range, and the text was generally more restrained than Ninian Niven's *Visitor's Companion*. A revised edition of Moore's *Handbook* appeared in 1859, with the text altered in minor ways and the map crudely redrawn; this edition was reissued in 1861. Four years later, a fourth edition appeared, with some further slight alterations.[50]

In his role as horticulturist and gardener, rather than field botanist, David Moore made numerous contributions. He reported on experiments carried out at Glasnevin on the formation of wood in flowering plants, on the cultivation of tussac grass from the Falkland Islands, on the effects of frost on exotic ornamental plants, and on grasses of agricultural importance, kohl-rabi, mangold-wurzel

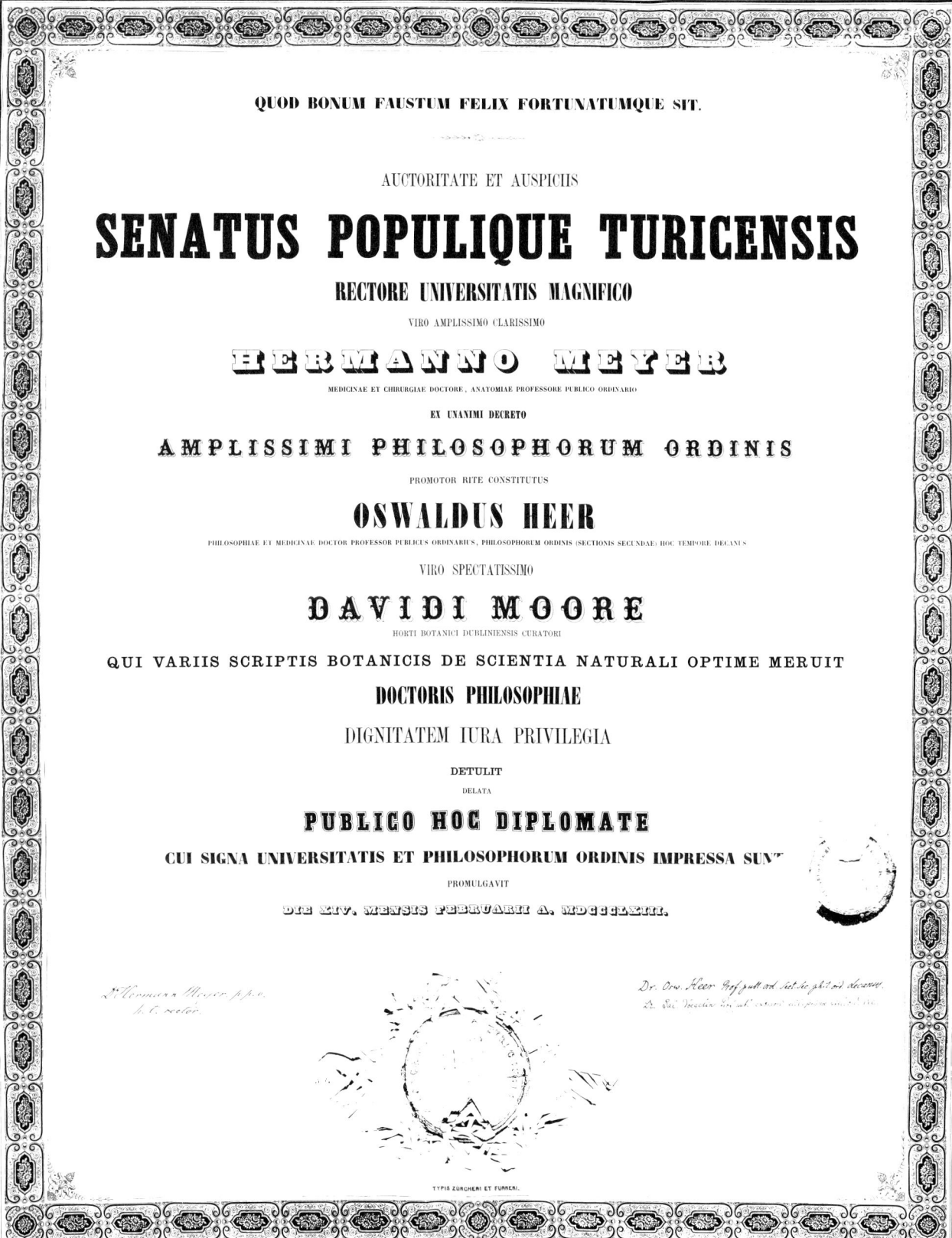

73. Doctoral scroll presented to David Moore by the University of Zurich, 14 February 1863

and potatoes.[51] He experimented with parasitic plants, and tried to establish several species in the Gardens. Two species of broomrape (*Orobanche*) and toothwort (*Lathraea*) were successfully cultivated. In one season there were six different dodders (*Cuscuta*) growing in Glasnevin, but none survived. They all had to be raised annually from seed. Dr Richard Whately, the Anglican Archbishop of Dublin, who was interested in botany, donated many packets of seeds to the Gardens in the 1840s, including several lots of mistletoe. From these seeds the common mistletoe (*Vicum album*) was established on six different trees, and by 1870 there were a couple of dozen plants thriving—this species still grows on trees in the Gardens. In that year, Moore was sent seeds of *Loranthus europaeus*, the only other mistletoe native to Europe, by Dr Edward Fenzl, Professor of Botany at the University of Vienna. This was the second time an attempt was made to germinate this species at Glasnevin, and it was successfully established on common oak (probably *Quercus petraea*) and on Turkey oak (*Quercus cerris*).[52]

Moore was helped in his work on mistletoe by Wilhelm Keit, who was also involved in Moore's experiments with the carnivorous pitcher plants. Insectivorous plants were one of David Moore's many interests, and he built up a fine collection. In June 1844, he received some plants of one of the North American pitcher plants (*Sarracenia purpurea*)—these are the first plants of the genus recorded at Glasnevin. By the late 1860s there were at least four species of *Sarracenia* in the collection, and attempts were then made to hybridize them. In the spring of 1868, pollen from *Sarracenia leucophylla* (formerly *S. drummondii*) was dusted onto the stigma of flowers of *S. flava*. The following spring, the resulting seeds were sown and they germinated. Special care was taken of the seedlings during their first winter. David Moore took plants of this exciting new hybrid (*Sarracenia* × *moorei*), which had pitchers two feet long, to Florence in 1874, for exhibition at the International Botanical Congress there. He also brought a plant of a second hybrid, which William Keit had raised by crossing *S. rubra* with *S. flava*. This hybrid, *Sarracenia* × *popei*, was named after William Pope, the Gardens' indoor foreman. In 1878, the original hybrid, *Sarracenia* × *moorei*, was recreated, and the following year a third hybrid, *S.* × *catesbaei*, was produced from crossing *S. flava* with *S. purpurea*. Since then, these hybrids have been recognized growing wild in eastern North America, where the parent species are indigenous. As well as raising *Sarracenia* species from seed, Moore succeeded in growing the rare and difficult West Australian pitcher plant (*Cephalotus follicularis*) from seed. According to a letter written to Dr Joseph Hooker, Moore intended publishing his observations on the formation of this plant's pitchers in the *Proceedings of the Royal Irish Academy*, but he did not do so.[53]

74. One of the two gold medals presented to David Moore by the Royal Horticultural Society of Tuscany for the exhibit of plants at the International Botanical Congress, Florence, 1874. By courtesy of Major-General F. Moore. (The second medal is in the National Museum of Ireland, Dublin, and is identical in all respects) (enlarged)

The Glasnevin display at the International Botanical Congress included the cobra lily (*Darlingtonia californica*), a close relative of *Sarracenia* from western North America. The particular plant, raised from seed sown in 1868, flowered in Glasnevin for the first time in 1874, and was described in the *Gardeners' Chronicle* as one of the finest living specimens. Killarney ferns, and the rare native orchid, *Spiranthes romanzoffiana*, were also exhibited in Florence. The Royal Horticultural Society of Tuscany awarded two gold medals (Fig. 74) to David Moore,[54] one for the exhibit of *Sarracenia* species and hybrids, and the other for a specimen of the Madagascar lace-plant (*Ouvirandra fenestralis*), an aquatic plant with peculiar lace-like leaves. Glasnevin's original plant of *Ouvirandra* was a gift from Veitch's Nursery at Chelsea in 1856, shortly after the plant had been introduced into Europe. Seedlings were later raised in Glasnevin.[55]

Growing insectivorous plants became a fashionable pastime during the latter half of the nineteenth century, when some of the splendid tropical species and new hybrids were as expensive as orchids: for example, a single plant of *Sarracenia* × *moorei* cost four guineas in 1877. At that time, Glasnevin possessed a substantial collection of tropical genera, including *Nepenthes* and *Heliamphora*. In 1864, the sun pitcher (*Heliamphora nutans*) from Venezuela, was growing in the Gardens, and there were fourteen different varieties of *Nepenthes*. The first of Glasnevin's *Nepenthes* species was raised from seed in 1848, and, in 1862, Veitch and Company donated plants of the hybrid *Nepenthes* × *dominiana*. *Nepenthes* are tropical climbers, and often occur as epiphytes in the wild; at Glasnevin, they were kept in the hottest of the propagating pits, and later were grown in the Orchid Wing of the Palm House, where William Robinson saw them in 1864.[24]

Moore's interest in insectivorous plants drew enquiries from Charles Darwin. Darwin's letters to David Moore were discovered only a few years ago. The first of these was dated 28 June 1874 (Fig. 75), and in it Darwin asked Moore to send plants of the bladderwort (*Utricularia vulgaris*) and the Portuguese sundew (*Drosophyllum lusitanicum*). In reply, David Moore commented that he had read about Darwin's recent work on the insect-trapping habit of butterworts (*Pinguicula*), noting that the greater butterwort (*Pinguicula grandiflora*), a native of Kerry and Clare, was the easiest species to keep in cultivation. Darwin responded by thanking Moore for his observations, and added that he had just received some plants of *P. grandiflora* from Cornwall 'where it is naturalized'.[56] Charles Darwin mentioned Moore's comments on the greater butterwort in his book on insectivorous plants published the following year.

David Moore also supplied material for Joseph Hooker, who read an address on insectivorous plants to the British Association for the Advancement of Science meeting in Belfast during August 1874. In a letter which contained a rare touch of humour, Moore told Hooker that

> Although I feel for my children especially the more distinguished among them and consequently don't like the idea of their bodies being subjected to the dissecting knife, I have sent away a netted hamper this afternoon...it contains a grand plant of *Sarracenia flava* in fine condition for the purpose you want it for...also good plants of *Sarracenia variolaris* and *Cephalotus follicularis*.[57]

As an intensely religious man, David Moore did not subscribe to Darwin's theory of evolution by natural selection, which had been published in 1859. He delivered an anti-evolution lecture in the Presbyterian Church, Rosemary Street, Belfast, in December 1874, as part of a series titled 'Science and Revelation'. These lectures were mounted as a riposte to the 1874 proceedings of the British Association for the Advancement of Science, at the opening session of which Dr Hooker had 'thought fit to assail some of the most important principles of religion'.[58] In the first lecture of the series, Dr J. L. Porter, the minister attached to Rosemary Street Presbyterian Church, attacked the theories put forward by Thomas Huxley, John Tyndall and Charles Darwin. The church was packed for Dr Moore's address; even the galleries were filled. The title chosen by Moore was 'Design in the structure and fertilization of plants: a proof of the existence of God'. The lecture was enlivened and enhanced by beautifully coloured diagrams of orchids, pitcher plants and the Venus fly-trap (*Dionaea muscipula*), as well as by living specimens, and a 'lamp of limelight...brought out the different diagrams admirably'.[59] Moore was more prudent than Dr Porter, and with innocuous words weaved his way between the shoals of contemporary theological thinking and the new scientific ideas. He was careful not to make a direct attack on the eminent scientists, some of whom were his friends, and instead tried to show 'preconcerted design, and infinite wisdom in the vegetable kingdom'. Among the things Moore suggested was that 'even in the almost universal colouring of plants, infinite wisdom is evinced, green being that colour on which the eye can look longest without tiring', but many of his ideas would now be considered naïve.

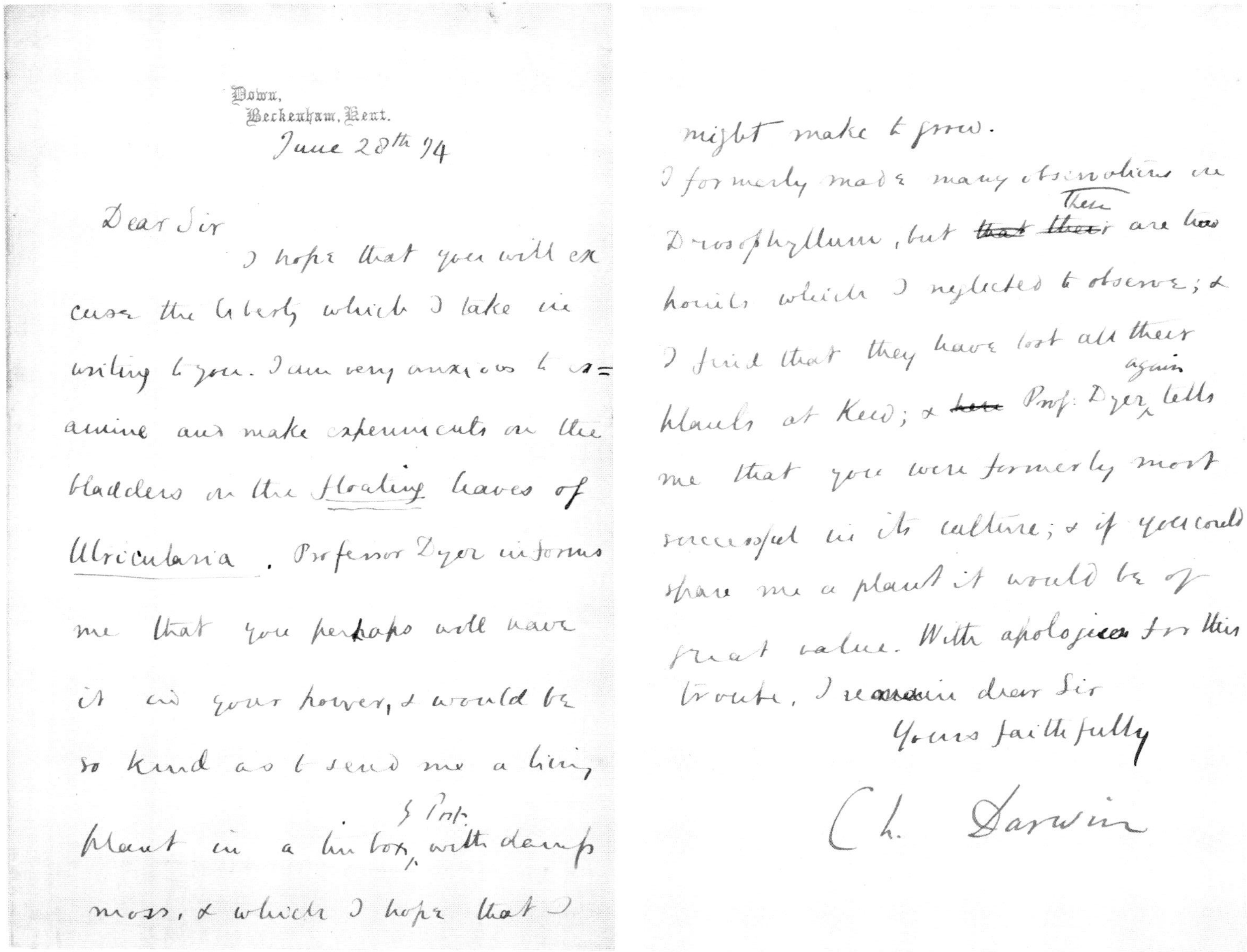

Down,
Beckenham, Kent.
June 28th /74

Dear Sir

I hope that you will excuse the liberty which I take in writing to you. I am very anxious to examine and make experiments on the bladders on the floating leaves of *Utricularia*. Professor Dyer informs me that you perhaps will have it in your power, & would be so kind as to send me a living plant in a tin box, with damp moss, & which I hope that I might make to grow. I formerly made many observations on the *Drosophyllum*, but the there are two houses which I neglected to observe; & I find that they have lost all their bloom at Kew; & Prof. Dyer tells me that you were formerly most successful in its culture; & if you could spare me a plant it would be of great value. With apologies for this trouble, I remain dear Sir

Yours faithfully

Ch. Darwin

75. Facsimile of Charles Darwin's letter to David Moore requesting insectivorous plants (reduced). The script is that of Francis Darwin, but the signature is that of Charles Darwin. (Reproduced by courtesy of Major-General F. Moore)

76. *Trachycarpus fortunei*; one of the two specimens planted by David Moore in 1870. This specimen still grows in front of the Curvilinear Range (photograph taken about 1900 — note the elaborate decoration on the guttering of the Range)

He spoke about pollination mechanisms, using as an illustration the work of Charles Darwin on the pollination of orchids, and concluded:

> I trust I have succeeded in showing that the simplicity of the works of nature is remarkable, and justly held forth for our admiration; yet, though simple, they are perfect, and, as we have seen, act harmoniously to accomplish whatever end they are designed for, without interruption in the stupendous and boundless works of the Creator.[60]

David Moore sought to assemble in Glasnevin as comprehensive a collection of plants as he could. He raised palms from seed supplied by his brother, and by other plant hunters. In 1870, two young Chusan palms (*Trachycarpus fortunei*) were planted in the open at Glasnevin to test their hardiness, and they survive today, one beside the Curvilinear Range (Fig. 76), and the other behind the Director's house, in the positions where Moore himself had them planted. The trial of plants in the open-air, without the protection of glasshouses, was one of the facets of horticulture that exercised David Moore. In his letters to the Hookers, he often recorded successful attempts to introduce supposedly tender plants into Irish gardens. One of his own successes in this quest was the beautiful pink-flushed lily from Natal, which Joseph Hooker named *Crinum moorei*. It was raised at Glasnevin from seeds sent home by Captain Webb, an army officer, who was serving in South Africa. As the lily would not flower in the glasshouse, Moore planted some of the bulbs in the border in front of the Curvilinear Range, where they survived and eventually blossomed. In 1874 specimens were sent to Kew for naming and for illustration in the *Botanical Magazine*.[61] The lily still grows where David Moore planted it over one hundred and ten years ago.

David Moore's work on hybridizing *Sarracenia* species has been noted, and he also produced at least one daffodil hybrid, which was illustrated by Frederick Burbidge in his monograph on *Narcissus*.[62] It is there described as a cross betweeen *Narcissus poeticus* and *N. pseudo-narcissus* var. *minor*.

Ferns and cycads were features in Glasnevin. 'Verily a good Tree fern is one of the most eternal of the verities of a good gardening establishment—especially a botanical one', wrote William Robinson,[24] and many adorned the Octagon House and the other Glasnevin glasshouses. Some came from Australia, from David Moore's brother, Charles, as living plants, but others, like the superb black-stemmed *Cyathea medullaris* from New Zealand, were raised from spores. The Killarney fern (*Trichomanes speciosum*) was cultivated with great success in a small, lean-to glasshouse built at the rear of the Curvilinear Range.

Australian plants were prominent in the Glasnevin collections throughout the latter decades of David Moore's curatorship, because he was able to obtain quantities of them from his brother in New South Wales, and from others, including James Drummond, one-time curator of the botanical garden in Cork, who had emigrated to the Swan River Colony (now Western Australia) in 1829. In 1864, William Robinson admired a fine specimen of the warratah (*Telopea speciosissima*) from New South Wales, growing in the Camellia wing of the Palm House Range at Glasnevin.[24] One of James Drummond's discoveries, *Hakea victoriae*, a native of the Barren Range in south-western Australia, was also in the Glasnevin collections then; it is a strange shrub with hard, variegated leaves, and looking like an elongated cabbage. Both of these plants belong to the family Proteaceae, which is one of the major components of the floras of Australia and South Africa. In a letter to Joseph Hooker, written in December 1870, David Moore made the following comments about growing these southern hemisphere plants:

> We still manage to keep a good many of the Cape and Australian Proteaceae which I am very fond of—but like you, I cannot get our people to take much interest in them. I have often to rescue them...when they are placed in some out of the way place. I have never found so much peat earth good for Proteaceae plants. Good rich loam with sharp sand and only a little peat best. Above all things, they abhor too much water. Those that are thick and fleshy about the root will only flower well when exposed to the warm sun in some sheltered nook and allowed to be well parched for a month or six weeks after they have made their young growth.[63]

77. A cycad (probably *Cycas revoluta*) in the Palm House, Glasnevin, photographed c. 1910

David Moore had a substantial interest in the primitive cone-bearing cycads (Fig. 77), which, though they resemble tree-ferns, are more closely related to coniferous trees. He assembled an outstanding collection of them in Glasnevin, again with the help of his brother, who sent Australian species, and of his contacts in South Africa. Many of the cycads growing in the National Botanic Gardens at the present time were acquired during David Moore's directorship, over a century ago. They are slow-growing and extremely long-lived plants, so their survival in the collections is easily understood. Moore's fondness for the cycads led him to study them in detail, and, in 1878, he decided that a plant, imported from Cuba about thirty years earlier, probably represented a new un-named species. Noting that it might be a form of *Ceratozamia longifolia*, according to Dr Edward von Regel, Director of the Imperial Botanic Garden in St Petersburg, Moore tentatively proposed for it the name *Ceratozamia fusca-viridis*. The Glasnevin plant, a female, was the only one then known in cultivation in Europe, and this same one is still growing in the National Botanic Gardens. The description and the new name were published in the *Scientific Proceedings of the Royal Dublin Society* in 1878, a short paper that was one of the last written by Dr Moore.[64]

But a botanical garden can never be exclusive, a botanist's 'ivory tower'. Even in a botanical garden the general public are not always impressed by quaint species of great scientific interest. Visitors expect to see pretty flowers. Glasnevin has always catered for these expectations, with beds of bright flowers at all seasons. Colourful displays were achieved in various ways; borders of hardy herbaceous perennials, and a rose garden, were created in Glasnevin at least by 1850. And thousands of annual plants were bedded out each year to provide great splashes of colour in high summer. By the 1860s the fashion for bedding-out vast numbers of half-hardy annuals was at its peak. But, William Robinson noted with warm approval, that Glasnevin was 'the very antithesis of bedder-out-aping botanic gardens'.[24] The public's expectations meant that David Moore did have to bow to this Victorian fashion, and he complained that, while the 30,000 plants bedded-out gratified the visitors who crowded the Gardens on Sundays, the summer bedding caused considerable extra work and expense. In 1871, Thomas Neill, who was working at Glasnevin in 1866, became manager of bedding-out plants, with a wage of nineteen shillings a week. The previous year Moore had expressed his dismay about the effects that this fashion was having on his staff, saying that '...the bedding-out affair has destroyed very much the interest young gardeners formerly took in the management and culture of exotic plants'.[65]

On occasions, the Botanic Gardens supplied plants as decorations for outside functions. Flowering shrubs were sent to Kingstown (Dun Laoghaire) to decorate the pavilion for the ceremony to receive Queen Victoria when she landed in Ireland in 1849, and, twenty years later, plants were sent to embellish the ballroom for the ball given for Prince Arthur when he visited Dublin. After that public entertainment, the Royal Dublin Society decided not to lend plants in future, although no absolute rule was made. It was difficult for the Director to satisfy the borrowers, and the Society's plants were liable to be damaged. Commercial nurserymen were unhappy about the interference caused to their trade.

No such problems arose over the supply of plants for other institutions. The School of Art, a sister institute formed under the aegis of the Royal Dublin Society, received vast numbers of specimens for use in classes. The Royal College of Surgeons, and the Apothecaries Hall, also obtained teaching materials from Glasnevin. Plants were sent to the university colleges in Cork and Galway, even after they established their own botanical gardens. When St Stephen's Green was improved, in 1850, the Commissioners appealed to the Royal Dublin Society for trees and help. They were invited to come to Glasnevin and select such trees and shrubs as were needed, but it was emphasized that David Moore did not have time to supervise their planting.

During Moore's tenure of office, although scientific work remained important, the Gardens became more and more an institution catering for the public demand for pleasure and relaxation. Left to itself, the Royal Dublin Society would have moved more slowly in this direction. But the Botanic Gardens at Glasnevin needed more money than the Society could afford. On average, it took £2,000 each year to finance the Gardens. Salaries and wages were the biggest item, and as the nineteenth century progressed, the proportion of expenditure on this item increased. The conservatories, erected between 1800 and 1860, cost about £9,000, of which the Society itself provided only about one tenth. So, in effect, the Society was dependent, in part, on an annual government grant to run the Gardens, and almost entirely dependent on government money for the provision of the necessary conservatories.

In these circumstances it was not surprising that the state increasingly demanded more say in the policy of the Royal Dublin Society towards the Gardens, which the Society had created originally with the help of a grant from the government.

Since the 1830s, the government had been unhappy about the way in which the Royal Dublin Society, an independent institution, received and spent public money in supporting activities such as the School of Art and Design, the Botanic Gardens, and the public lectures. The government tried to persuade the Society to spend the money in ways which the government thought were wise and efficient, but this led inevitably to disagreement and clashes with the Society. The swing of the pendulum of power between the Whigs and Tories also played a part. As has been mentioned, the disagreement in 1841 might have precipitated the closure of the Gardens. In 1851, the Department of Science and Art was established. It took over responsibility for disbursing grants to the Society, and, incidentally, asserted its authority in other ways. In 1854, the Royal Dublin Society lost its role in scientific education when its professors were transferred to the Museum of Irish Industry, which was established under the Department of Science and Art. However, the Society continued to organize popular science lectures for the general public, and its Professor of Botany, Dr Harvey, assisted with these.

Following the resignation of Professor George Allman from the chair of botany in Trinity College, Dublin, William Harvey was appointed Professor of Botany there in 1856, and he also retained the professorship in the Museum of Irish Industry. In 1861, at the age of fifty, Harvey married Miss Phelps, a childhood friend from Limerick. A few weeks after his marriage he suffered a serious haemorrhage in his lungs, caused by tuberculosis which had dogged him since his thirties. His health declined steadily after this, and in February 1866 he went to Devon in the hope of recovering. But, on 15 May 1866, William Harvey died in Torquay, at the home of his old friend Lady Hooker, widow of Sir William Hooker.

William Henry Harvey possessed the gift, not always allied to scientific ability, of presenting his work in beautiful, lucid and pleasing English. The elegance of his annual reports on the state of Glasnevin suggests that he might have become a distinguished literary figure if he had taken to writing professionally. His mastery of language was recognized by his contemporaries. Asa Gray, the distinguished American botanist, wrote of Harvey:

> He was a keen observer and a capital describer...[he] wrote perspicuously, and where the subject permitted, with captivating grace; affording, in his lighter productions, mere glimpses of the warm and poetical imagination, delicate humour, refined feeling, and sincere goodness which were charmingly revealed in intimate discourse and correspondence, and which won the admiration and the love of all who knew him well. Handsome in person, gentle and fascinating in manners, genial and warm-hearted but of a very retiring disposition, simple in his tastes and unaffectedly devout, it is not surprising that he attracted friends wherever he went, so that his death will be sensibly felt on every continent and in the islands of the sea.[66]

The Royal Dublin Society was unable to replace Professor Harvey. The continuing skirmishes between the Society and the government had resulted in the erosion of the Society's control over the institutions it had established. In 1865, the Committee of Council on Education took the first steps towards the creation of a College of Science in Dublin, to which the various professors in the Museum of Irish Industry would be attached. Two years later, the Museum was abolished and the Royal College of Science was established; the first meeting of the College Council took place on 10 September 1867.[67]

The chair of botany in the new college was filled by Dr Alexander Dickson, who was already Professor of Botany in Trinity College, Dublin. This was acceptable, as the duties of the professor at the Royal College of Science were part-time. On Dickson's departure to the University of Glasgow, Charles Wyville Thomson was appointed Professor of Botany in the Royal College, but he vacated the chair after about one year, and moved to the University of Edinburgh. In June 1870, William Thiselton Dyer became Professor of Botany in the Royal College of Science. Like his predecessors, Dyer had no formal connection with Glasnevin Botanic Gardens, but he was provided with botanical specimens by the Gardens for his lectures and practical classes. Dr Thiselton Dyer resigned the chair in January 1872, and moved to the Natural History Department of the School of Mines, in South Kensington, London, being succeeded in Dublin by William Ramsay McNab.

78. William Ramsay McNab, a photograph taken c. 1880. (Reproduced by courtesy of Mrs Cowie and James White)

Dr McNab was the grandson of William McNab (1780-1848), curator of the Royal Botanic Garden, Edinburgh, and son of James McNab (1810-1878) who had also held that position. W. R. McNab was born in Edinburgh on 9 November 1844, and was educated at Edinburgh Academy. He studied medicine and botany at the University of Edinburgh, graduating at the age of 22. While an undergraduate, he was appointed assistant to the Professor of Botany, James Balfour, and later spent some time in Berlin studying botany, pathological anatomy and histology. He practised medicine for a short time in Dumfries, but, in 1870, he was appointed Professor of Natural History at the Royal Agricultural College in Cirencester, succeeding William Thiselton Dyer who had, by then, moved to Dublin. Two years later, in March 1872, McNab again followed in Thiselton Dyer's footsteps, becoming Professor of Botany at the Royal College of Science in Dublin.

William Ramsay McNab (Fig. 78) was a prolific author on plant morphology and physiology. He published the results of experiments on the movement of water in plants in the *Transactions of the Royal Irish Academy* in 1874. Material for his classes and research was provided by the Glasnevin Gardens, following the established custom. McNab was a thorough and precise demonstrator in the laboratory, and his lectures were fluent and entertaining. He had a great appetite for work, and it is probable this led to the uneasy relationship that gradually developed between himself and David Moore.

1 April 1878 was a momentous day for the Royal Dublin Society's Botanical Gardens at Glasnevin. After eighty-three years, the Society relinquished control of this 'brightest jewel', and the Royal Botanic Gardens, Glasnevin, along with the Society's Library and Museum, passed into the direct control of the Department of Science and Art. Within the limits imposed by its resources and the enthusiasm of its members, the Society had managed the Gardens satisfactorily, and, under Dr David Moore, the plant collections, and the Gardens' reputation world-wide, had grown immensely. Moore was,

by this time, seventy years old, and although it was suggested that he might retire, he had no intention of doing so, and was retained as Director by the Department of Science and Art. The editor of *The Garden*, William Robinson, paid this tribute to Dr Moore on 29 June 1878:

> Often sorely hindered for want of funds, with but few sympathetic spirits to cheer him on, the Doctor laboured long and hard, until he has, as now, made these gardens to hold their own among the gardens of Europe...Still in the prime of life, now surrounded by many grateful and attached friends, we trust he may long be spared to preside over the beautiful gardens which owe so much to his labours.

To a casual visitor, the change in the administration of the Gardens was barely noticeable—the Department's insignia did replace that of the Royal Dublin Society on notice-boards and letter-heads. But outwardly that was all that changed; the Botanic Gardens continued to blossom.

But, change was inevitable following the removal of the Gardens from the control of the Royal Dublin Society, especially in the Gardens' relationships with other institutions. The Royal College of Science, in its early years, only had an informal connection with the Botanic Gardens, but after 1 April 1878 the Department of Science and Art administered the grants which maintained both College and Gardens. Professor McNab seized upon the opportunity which now presented itself to obtain better facilities at the Botanic Gardens for his own research and teaching. And his personal ambitions extended further.

Before the Royal Dublin Society Bill received the Royal Assent and came into force on 1 April 1878, Dr William Steele, who was appointed to administer the Library, Museum and Botanic Gardens, was sent a memorandum[68] asking that the Professor of Botany in the Royal College of Science be afforded 'greater facilities for study and research' in the Botanic Gardens, and that the Gardens should be more readily available for 'the instruction of students in the College'. It was stated that the Lords of the Committee of Council on Education believed that special arrangements should be made between these separate branches of the same government department.

Dr Moore responded with a lengthy memorandum[69] setting out what facilities were already provided for Professor McNab. He was allowed free access to collect whatever plant specimens he required, and these he selected himself and brought to the College. The Professor could use any of the books in the Gardens' library, as long as the books were not removed. As for providing a room for Dr McNab, Moore stated that there was no room available. He agreed that the Professor could have whatever additional specimens he required, provided that the material could be removed without detriment to the living collection in the Gardens. The Professor's other request, to be allowed unlimited access at all times to all the propagating houses and the nursery, did not find favour with Moore. However, Dr Moore stated that he would issue directions to the foremen to open the doors of the glasshouse for Professor McNab whenever the professor wished, and to give him every possible assistance. Dr McNab would also be given keys to the public glasshouses when he called at the Gardens.

David Moore's answer did not satisfy Dr Steele and the senior officials of the Department of Science and Art in London. They did not like the tone of the memorandum, complaining that it has a 'stand off flavour about it...and not a cordial recognition of the fact that the College of Science, as part of the same Department, stands in a different position to the general public'.[70] The Department suggested that a table should be placed in the Professor's room in the Botanic Gardens for experimental work, and that a small glasshouse should be erected outside his window. As for giving keys to the glasshouses to Professor McNab, the Department felt that this was not unreasonable, and that the Professor would understand and respect the special and valuable plants. 'His merely going into the house won't, I suppose, damage them', wrote Colonel Donnelly, 'however, if there are any such modest and sensitive plants—from Turkish seraglios I presume—why the Professor must be warned off'.[71] Dr Steele was instructed to speak to Dr Moore, and a new memorandum was drafted in which all William McNab's requests were conceded. David Moore signed this new memorandum,[72] but in a private letter to Dr Steele he clearly indicated that he did this against his better judgement.[73] Moore commented that '...the possession of keys to the departments has never been allowed or indeed asked for by any previous professor...', and he stated that, in his opinion, a precedent was being created that could have unfortunate repercussions in later years. Under the new regulations, plants would be provided for experiments and teaching as far as possible within the limitations of the Gardens, and the room adjoining the Director's office and the library in the right-hand gate-lodge, previously used as a paint store by the Board of Works, would be fitted out for the Professor. The Department was, it seems, satisfied with the capitulation of the Director of the Botanic Gardens, and arranged for a small glasshouse to be added to the Professor's new office at the Gardens.

David Moore's relationship with his new superiors was not at all easy. He found the bureaucrats difficult to deal with, and resented their interference which, at times, was petty. Professor McNab also became exasperated. Work on converting the room into a laboratory for his own use went slowly, and the glasshouse when completed had no window that could be opened.[74] There were even arguments about the wording of a notice board at the entrance to the Gardens.[75]

All these problems, caused by Professor McNab, convinced David Moore that there was a campaign to remove him from the directorship of the Botanic Gardens. He wrote to Dr William Thiselton Dyer, saying that

> the whole affair seems to have been arranged between Dr Steele, Colonel Donnelly and McNab, that I was asked to retire on pension, having served 40 years and [McNab] was to step into my shoes, become Director of the Glasnevin Gardens and Professor of Botany at the College of Science. This accounts for his eagerness to get the keys of the conservatories. He is certainly very innocent regarding the management of a botanic garden, especially one of the magnitude of Glasnevin. It will be a sad thing if this fine establishment be allowed to fall into the hands of an incompetent person, after all I have done to raise it to the position it occupies among similar establishments in Europe.[76]

Moore continued: 'I am not yet gone and it may please the Lord to enable me to perform my duties alike creditable to myself, my friends and the public some years longer if no particular pressure be put on me to retire which has not yet been done'.[76]

A few days later, on 21 November 1878, presumably in reply to a letter expressing disbelief in the plot, David Moore told Dr Thiselton Dyer that 'however ridiculous it may appear...there is not the slightest doubt about the connection of the plot. Colonel Donnelly stated openly that he would be a great loss to them as it had been arranged if I would retire they had [McNab] ready to put in my place'.[77]

In March 1879, William McNab applied for the post of Professor of Botany in the University of Glasgow, but Isaac Bayley Balfour was appointed. With a certain sarcasm, Moore wrote to Thiselton Dyer:

> I don't know what we shall do here if we lose McNab as the Board of Works people are busy preparing a lab for him and a plant house on the top of it to keep his plants under experiments. It is part of one of the lodges at the entrance gate and will look *abominably* bad when a glass house is perched up on the top of it. Dr Steele and Colonel Donnelly think otherwise. We shall see.[78]

But David Moore was not to see. He had for some years suffered from a diabetic condition which suddenly worsened and, after a short illness, he died on Monday, 9 June 1879. His son, Frederick, wrote next day to Sir Joseph Hooker: 'We have not as yet ascertained the exact cause of death; he was for the last fortnight suffering slightly from irritation of the bladder; on Thursday last Surgeon Butcher performed an experimental operation on him, and the pain and nervous excitement were, we fear, too much even for his constitution, and he died very quietly having been unconscious for a considerable time'.[79] David Moore was buried in Mount Jerome Cemetery.

Numerous tributes were paid to Dr Moore. The *Dundee Advertiser* of 9 June, noting his death, described him as 'a distinguished Dundonian...who [was a] corresponding member of nearly all the leading botanical societies in Europe. His death will be deeply regretted by his numerous friends, to whom he has endeared himself by his upright sincerity and genial disposition'.[80] His memorialists agreed that the Botanic Gardens at Glasnevin was his 'best monument', and that he had raised the Gardens from a lesser state until it was 'excelled only by the world-famous gardens at Kew, and equalled by only one other public garden, that of Edinburgh'.[81] The editor of the *Gardeners' Chronicle* summed up David Moore as 'one of a brilliant galaxy of botanists and gardeners who have constituted an era in gardening'.[82]

Botanical explorers and the supply of plants
1800-1920

ONE function of a botanical garden is to cultivate as many different plants as possible within the limitations of its situation, and gardens need to obtain plants from far afield. This can be achieved in many ways: by getting other botanical gardens to send propagation material, by encouraging enthusiastic amateurs to collect seeds in foreign lands, or by employing someone to travel to distant places to bring back seeds and plants. Today, exchange between botanical gardens throughout the world provides the majority of the new species grown each year at Glasnevin. However, in the nineteenth century, the Gardens received most of its new plants from travellers and explorers, both amateur and professional. Those collectors sent exciting unnamed plants, and known species which had not previously been cultivated, to Dublin's two botanical gardens. Later, some of these new plants were despatched to other gardens, and some are now commonly cultivated in Ireland and elsewhere.

In the 1800s, the vitality of a botanical garden could have been judged by the number of new plants which were cultivated, as well as by the size of the collections. Botanical gardens vied with one another, and with private collectors, to display their latest treasures. This gentlemanly competition took place at horticultural shows. The two Dublin botanical gardens both provided non-competitive exhibits for the shows of the Royal Horticultural Society of Ireland during the mid-1800s. The new plants were also vividly displayed in the numerous horticultural and botanical periodicals which flourished in the Victorian era. Using hand-coloured engravings accompanied by scientific descriptions, these magazines provided species-by-species accounts of the progress of botanical exploration and of botanical gardens.

In 1813, a Glasnevin plant was figured for the first time in the most famous of these periodicals, William Curtis's *Botanical Magazine*. The subject was *Leptospermum lanigerum*[1] (Plate 1), a native of Australia. It was probably one of the plants purchased from Lee and Kennedy. In 1821, Dr Wade sent specimens of *Ulex europaeus* 'Strictus' (Plate 4) to John Sims who edited the *Botanical Register*, and to Sydenham Edwards who edited the *Botanical Magazine*, hoping that they would have the plant sketched for their journals. 'I trust both will let it have a place in their respective publications', he told John Foster, '...for the honour of poor old Ireland!!'[2] Neither editor complied with Walter Wade's request.

Ninian Niven and David Moore were much more successful in placing species from Glasnevin in the *Botanical Magazine*. A steady flow of living specimens was sent from Dublin to London—in those days it was possible to get a flower from one city to the other in less than twenty-four hours! Sometimes drawings were sent—Ninian Niven sent watercolours to Dr William Hooker on at least one occasion, including portraits of *Oxypetalum caeruleum* and *Verbena tweediana* (Plate 2). Through the *Botanical Magazine* in particular, plants introduced into cultivation from Glasnevin became widely known.

Occasionally, botanical gardens sold new plants and so made a profit, but more usually the public gardens and private gardeners swapped their treasures. They followed the maxim that the best way to keep a plant is to give it away! Thus new species soon became established in gardens, and the number of plants available to gardeners in Ireland and Britain was greatly increased.

In its first decades Glasnevin obtained most of its plants of botanical significance from nurserymen who had benefited from introductions made by the great explorers of the last decades of the eighteenth century, such as James Cook, Joseph Banks and Tobias Furneaux. Lee and Kennedy received seeds from South Africa and South America, and they were particularly noted for their Australian plants, which were raised from seeds sent back by new residents of the colonies at Port Jackson and in Tasmania. Wade's first catalogue of the glasshouse collections indicates that, in 1802, there were eighty different species of *Erica* (heather) from the Cape of Good Hope, as well as many genera and species of the southern hemisphere family Proteaceae, including *Protea* from southern Africa and *Banksia, Lambertia, Hakea* and *Persoonia* from New Holland (Australia).[3] These would have come through Lee and Kennedy, as no Irish nurseryman was capable of supplying these plants in the 1790s or 1800s.

The appetite of Glasnevin for plants in its formative years was well-known. An American collector and nurseryman, John Lyon, made a special visit to Dublin with plants in the hope of selling some to the Dublin Society.[4] John Lyon was born in Scotland, and emigrated to Pennsylvania in the latter years of the eighteenth century. He became gardener to William Hamilton, a wealthy gentleman and art collector, who had a three hundred acre garden in Philadelphia. It is said that Hamilton's garden reached its peak under Lyon, who cared for the collection of ten thousand plants. John Lyon made several plant collecting trips into the Allegheny and Appalachian Mountains, to the 'land of the Cherokee Indians' and to islands off the coast of Georgia. He grew and propagated the plants he collected in a nursery in Philadelphia, before shipping plants and seeds to London for sale.

On 8 December 1805, after returning from the Pennsylvania Alleghenies, John Lyon embarked on the ship SALLY for Dublin. His journal records the cost of the journey including the freight charges for his plants. On 17 January 1806, the ship was off Cape Clear, but encountered bad weather, and on 25 January put into the port of Liverpool in distress. On 28 January she docked in Dublin after 'a very boistrous passage of 51 days from Philadelphia'. John Lyon remained in Dublin until 23 February, going around the gardens and nurseries trying to dispose of his cargo. He probably called on Dr Walter Wade and John Underwood at Glasnevin, but there is no record that they purchased plants. Lyon may also have gone to Trinity College to see James Mackay. Messrs Simpson and Sons, who had a nursery at Inchicore, Dublin, purchased £47 16s. 9d worth of plants, but the only other known sale he made in Dublin was to Commissioner Rowley. John Lyon travelled to Bristol and to London, where he set up premises at Parsons Green and held a public sale of plants. Lee and Kennedy were among the purchasers, and so was the Right Honourable John Foster, who paid £52 6s. 6d for thirty lots of plants specially for Glasnevin. Foster was probably in London when Lyon visited Dublin, and as he had the final word in all matters relating to the running of Glasnevin, a direct purchase without Foster's approval was unlikely. There is, alas, no record of the plants Foster bought from John Lyon.

John Foster had a considerable interest in plants from North America, and at Collon he had assembled a substantial collection of trees and shrubs from that region. Indeed, at that time Foster's arboretum contained a better collection of plants than Glasnevin. He would have been influential in persuading the Dublin Society to grant ten pounds towards an expedition by John Bradbury, which was mounted in 1809 by Liverpool Botanic Gardens. John Underwood compiled a list of 525 'sorts' of plants required, 'two plants of some sorts, four and six of others, but perhaps it will be a long time before any of them come to the Garden'.[5] Bradbury set sail for North America in April 1809, bound ultimately for Missouri Territory. He travelled by way of New Orleans to St Louis, over 600 miles from the mouth of the Mississippi River. At St Louis he acquired land and established a garden. In March 1811, he began a strenuous and eventful trip by boat up the river. On 31 May, he and his companions were attacked by about six hundred Indians. A volley of shots checked the assault, and the chiefs came forward to parley. After giving the Indians presents, and explaining the purpose of the expedition, the explorers were allowed to go on. By late June, Bradbury reached Fort Mandan, over 2,000 miles up river from St Louis. He had collected plants at various places along the river, and he brought these back to St Louis. Aided by the rapid flow of the river, the return journey took only three weeks.[6]

John Bradbury packed his plants, journeyed to New Orleans, and thence to New York, where he was delayed for many months by the war in Europe. He eventually left America and reached England about two years after his plants. It is not recorded whether Glasnevin received any material in return for the grant of ten pounds, but on 13 June 1811, Walter Wade wrote to John Foster enclosing a letter written by Bradbury from America.[7] In this, Bradbury had given Wade a list of plants sent to Liverpool Botanic Garden. Foster's only comment on the letter was that John Bradbury 'does not seem...very contented...but he may do much good. North America is not nearly explored...'.[8] Two years later, in January 1813, the Dublin Society paid for some seeds from Louisiana, which may have been some of those collected by Bradbury.

Nurserymen and specially commissioned collectors were not the sole suppliers of seeds and plants to Glasnevin. Over the years private individuals gave material to the Botanic Gardens, and these gifts came from diverse sources. A vice-president of the Dublin Society, the eccentric General Charles Vallency, donated some seeds that had been washed ashore on the coast of Kerry. These greyish seeds, called nickar nuts, had been carried to Ireland from the West Indies by the currents of the Atlantic Ocean. They germinated, and Vallency sent a drawing of one of the seedlings to Sir Joseph Banks, who identified the plant as *Caesalpinia bonduc*.[9]

People serving abroad in colonial administrations often sent seeds to botanical gardens in Europe. The Marquis of Hastings, a descendant of Sir Arthur Rawdon of Moira, was Governor-General in India from 1813 to 1823, and sent several packages of seeds to Glasnevin. The curator of the Calcutta Botanic Gardens, Dr Nathaniel Wallich, also supplied Indian plants, and over one thousand packets of seeds were received from Calcutta before Wallich's death in 1854. Plants came from St Helena through William Burchall, who was 'school-master and acting botanist' and an employee of the East India Company. In 1819, the Government botanist in New South Wales, Charles Frazer, despatched a parcel of seeds from Sydney, whence many plants came to Glasnevin in the latter decades of the nineteenth century.[10]

North and South America were the major hunting grounds from which Glasnevin received seeds during the first half of the nineteenth century. In 1824, the Dundalk-born botanist, Thomas Coulter, went to Mexico as a doctor to a mining company.[11] He had graduated in medicine from Trinity College, Dublin, but his passions were botany, fly-fishing and catching moths and butterflies. For ten years he lived and worked in Mexico, sometimes managing silver mines, sometimes trying to establish his own businesses, and sometimes collecting plants, birds and insects. In 1831, he went to California, which was a province of Mexico at that time. He collected a number of pines and other plants, which were later introduced into cultivation in Ireland. He returned home in 1834. Shortly afterwards Coulter gave some seeds to Glasnevin Botanic Gardens, among which may have been some of the big-cone pine, named *Pinus coulteri* after him. In 1828 he had sent a collection of living cacti from Mexico to Trinity College Botanic Garden, which included many new species. In California Thomas Coulter discovered the white-flowered matilija poppy (*Romneya coulteri*) which first flowered in Glasnevin in the 1870s (Fig. 79).

In 1829 Thomas Drummond was appointed curator of the newly established botanical garden in Belfast. He had recently returned from the Rocky Mountains in British Columbia. He brought back seeds, some of which were germinated in Belfast, and he shared the young plants with Glasnevin; forty species from this collection were among the first plants exchanged by the two gardens. Drummond resigned the curatorship in Belfast in 1831, and returned to North America as a collector for Glasgow Botanic Garden. He sent seeds of various plants, including the annual *Phlox drummondii* from Texas, to Glasgow.

The records of plant introductions to Glasnevin before 1834 are incomplete, although John Underwood had made a half-hearted attempt to keep such records from 1830 to 1832. Ninian Niven started an accurate register of acquisitions and gifts in April 1834 shortly after he became curator[12] and he also recorded the names of those who received plants from the Gardens. However, some earlier donations are recorded in the minutes of the Royal Dublin Society.

Niven was anxious to increase the range of species cultivated and during his time as curator he encouraged many people to donate seeds and plants to the Gardens. Shortly after assuming the curatorship Ninian Niven published a circular which contained hints 'for the gathering of curious plants and seeds'.[13] This was intended for forwarding to people living abroad. Thus Niven laid

79. *Romneya coulteri*, a coloured plate, prepared from specimens grown at Glasnevin by Dr David Moore in 1877, and published in *The Garden* 25 May 1878. (National Botanic Gardens, Dublin)

foundations which were useful to his successor, David Moore. Numerous members of the Royal Dublin Society donated seeds, sent to Ireland by relatives who were living or travelling overseas. J. D. Croker, a wine merchant of Great George's Street, Dublin, gave seeds of *Rhododendron arboreum* which he had received from J. Cavit in Ceylon. Captain Edward Cottingham presented twelve different greenhouse plants in March 1835, which may have included *Abutilon vitifolium*, a plant he introduced from South America. Mr and Mrs John Gibson, of Straffan, County Kildare, gave a few seeds of hardy American trees in April 1836, and in July some packets of seeds 'such as are generally made up and sent to Europe from Calcutta'. There are many other examples of small private gifts, for it soon became known that the curator of the Society's Botanic Gardens would accept such souvenirs.

Ninian Niven's main source of South American plants was a fellow Scot, John Tweedie (Fig. 80), who had emigrated to Buenos Aires in 1825. Tweedie had been a gardener in the Botanic Garden in Edinburgh before resigning in 1800, to become head gardener at Castle Hill in Ayrshire.[14] He developed an aptitude for landscape gardening and worked on a number of large estates in Scotland, improving them and thereby making a name for himself in that profession. At the age of fifty, John Tweedie decided to move to the Argentine, where it is said he practised as a landscape gardener. However, in 1840 he was being supported by his family, who kept a store in Buenos Aires.[15] He

80. John Tweedie (Reproduced by courtesy of the Hunt Institute for Botanical Documentation, Pittsburgh, USA)

undertook several arduous trips during which he collected seeds, living plants and dried specimens. The latter he sent to Dr William Hooker. Hooker's accounts of the flora of South America contain many references to John Tweedie's collections.

Tweedie became friendly with other British residents of Buenos Aires, including consular officials. As far as is known he made his first expedition in 1832, when he travelled with Henry Fox, British Consul in Brazil, from Buenos Aires to Rio de Janeiro. In later years, John Tweedie went to Patagonia (his first trip to that region ended in disaster), to Tucuman (about five hundred miles north-west of the capital, in the foothills of the Andes), and to the Sierra de Tandil which lie almost 150 miles south of Buenos Aires. John Tweedie also became friendly with the Earl of Arran, who was in the diplomatic corps. It was through Lord Arran that Ninian Niven originally obtained seeds from John Tweedie. Lord Arran was a keen gardener and a 'kind open-hearted man'.[16] His sister, Lady Mary Lindsay, was married to the son of the Bishop of Kildare, and lived in the bishop's house on the north bank of the Tolka, opposite the Botanic Gardens. Lord Arran visited Glasnevin occasionally, and generously gave seeds and plants to the Gardens. The first consignment of seeds from Tweedie reached Dublin in 1834, but it was not recorded in the accessions book.[17] The earliest consignment listed by Niven was 'a box of roots' received in April 1836, for which the Royal Dublin Society paid £8 15s. 0d in freight charges from the Argentine. In March of the following year, a box of seeds and epiphytes arrived in Glasnevin. In September 1837, John Tweedie's consignment included thirty species of cacti from southern Brazil.

Among the plants raised from John Tweedie's first batch of seeds was a species of *Verbena* which Ninian Niven later offered for sale. It was named *Verbena tweediana* after the collector, but it is now regarded as a form of *V. phlogiflora* (Plate 2). On 24 July 1837, Ninian Niven sent seedlings of several other species and two drawings to Sir William Hooker in Glasgow, with a note saying that these plants were all

> new to us gentlefolk on this side of the Channel. They have been raised by your humble servant from South American seeds, collected by our indefatigable friend Mr Tweedie. The Verbena I only succeeded in raising one plant of last year. It is a *free* grower and altogether a very splendid plant. It is...a much more distinct plant than 'Tweediana' is...If it is agreeable to Sir William I would be glad to have it named 'Arraniana' in honour of Lord Arran...who is, besides, a nobleman of high attainments, and likely to be a zealous encourager of Botanical science.[18]

Today, this is also regarded as a mere form of *V. phlogiflora*, but at the time was illustrated in botanical journals as a new and distinct species. John Tweedie sent three other new species of *Verbena*.

Not all of Tweedie's consignments reached Glasnevin in perfect condition. At this period, sending living plants by ship was hazardous; even a small amount of salt spray will kill plants, and delays often meant the plants rotted before they were received. In 1841, a small box of seeds of *Araucaria araucana* (the monkey puzzle) arrived in Glasnevin, but they did 'not appear fresh'. Cacti received in September 1842 were 'in very bad order'.

81. *Cortaderia selloana*, pampas grass, as illustrated in *L'Illustration Horticole*, January 1855

John Tweedie lived for thirty-seven years in the Argentine, and continued to send packets of seeds to Glasnevin until 1857. Thus David Moore, who succeeded Ninian Niven, received the majority of Tweedie's collections. A passion-flower sent from Buenos Aires was named *Passiflora mooreana* by William Hooker in honour of David Moore; it was figured in the *Botanical Magazine* in 1840.[19] Tweedie introduced the tree tomato (*Cyphomandra betacea*) and the cruel plant (*Araujia sericofera*) to Europe through Glasnevin. He also sent *Mandevillea sauveolens* (Chilean jasmine), which was named after the British Minister in Buenos Aires, H. J. Mandeville, who encouraged Tweedie's botanical work. However, the most enduring introduction by Tweedie was of the pampas grass (*Cortaderia selloana*) (Fig. 81).

Seeds of this plant were included in a package received in Dublin in 1840. Seedlings were planted out in a nursery bed at Glasnevin and survived the winter without damage. Two years later, the plants flowered for the first time in Europe. Their elegant, tall, white plumes thrilled David Moore. John Tweedie had described the grass as 'the most showy plants of any class' in the Argentine, with 'the flower stem six to twelve feet high, [and] large spikes of twelve to eighteen inches appearing like sheets hung on poles'.[20] He said that the tall plumes could be seen from several miles away on the flat plains, the pampas. Moore predicted that the grass would be suited perfectly to gardens in these islands, and it was soon despatched from Glasnevin to other gardens.

In 1855, a French botanist, Charles Lemaire, suggested that the pampas grass should be called *Moorea argentea* after David Moore, 'botaniste distingué et directeur du Jardin botanique de Glasnevin, à qui...on en doit l'introduction dans nos jardins'. Unfortunately, Lemaire's suggestion was not taken up by botanists, and another name was coined for the grass.[21] Today, the correct scientific name is *Cortaderia selloana*, which commemorates neither John Tweedie nor David Moore. However, Tweedie was remembered in the genus *Tweedia*; one species (*T. caerulea*) introduced by him about 1832, is an attractive greenhouse plant with blue flowers, but is it now placed in the genus *Oxypetalum*.

While John Tweedie was collecting in South America, Professor John Scouler, Professor of Geology and Mineralogy to the Royal Dublin Society, attempted to persuade the Society to send out its own collector.[22] He had in mind the young botanist William McCalla, from Roundstone in Connemara, and decided that his destination should be New Zealand. McCalla had worked under David Moore in the Ordnance Survey, but was dismissed for passing information about Moore's collections to outside naturalists. Despite this, William McCalla continued to collect plants, especially seaweeds, from the coasts of Ireland. While Moore did not easily forgive McCalla's misdeed, he considered that the young man was a competent and capable naturalist, an opinion shared by John Scouler and the botanists in Trinity College, Dr. Thomas Coulter and James Mackay. Dr Scouler began to make arrangements for McCalla's passage to New Zealand. McCalla himself wrote to Sir William Hooker at Kew seeking advice and help. On the recommendation of the Professor of Botany, Dr Samuel Litton, the Royal Dublin Society advanced ten pounds for plants to be collected by McCalla. Scouler and Coulter personally provided additional financial support. However, after many delays and much procrastination on the part of William McCalla, John Scouler cancelled the plans, and McCalla returned to Roundstone. Dr Scouler wrote to Sir William Hooker saying that McCalla was cowardly, and that he would do 'nothing today which could be deferred until tomorrow and to do nothing for himself while there was a chance of someone else doing for him. I am sorry for it, for with all his faults he is the most general naturalist and algologist in this place and also of a literary and philosophical turn of mind'. David Moore had expressed a similar opinion earlier, saying that while McCalla was 'an indefatigable collector possessing of very considerable degree of natural talent...he wants almost every other acquirement necessary for crowning his labour with honour and success. He wants industry, taste and a due sense of honourable and faithful motives'.

McCalla went on collecting seaweeds and built up a reputation as a competent phycologist. He published two collections of Irish marine algae, and was awarded a silver medal by the Royal Dublin Society in 1845 for this work. He died of cholera four years later.

It had been proposed that William McCalla should collect dried specimens and seeds in New Zealand. This was the only time an Irish institution attempted to initiate its own plant-collecting expedition and to sponsor its own collector. Generally, the botanical gardens in Belfast and Dublin relied on collectors sent out by British gardens and nurseries, or on the goodwill of Irish people travelling and residing abroad. Quite often the botanical gardens received seeds from men serving in the armed forces in colonial outposts.

One such benefactor of Glasnevin was Edward Madden, an officer in the Bengal Artillery. He served in India from 1830 to 1850. His family came from Kilkenny, and when on leave Edward Madden often returned to Ireland. His hobby was exploring, and during his Indian expeditions he collected seeds and plants. Some of the seeds were sent to Glasnevin, but the packages were often shared between several botanical gardens; one lot, sent in April 1847, was divided between Glasnevin, Belfast and Kew, Trinity College Botanic Garden, Dr Royle of the East India Company in London, and the nurserymen Conrad Loddiges of London and Patrick Fennessy of Waterford. Edward Madden's

82. *Cardiocrinum giganteum*, Edward Madden's introduction from the Himalayas, photographed at Glasnevin by J. W. Besant in 1932

consignments of seeds were usually accompanied by long and detailed letters. In these he described the problems of collecting and the localities visited in his quest for new plants. Madden was well-organized; his military training and experience clearly influenced his attitudes towards his hobby. In 1846, he set down guide-lines for collectors working in the Himalayas. Madden suggested that the expedition should contain not more than three people, and that the collectors should engage new porters each day.

> Let your cups, jugs, plates and dishes be of metal; with these only may you defy fate and falls; and as for provender to adorn them, an ample supply of tea, sugar, Carrs biscuits, hermetically sealed soup and bouilli, fowls, sliced bread rebaked into everlasting rusks, with a liberal allowance of beer, wine and brandy, the latter precious article insured against damage by being decanted into stone bottles.[24]

He did not agree with the usual practice of employing large numbers of servants, suggesting that a few able-bodied men 'in sound health, and warmly clothed, provided with a small tent for their own use, would save much trouble'.[24]

Madden's expeditions into the mountains of northern India provided Glasnevin with many valuable plants. His first consignment reached Dublin in January 1841, and was entered in the accessions book as 'a very valuable package of seeds from the Himalayan Mountains [presented] by Mrs Madden Senr. of Kilkenny, collected by her son Captn. Madden of the Royal Artillery'. The last lot, consisting of '20 papers of Nepal seeds, 20 rare orchidaceous epiphytes' was received on 25 April 1850, and was the thirty-seventh batch sent by Madden.

The most exciting plant introduced by Edward Madden was the giant Himalayan lily, which is now called *Cardiocrinum giganteum* (Fig. 82). It has large, heart-shaped leaves and a flowering spike up to three metres tall. The flowers are large white trumpets, and they are succeeded by enormous seed-pods. Madden's seeds took many months to germinate, but sufficient plants were raised for Moore

83. *Abelia triflora* from the Himalayas—this is one of the original seedlings raised from Madden's seed.
The photograph was taken in November 1981 by D. Davison (see also Plate 7)

to distribute seedlings to other gardens. In 1851 the lily flowered for the first time in the garden of the Reverend John Townsend Boscawen at Truro in Cornwall. In June 1852, Madden saw his giant lily in bloom at James Cunningham's nursery near Edinburgh, and wrote to Sir William Hooker at Kew saying that 'they forced it here to 10 feet...with 12 flowers not quite as large as in its proper temperature'. Another lily, *Lilium wallichianum*, which was introduced by Madden as bulbs, flowered in August 1850 within four months of arrival at Glasnevin.

Madden was familiar with many Himalayan rhododendrons, and one species (*Rhododendron maddenii*) was named after him by Joseph Hooker. Writing to Joseph's father, Sir William, Madden asked him to convey 'my best acknowledgement for the honour [Joseph] has done me of naming one of his Rhododendrons after so humble an explorer as myself'. Madden introduced the dwarf ericaceous shrub, *Cassiope fastigata*, into cultivation through Glasnevin, as well as *Buddleja crispa*, which has large woolly leaves and panicles of pale lilac and orange flowers. A plant of *Abelia triflora* (Plate 7), raised from Madden's seed in 1849, still flourishes in the National Botanic Gardens, beside the entrance to the Fern House (Fig. 83). It is a most charming shrub, with bright pink buds which open to pure white, star-shaped flowers that are sweetly perfumed. Madden was not the first to find this plant, but the seeds he sent to David Moore resulted in its establishment in European gardens.

Edward Madden retired from the army in 1849 and came back to Europe. He and his wife settled in Edinburgh, where he was able to work on Himalayan plants. He became president of the Botanical

Society of Edinburgh, and in 1853 delivered a presidential address on 'The occurrence of palms and bamboos, with pines and other forms considered northern at considerable elevations in the Himalaya'. He died in Edinburgh in 1856.

Edward Madden and John Tweedie were the major donors of plants and seed to Glasnevin in the first half of the nineteenth century. However, Tweedie was not the only person sending specimens from South America. Dr A. Gogarty, a licentiate of the Royal College of Surgeons in Ireland, lived for a number of years in Rio de Janeiro. Between 1839 and 1843 he sent several cases of plants to Glasnevin. He used the newly invented Wardian case, a hermetically sealed miniature glasshouse, which prevented plants from being damaged and killed by salt-spray during long sea passages.

The transportation of plants and seeds over long distances in sailing ships presented many problems. If not kept perfectly dry, seeds will germinate, and in the heat and humidity of the tropics, those seeds which survived the ships' rats often rotted after germinating. Living plants, if kept in the ship's hold, succumbed to lack of light and excessive heat, and, if kept on deck, were adversely affected by salt spray. The provision of fresh water on a long voyage for the crew and passengers could be difficult, and there was often little to spare for keeping plants alive. With such hazards, it is not surprising that very few plants reached their destinations in perfect condition—it has been estimated that on some voyages from China only one plant in a thousand reached Europe alive. These problems were partially solved by the Wardian case in the early 1830s. It worked on the principle of recycling water, just like a bottle-garden. The plants were placed in a container in well-watered soil, and were covered by a glazed dome. The water, transpired by the plants, condensed on the inside of the dome and trickled down into the soil to be reabsorbed by the roots. Dr Nathaniel Ward, the case's inventor, successfully sent plants from England to Tasmania in 1833, using two of his cases. The nurseryman, George Loddiges of Hackney, employed Wardian cases on over 500 occasions between 1835 and 1842 to import plants for his nursery. David Moore used these cases in Glasnevin before 1842 for growing ferns, mosses and liverworts, and he was particularly pleased with the progress of the Killarney fern when grown in such a container, for it produced spores and 'larger fronds than it usually does in its native habitat'.[25]

Wardian cases containing plants from overseas first reached Glasnevin in the late 1830s. Dr Gogarty's cases, despatched from Rio de Janeiro, contained orchids and ferns, as well as bulbs. Two orchids collected by Gogarty flowered at Glasnevin in 1841, and were figured in the *Botanical Magazine*—*Catasetum globiflorum* and *C. abruptum*. He was an enterprising gentleman, a keen supporter of the Royal Dublin Society and other Dublin institutions—the Royal Zoological Society of Ireland benefited from his gift of a jaguar.[26]

In the 1850s, Richard O'Reilly, a barrister from Upper Dorset Street, Dublin, a judge in Jamaica, despatched numerous parcels of seeds and Wardian cases to Moore at Glasnevin. His first consignment, received in December 1846, consisted of some seedling limes (*Citrus aurantiifolia*) and one grafted West Indian lime. In 1853, he sent a box of yams, but most of the cases contained ferns and orchids.

Men like Madden, Gogarty and O'Reilly had the advantage of being based in one place for several years at a time, and were able to spend much their spare time collecting; it was their hobby. The professional explorers, who undertook some of the most daring exploits of the Victorian era, managed somehow to collect seeds as they travelled. Glasnevin did not benefit as much from these collectors as, for example, the Royal Botanic Gardens, Kew, but some names are worth recording. David Livingstone's travels in Africa are famous. In 1856, Dr Richard Whately, Archbishop of Dublin, presented Moore with a package of seeds collected by Livingstone. Their precise source is not recorded, but Livingstone possibly collected them during his eastwards crossing of Africa, when he saw the Victoria Falls for the first time. Sir Thomas Mitchell, who made several journeys in eastern Australia, gave seventy packets of seed to Glasnevin in 1847. He was Deputy Surveyor-General of New South Wales, a soldier, a scientist and a skilled artist. His final expedition left Sydney in November 1845; for fourteen months he explored central New South Wales and reached as far as northern Queensland. Thomas Mitchell paid a short visit to England in 1847, and probably brought with him the seeds donated to Glasnevin.

Australia became one of the main sources of new species in the last half of the nineteenth century, mainly because David Moore's brother, Charles, was director of the Royal Botanic Garden in Sydney. Charles was thirteen years younger than David. In 1832 their mother died. Not wishing to be a burden to his ailing father, Charles left Scotland and came to Dublin, where he joined his brother at Trinity College Botanic Garden. There Charles was trained as a gardener. He was bright and was awarded a number of prizes. In 1835, he won the 'first premium' in the Horticultural Society of Ireland's annual examination for journeymen gardeners, although he was only fifteen years old, displaying 'evidence of a deeper acquaintance with the structure and physiology of plants than his age would lead us to expect'.[27] In May 1837, after the dismissal of William McCalla, Charles joined his brother in the Ordnance Survey. When David left to take up the curatorship at Glasnevin, Charles was promoted and became the Survey's botanist.[28] For a while he collected plants in Donegal, but he resigned two years later. He went to England and moved from position to position, eventually becoming a gardener in Kew in 1847. Amid embarrassing confusion—'a grand bungle'—for the government and the botanical establishment,[29] Charles was appointed superintendent of the Botanic Garden in Sydney in July 1847. He was recommended by Dr John Lindley and Professor John Henslow, and appointed by Earl Grey. Sir William Hooker was not consulted and commented to Charles Moore:

> I can scarcely congratulate you upon that inasmuch as the appointment has been in my hands for the last three years, and you appear to have stepped in and taken it away from me...had it been almost any other person, I should have felt it my duty to have expostulated with the Secretary of State.[29]

The Governor of New South Wales, Sir Charles FitzRoy, about the same time appointed John Carne Bidwill as government botanist and director of the Botanical Garden, but, because of the time taken for dispatches to travel between London and Sydney, Earl Grey did not learn of this decision until after Moore's appointment. Charles Moore left England on 18 September 1847, on 'the most friendly terms with the government, Dr Lindley and Sir Wm. Hooker', having been provided with letters of introduction to various people, including Bidwill. Moore was given free passage to New South Wales, and landed in Sydney on 14 January 1848. Charles Moore (Fig. 84) served as director of the Botanic Garden from 1 February 1848 until he retired, aged seventy-six, on 5 May 1896. He returned to Europe on three occasions; in 1867 to attend the Paris International Exhibition, in 1874 to attend the International Botanical Congress in Florence, and on his retirement in 1896, when he visited Dublin and the botanical gardens in Glasnevin and Ballsbridge. He died in Sydney on Sunday, 30 April 1905.

Once he was established in his new post, Charles Moore began to send Wardian cases of Australian plants to Dublin. These were then filled with other plants at Glasnevin and returned to Sydney. Seeds were also exchanged. Charles Moore travelled and collected in eastern Australia. He toured the South Pacific islands on board HMS HAVANNAH in 1850, and, in 1869, visited Lord Howe Island. Among the plant consignments which he sent to Glasnevin was one consisting of ten tree-ferns, each between four and eight feet in height. These reached Dublin in July 1863, and were the forerunners of the Australian tree-ferns which now flourish in the gardens of counties Cork and Kerry.

Although not always the closest of brothers, Charles and David Moore both sought to improve the collections in their separate botanical gardens. There are also several ironic parallels in their different situations, and in the battles which each had to fight to maintain the standards of their gardens. In one instance, they acted as joint sponsors of a botanical explorer. In a letter to Sir William Hooker, David Moore explained that a bill handed to Sir William, endorsed by 'Captain Denham of HMS HERALD, my brother and myself is genuine and correct. My brother advanced the money to Milne at Sydney on his getting Captain Denham to endorse the bill as part payment for books'.[30] The person concerned was William Grant Milne, who had been assigned as a botanist to an expedition to the South Pacific commanded by Denham. Milne trained at the Royal Botanic Garden, Edinburgh, and had been recommended for the expedition by Professor John Balfour. After returning from the voyage, he went to Glasgow Botanic Garden, but soon decided to go abroad again. The curator of Glasgow Botanic Garden arranged a free passage for him to Old Calabar on the coast of the Cameroons, and David Moore was persuaded to act as one of Milne's sponsors.

84. Charles Moore, 24 February 1887 (By courtesy of Major-General F. D. Moore)

William Milne was a professional collector. He had to satisfy his patrons and clients by obtaining sufficient specimens of interesting new species to keep them happy, and he collected shells as well as plants to supplement his income. David Moore corresponded with him during this African adventure, and Milne's letters reveal some of the hardships of this profession. On 26 November 1865, William wrote from Fernando Po to Moore:

> My dear Sir,
> I got your letter at Calabar on the first of the month and the watch which is in a fine going condition. I do not know how to thank you for all this kindness. I will try to pay it back with gratitude. But I must tell you that your letter came under a misfortune on board of the Mail Steamer. After glancing over the letter I laid it down on the table, and while looking over several other letters a young gentleman lifted your letter divided into too, and lighted his *cigar* to my great mortification. The part which is left I see that you speak about £5 sent to the Consul in Fernando Po by Mr Lauch. Such an order has never reached the English consulate. Perhaps it will be here on Monday night with the mail. And many thanks for your kindness in telling Mr Lauch to allow me a little more. It is kind, you are the only gentleman of consideration amongst my Botanic friends.[31]

Milne asked Moore to try to find other people willing to purchase land and marine shells from him, and indicated that he was also going to collect birds and quadrupeds. Milne's health had been very poor; he explained that he had gone to Calabar for medical advice.

> Calabar is certainly not a healthy place and as for Fernando Po it is the worst place on the west coast. One Scotch firm within the last two years has lost nine young Scotsmen out of fifteen at this place. All with fever. Few whites can exist here, and those that are here is never well. Fever, fever, day after day. Teatottleism is the best medicine...month after month passes away without every tasting spirits or strong drinks of any kind. Dr Horsen at Calabar ordered me to take a glass of hott Brandy every night. I took it for a few weeks and now I have giving it up: it is true I was never in better health and if I should be attacked with fevers again which there is no doubt I will take a glass of hott Brandy every night until I am better. Dr Mullings as well as Dr Howston orderd me to take stimulants. Dr Mullings told me if I did not use a active spirits I would not live.[31]

Edward Madden had also regarded brandy as a necessary item in an explorer's luggage, but it can hardly be reconciled with 'tea-totalism'!

Milne's letters are highly idiosyncratic, as can be seen, and colourful. His life was hard and his pleasures were few. He provided a vivid picture of life in tropical Africa and the perils of plant collecting. His journeys were not 'picnics'. He described one of his camps on the slopes of the Cameroon Mountains at nine thousand feet above sea level, in the summer of 1865. At that altitude Milne stated that there was nothing

> but long grass and a scant scruby vegetation and my boys had to carry water 5 miles. I shifted my camp to 6600 feet close to a fresh water spring. Let me tell you something about a collectors camp. Fancy that you see a Temporary hut covered with leaves on one side towards the mountains and that you see a sort of a sleeping place under the shelter composed of saplings to allow the rain to run clear of your person and a block of wood for a pillow and a wet blanket to keep you warm. Let us have a peep outside of the hut and here are met dirty blankets rolled round the sholders of dirty nigers. Bundles of dirty socks inlivens the sceune. Tin plates and cooking pans, old boots & insect nets, wet cloths hanging at the fire. Also plants placed to dry on a sapling grating over the fire, bottles of insects and a copy of the vegetable Kingdom &c. Such was my camp on the mountain of Fernando Po.[32]

After collecting and drying his plants, Milne found time to paint some of the flowers. He sent several sketches to England but their fate is not known. One crude watercolour of a plant belonging to the family Melastomataceae survives among David Moore's papers. On the back is scrawled one of Milne's letters, written just before he went to Calabar for medical advice (Fig. 85). 'Upon the whole my system has got a fearful shake', he wrote, 'my nerves are such that it is with a struggle that I can write my name or hold a brush. I am not able to write to your brother from the state of my health'.[33]

Milne got the medicines he needed in Calabar and returned to the Cameroon Mountains early in 1866, but again he contracted fever. On 4 April 1866, he was found by Howard Lauche, a Dublin man who was resident in Calabar. In a letter to David Moore, Lauche said that he had 'found little Milne nearly dead of dysentery, and very hard up. The doctor said that if he did not get a change of air he would not live another week, so I took him round in the mail to Calabar, where he is sure of comfort and good medical advice'.[34] William Milne did not recover and died in Calabar in July. David Moore wrote an obituary for the *Gardeners' Chronicle*, concluding with the remark that William Grant Milne was 'another victim...fallen while prosecuting scientific researches in Western Africa'.[35] One of Milne's introductions, *Lankesteria barteri* flowered at Glasnevin and was depicted in the *Botanical Magazine* (Plate 8). All the plants Milne sent to Glasnevin were destined for the glasshouses. David Moore wanted palms and Milne tried to help. In June 1865 he had sent ferns and some flowering plants including balsams (*Impatiens* species). Howard Lauche, who supervised Milne's financial arrangements with David Moore, also sent boxes of plants and seeds.[36] In May 1864 Lauche despatched two cases with 'palms for your large house also a large orchid nearly a fathom in diameter'. Moore filled Wardian cases with plants before returning them to their original owner, and Howard Lauche thanked him for plants sent to Africa.

After the death of David Moore in 1879 and the appointment of his son, Frederick, as curator in Glasnevin, a change in policy is apparent in the acquisition of plants. More and more material was received from other botanic gardens and from specialist nurserymen who stocked the plants that interested Frederick Moore. Orchids were his passion, and many new species came to Glasnevin from nurserymen who employed their own collectors. Indeed, the pattern of collecting in the wild was changing. Individuals sponsored by a few gardens or private persons, and working on derisory budgets, were no longer able to compete with the well-organised work of professional collectors sent out by large nurseries. The professional collectors like Milne often had to collect herbarium specimens, seeds, living plants, birds, insects and shells during their time abroad, but their successors were sent out to collect nothing but seeds. This did not make their task any easier, for they had to visit areas twice, first when the plants were in flower so that they could be identified, and again when they were in seed. Some of the professional plant-hunters were seeking specific prizes; some concentrated on orchids, others collected only plants likely to be hardy in European gardens.

Glasnevin benefited greatly from these expeditions, especially the ones arranged in the first few decades of the present century. Large quantities of seeds were imported from the Far East in particular.

85a. Letter from William Milne to David Moore from Mount Cameroon, 26 July 1865. (Reproduced by courtesy of Major-General F. Moore) On the back of this letter is a crude watercolour (see overleaf)

Collectors, including Ernest Wilson, George Forrest, Frank Kingdon Ward, Reginald Farrer and William Purdom, introduced many of the garden plants which are commonly grown today. Sometimes seeds came directly to Glasnevin, but if not, Glasnevin soon acquired duplicate seedlings through the network of exchange that flourished between private and public gardens.

However, the day of the enterprising amateur was not, and is not, past. In the first decade of this century, Glasnevin received many fine garden plants from New Zealand, collected by a retired lawyer, Henry Hammersley Travers. Henry Travers was born in Ireland in 1844, and arrived in New Zealand at the age of five when his parents, William and Jane Travers, emigrated from Europe. Henry spent several years travelling about New Zealand with his father, who was a keen explorer and amateur

85b. Unidentified species of Melastomataceae from Fernando Po (Watercolour by W. Milne)

naturalist, and many New Zealand plants bear names commemorating them. In 1864, Henry Travers was sent by his father to the Chatham Islands, which lie 600 miles east of South Island. In November 1864, he landed on Chatham Island itself, a low, gently undulating island covered with scrub of *Coprosma, Dracophyllum, Olearia* and tree-ferns, including the beautiful silver-backed *Cyathea dealbata*. On the beach grew the island's best-known native plant, a forget-me-not with cabbage-like leaves, *Myosotidium hortensia*. Among the new species Henry Travers collected were *Olearia traversii* and *Olearia semidentata*.[37]

In August 1906 Henry Travers (Fig. 86) sent some seeds, including *Myosotidium* and *Olearia semidentata*, which he had collected on the Chatham Islands earlier in the year, to the Botanic Gardens at Glasnevin. He asked for plants in exchange, especially *Anthurium, Caladium, Alocasia*, and the rarer saxifrages

86. Henry Hammersley Travers. (Photograph by courtesy of Mr and Mrs W. T. L. Travers, Taupo, New Zealand)

and campanulas. Frederick Moore sent *Caladium* plants in return, and Travers acknowledged receipt of these in March 1907. He had just returned from another excursion, to mountains in the Nelson region, and was able to send Moore seeds of *Pseudowintera* (= *Drimys*) *traversii*, noting that 'this plant has seldom been obtained. I was the discoverer over 23 years ago'. Henry Travers was keen to exchange plants with Glasnevin, but he told Moore that he could not afford the cost of sending living plants, although he was prepared to collect whatever seeds Moore desired. In return he asked for plants such as 'any really good tuberous Begonias both single and double'. Frederick Moore, however, wanted living plants as the seeds proved difficult to germinate, and asked Travers to do his best. Henry Travers replied in May 1907

> I intend sending you a small lot of living plants—if they die I shall much regret it for your sake. I have been so unlucky in sending plants to Europe—they are all right until the tropics. I have tried to send them in the cool chambers, but the shipping people will not take them.

Henry Travers offered Moore tree-ferns, 'the price to be 1s/6d per foot, but not more than 5ft long to be sent', and other fern plants at the same price, one shilling and six pence each. In June he posted a list of native flowering plants to Glasnevin, and set out his terms—Glasnevin had to order a minimum of three plants of each species, but only those that reached Dublin alive would be paid for, and he asked two shillings and six pence for each plant. With the list he enclosed a pressed specimen of the New Zealand eidelweiss, *Leucogenes leontopodium*.

In April 1908 Henry Travers had a consignment ready for Glasnevin; the plants were packed and loaded on a trolley to be sent to the railway station, when

> 'through some stupidity on the part of one of the men, the trolley was started down an incline of 1 in 20, without any sprag in the wheels or break [sic] on—the trolley collided with another at the bottom and the cases of plants were distributed over the river bed. One case containing my choicest plants was smashed to pieces, but marvellous to say not a plant was injured. It took me 2 & ⅓ hours to repair damage'.

In October 1908 a major consignment of New Zealand plants arrived from Wellington, including many species not previously introduced into cultivation in Europe. Plants of *Olearia semidentata* (Fig. 87) were included—these arrived two years before the same species was brought to Tresco by Captain Arthur Dorrien-Smith. Dorrien-Smith had met Henry Travers in New Zealand in March 1908, and had brought some plants to England on the ARAWA. Travers told Moore that Dorrien-Smith's plants 'are certainly a splendid lot—there is the advantage of money, so being able to travel where and when. I am intending to wait until the "Arawa" leaves again...before I send your plants as one of the officers of the ship will look after the man who will attend to them'.

Henry Travers loved the work of collecting and was keen to get some return for his labours. On one occasion he asked Moore if there was anyone in Dublin who would auction living plants of some of the New Zealand buttercups. Plant collecting was not easy; the weather was often wet and some of the plants he required were shy flowering and produced very little seed. Travers even employed a man to collect for him, but he had dragged plants out of the ground and damaged the roots.

87. *Olearia semidentata* on Chatham Island; photographed by Capt. Arthur Dorrien Smith, 1910 (from *The Gardeners' Chronicle*). It is possible that this is not the true species but hybrid.

Captain Arthur Dorrien-Smith returned to New Zealand in 1909, and again called on Henry Travers who had been collecting plants for him. These were well established in pots, and Travers was confident that they would reach England safely. Dorrien-Smith also visited the Chatham Islands specially to collect *Olearia semidentata*. He left New Zealand in December 1909 and reached England in March 1910, having lost some plants during the voyage through the tropics.[38] By this stage Travers was frustrated with the difficulties of getting plants to Europe and at the end of 1910 stopped shipping living specimens of plants that were difficult to establish after the journey. He told Moore he would concentrate on woody subjects. However, after 1911 he did not despatch any more plants or seeds to Glasnevin. The plants he had sent greatly enriched the collections, and *Olearia semidentata* was certainly his finest discovery and introduction.[39]

In the 1910s, Glasnevin benefited from the work of another amateur botanist who is unique in several respects, Lady Charlotte Wheeler Cuffe (Fig. 88). She was the wife of Sir Otway Wheeler Cuffe, who was employed by the Public Works Department in Burma and who held various official positions, including that of aide-de-camp to the Viceroy of India. Lady Cuffe was a small but formidable lady, an excellent artist, and fond of gardening. During her years in Burma and India she did some superb watercolours of rhododendrons and orchids.[40]

88. Lady Charlotte Wheeler Cuffe, a photograph taken in Rangoon about 1911
(Reproduced by courtesy of Capt. A. Tupper R.N., Kilkenny)

In 1911, Charlotte Wheeler Cuffe was invited to spend a few weeks travelling with Mrs Winifred McNabb. The two ladies journeyed to Mount Victoria, which is situated about 600 miles north of Rangoon and rises to over 10,000 feet. To get to the summit, they had to stay at Kampetlet village, whose only European inhabitants were two young men. Winifred McNabb would not go by herself, so she invited Charlotte Cuffe to accompany her. Bringing paints and sketchpads, the two women set off from Kampetlet at a leisurely pace, and stayed overnight in a hut on the lower slopes of Mount Victoria. Next day they walked through the oak and rhododendron forest which covered the cool, upper parts of the mountain. Charlotte Cuffe noted a fine yellow-flowered rhododendron growing near the summit. Another species with large white flowers grew as an epiphyte, its roots embedded in the mossy covering that clothed the tree trunks. Near the summit they strolled through grassy meadows where blue buttercups and eidelweiss bloomed. From the summit they could see the snow-capped peaks of the eastern Himalayas stretching northwards towards the frontier with Tibet and China. On the summit there was a single gnarled rhododendron with crimson flowers 'brandishing defiance to the four winds of heaven' (Fig. 89).

This beautiful mountain impressed the two women, and they decided to return again the following year. This time they brought a camera. A few rather fuzzy photographs taken with their old-fashioned equipment survive, along with a series of watercolour sketches. On the second ascent Charlotte Cuffe collected seedlings of the yellow and the white rhododendrons, and also of the blue buttercup. She sent these to Glasnevin. The shrubs proved to be new species. The yellow one was named *Rhododendron burmanicum* and is hardy enough to grow out-of-doors in Ireland in gardens near the coast. The white one only thrives indoors, and is extremely rare in cultivation; indeed, it has not been collected by any botanist since Lady Cuffe found it in 1911. Appropriately it bears her name, *Rhododendron cuffeanum* (Fig. 90). The blue buttercup impressed Charlotte's friends who nicknamed it 'Shadow's buttercup', for her pet-name was Shadow. It was also a new plant, but only a local form of a widespread Himalayan species of *Anemone obtusiloba*.

89. *Rhododendron arboreum* on the summit of Mount Victoria, Burma; a watercolour by Charlotte Wheeler Cuffe. (Reproduced by courtesy of Capt. A. Tupper R.N., Kilkenny)

90. *Rhododendron cuffeanum* — a photograph taken about 1920 of the original plant growing in Glasnevin

Charlotte Wheeler Cuffe had not met Sir Frederick Moore when she sent the plants from Mount Victoria, but on her return to Ireland in 1913 she made a point of going to Glasnevin to see her plants. A friendship was formed which lasted many years. More plants came from Burma to Glasnevin between 1913 and 1921. In her delightful letters, especially those to her cousin Baroness Pauline Prochazka, Lady Cuffe tells of her other adventures in Burma, travelling on mules along rough tracks, accompanying her husband as he inspected the work of his department. She travelled along river valleys clothed with rhododendrons, and entered areas which no botanist had visited. Indeed, she blazed trails that were later followed by George Forrest and Frank Kingdon Ward. Charlotte Wheeler Cuffe met both men and entertained them in her home at Maymyo, swapping stories about travels in search of botanical treasures. She discussed her rhododendrons from Mount Victoria with George Forrest, and had 'great talks about the plants and trees of the Irrawaddy-Salween Divide' with Frank Kingdon Ward as he rested in Maymyo in 1915.

Lady Cuffe is unique in one other respect. She founded, designed and planted a botanical garden. In 1917, she was invited by the Burmese Secretary, William Keith, and the head of the Forestry Department, Charles Rogers, to undertake the job. 'I couldn't sleep last night with excitement over it (which was very silly of me)', she told Pauline Prochazka. She was given complete freedom and as many labourers as she needed. In a letter to Sir Frederick Moore, Lady Cuffe said that

> the idea is to have a garden of all the beautiful indigenous flowers, trees and shrubs, with just a few imported things, but very few. There are a lot of beautiful wild things in the area now, including a small patch of primeval forest, a marsh, some rocks, a little lake, and a wide stretch of open valley covered with bracken fern, wild raspberries—and weeds!

She set about her new job with joy and enthusiasm. By November 1917 she was managing 170 acres. Scrub had been cleared, and the large areas had been planted. Springs of water had been found and tapped. 'Gardening has its excitements in this country', she told Sir Frederick, 'a leopard has been prowling around the garden and a wild boar rooting...wild duck have already discovered the pond, and there are woodcock and snipe as well, but I am going to prohibit shooting (except the leopard and the pig)...' She sought Moore's advice, which he gave freely, and plants suitable for this tropical paradise were sent from Glasnevin in Wardian cases.

Lady Cuffe continued her botanical adventures while supervising the new garden. In May 1918, she went to north-eastern Burma with Sir Otway, who had to inspect the boundary pillars that marked the frontier with China. It was their second visit to the region where Charlotte had originally collected seeds of a rose with large fragrant, white flowers. She planted the rose at Maymyo, where it flourished.

In 1920, Reginald Farrer visited Maymyo and saw Lady Cuffe's handiwork. He praised the garden in an article in the *Gardeners' Chronicle*:

> All the natural advantages are being made the very best of by Lady Cuffe and Mr Rogers, who, bit by bit, are laying out the garden with a special eye to aesthetic as well as cultural effects...I have no doubt that the Maymyo Botanical Garden will be a paradise.[41]

The Cuffes left Burma in April 1921. Lady Cuffe handed over charge of the botanical garden to Roland Cooper, who had been appointed by the India Office on the recommendation of the director of the Royal Botanic Gardens at Kew. She was pleased with him, and he promised to collect seeds from the white rose and send them to Glasnevin. The rose was introduced to the Botanic Gardens; it bears the name *Rosa laevigata* 'Cooperi'. Sir Otway and Lady Cuffe returned to Ireland and took up residence in the Cuffe family home, Leyrath, outside Kilkenny. There they frequently entertained Sir Frederick and Lady Moore.

91. Frederick William Moore; a photograph taken in the 1880s.
(Reproduced by courtesy of Major-General F. D. Moore)

Frederick William Moore
1879-1922

IN 1874, Dr David Moore sought the advice of several colleagues and friends about the future career of his eldest son, Frederick. It was suggested that young Moore should be prepared for the position of superintendent of a botanical garden. David Moore accepted this recommendation because his son did show a fondness and aptitude for botany and natural history. Frederick William Moore (Fig. 91) was informed confidentially on 23 June 1879, a fortnight after his father's death, that he had been appointed Curator of the Botanic Gardens[1].

After the death of his second wife in 1847, David Moore remained a widower until 7 December 1854, when he married Margaret Baker, daughter of a Dublin builder, Thomas Baker[2]. David and Margaret Moore had five children, Helen, Norah, Frederick, David and Malcolm. Helen was the oldest child, and Frederick, born on 3 September 1857 at Glasnevin, was the eldest of the boys.

Many years later, Frederick Moore recalled his childhood days in the Botanic Gardens[3]. His earliest memory was of an incident when he was almost seven years old. At the time, builders' laths were purchased for use as plant stakes, and on wet days the men would split them to make the thinner rods needed for the pot-plants. Frederick and his younger brother, David, discovered that these laths made fine toy swords. At the back of one of the glasshouses, in a specially-made border, Edward Madden's giant Himalayan lilies (*Cardiocrinum giganteum*: see Fig. 82) grew; they were given special attention, and year after year rewarded the care bestowed on them by producing tall stems bearing huge, white, trumpet-shaped flowers. The two boys thought these plants would make excellent foes. So, on one occasion, careless of the consequences, they slashed down the lilies with their wooden swords. Frederick Moore recalled that while 'we did not break our swords upon the lilies...later our swords were broken on us, and my brother and I had good cause to remember' Edward Madden's lovely lily[3].

Frederick received his early schooling in Dublin, but in the autumn of 1869, at the age of twelve, he and his brother, David, were sent to school in Hanover. While there, Frederick boldly went to the Royal Garden of Herrenhausen and asked the director, Herman Wendland, for a little prepared soil in which to grow plants. He had expected a rebuff, but was given soil and some bulbs, and he returned to school delighted. After that, young Moore paid regular visits to the Royal Garden to chat with Herr Wendland. The friendship between the boy and the gardener developed into a fellowship that lasted many years, and Glasnevin Botanic Gardens received rare palms from Herman Wendland during Frederick Moore's curatorship.

Frederick and David returned to Ireland in August 1872, fluent in German and competent in French, but their English needed improving. Their father thought that they had 'the appearance of being large strong men', and wondered whether a career in the Indian Forest Service might be appropriate for Frederick[4]. However, Dr Moore was advised against this, and so began to encourage his son to follow him as a botanical gardener[5]. Years later, Frederick remembered being solemnly interviewed by his father, and being asked if he had any taste for plants. He was instructed to go to the Gardens' library, borrow copies of Robert Sweet's *Hortus Britannicus* and John Loudon's *Encyclopaedia of Plants*, and to go out and begin to learn the names of plants. 'After five days' work in the conservatories', Moore recalled, 'I discovered that an *Aloe* and a *Gasteria* were very much like each other and further, with the audacity of youth, I went to my father's office and told him there were two plants which were different but had the same labels. On inspection he found this to be true, and told me to continue learning plants and to go to the College of Science, both of which I did'[3]. At the Royal College of Science Moore attended lectures given by, among others, Professor William McNab. In 1874 he sat the examination for the Indian Forest Service, but although he passed with credit, he was not included among the candidates selected for training.[6] In February 1875, his father sent him to the nursery of Louis van Houtte at Ghent, as a 'Volentaire'. Lodgings were found for him with a cobbler in a street adjoining the nursery. The food was wholesome and good; Moore remembered especially the treat on Fridays, 'a red cabbage with a hole made in the middle into which a Blenheim orange apple was placed and boiled with the cabbage...it was delicious but I never met it anywhere else'.[3]

Apprenticeship at van Houtte's nursery provided an excellent training, but it entailed hard work and there was no pay. Work began at 6 o'clock in the morning. Breakfast was at 8 o'clock, and at noon an hour was allowed for dinner. There was a half-hour coffee break at 4 o'clock in the afternoon, and work continued until 8 o'clock in the evening. On Sundays, during the busy season, the men worked from 5 o'clock in the morning until 1 o'clock, with only one hour's break.

At first Frederick had to prepare thousands of pots, filling each with compost. After a few weeks he was allowed to prick off seedlings. The daily routine included removing the heavy wooden panels which protected the glasshouses at night and then, wearing wooden clogs on bare feet, damping down the houses by pouring water on to the hot pipes. Each evening in the winter, at about 4 o'clock, the wooden panels had to be replaced on the outside of the glasshouses. In the summer, the young azalea, *Camellia* and *Begonia* plants, for which the Ghent nursery was famous, had to be watered by hand starting about 5 o'clock. This work was tiresome, monotonous and uninteresting. However, three mornings each week the 'Volentaires' went to Ghent Botanic Garden School of Horticulture for lectures and practical classes. Moore found these instructive and enjoyed best of all the demonstrations of budding, grafting and pruning.

There were two English apprentices at van Houtte's during Frederick Moore's time, and on Sunday afternoons after work was finished, the three young men visited other nurseries in the area. In the process they learnt a great amount about the Belgian nurseries and the competition that existed between the nurserymen. This rivalry came to a climax at the famous Ghent Quinquennial Horticultural Show, which Frederick Moore attended many times while Keeper of the Glasnevin Botanic Gardens; on these later occasions he was a juror, judging the exhibited plants.

On 1 May 1876, after fifteen months at Ghent, Frederick Moore moved to the important and long-established botanical garden attached to the University of Leiden.[7] It had a fine collection of plants, especially the larger species of *Lycopodium* (club mosses), many rare ferns and unusual palms from the Dutch East Indies. Moore found the work less exacting than in Ghent; he had more leisure time and became friendly with other young men, many of whom were from South Africa. While he did not speak Dutch, his knowledge of German and the Flemish he had picked up in Ghent allowed him to follow the botanical lectures with ease. He attended classes conducted by Professor Willem Suringar on botany, horticulture and the use of the microscope, and went on field excursions. Moore also visited the local nurseries of de Graff, van Tubergen and others. In his spare time he went to the University Rowing Club, of which he was made an honorary member, and rowed in two winning crews.[3]

Frederick received an allowance of six pounds a month from his father, and in June 1876 'a Bank Post bill for no less than seven pounds! This must fairly put *you square in every way*, for the extras you had to pay on your microscope and other matters'.[8] With the money Dr Moore sent a set of clear instructions about the procedure his son was expected to follow while at Leiden Botanic Garden. He was to go to the sand dunes once a week with Professor Suringar and to become well acquainted with the Dutch flora; to visit the Leiden Herbarium; to get a working knowledge of Dutch, and to take lessons in Latin, for which his father would pay. Frederick was told that

> such work requires close attention and above all *system* in working. You appeared to me to want method sadly in everything. Even the throwing of your clothes about in your bedroom and on the floor, shewed you had no system in your arrangements. I pray you look to this before it is too late and think how you are to get on through the world if you are spared.[8]

However, the rigorous programme laid down for Frederick by his father only lasted until the autumn. The Board of Trinity College, Dublin, was seeking a head gardener for its botanical garden to replace

Michael Dowd, who, having been appointed curator in October 1875, had resigned because he had 'the good sense to see he knew nothing'[9] about curating a botanical garden. In April 1877, Dr David Moore was asked to advise the University's Board about the duties of a new head gardener. During that summer his comments were carefully considered, and in early October, just before he departed on a visit to Glasgow, Dr Moore was asked if his son might be considered for the appointment. David Moore wrote to Frederick, who immediately submitted a letter of application.[10] On 28 October the Registrar of Trinity College wrote to Dr Moore saying that Frederick had been appointed.[11]

Frederick Moore took up the post of head gardener in November, after certain misunderstandings about the terms of his appointment had been rectified. His salary was set at £120 a year, rising by increments of ten pounds to a maximum of £160 after two years. He was given an allowance of forty pounds for his accommodation and was instructed to find a house near the Botanic Garden in Ballsbridge.[12] Moore was made responsible to the Professor of Botany, Dr E. Perceval Wright, and had to ensure a supply of specimens for botany lectures.

A former curator, John Bain, had officially retired on 28 March 1868, but he maintained a considerable interest in the College Botanic Garden.[13] On 29 October 1877, Bain was in the Garden with Moore, the Registrar Dr Hart, and his son Henry C. Hart. Moore and Bain exchanged words — at the beginning Frederick Moore 'assumed a different tone and manner from which Mr Bain's age and experience entitled him to expect', but 'nothing...could justify some of the words which Bain's hot temper led him to use'. Dr Hart was upset by a confrontation which appears to have resulted from John Bain's forcefully expressed opinions about some of the changes which had taken place during the eleven months Moore had been in charge.[14]

Frederick Moore was employed in the same garden in which his father and his uncle had received some of their early training, yet he was 'still a youth'.[15] His father hoped that a better position would eventually come Frederick's way, and when he was asked to give lessons in botany to Dr Nevill's son, David Moore advised Frederick to undertake the tuition and to charge five pounds for twenty lessons. Dr Moore told Frederick that he 'considered this proposal...probably of a providential nature because it will make you known to the very men above all others who may yet assist you to something better, so lay hold of it'.[16]

Trinity College Botanic Garden was a private teaching garden to which the general public was admitted only by special arrangement. Its eight acres were enclosed by high stone walls, abundantly clothed with shrubs and climbing plants, and the glasshouses contained a fine general collection of species, including orchids and ferns. Moore took a special interest in the herbaceous plants. He began to improve the collections, and, for the first time in many years, plants were exchanged between Glasnevin and the College Garden. A visitor to the Botanic Garden at Ballsbridge in 1878 noted that the Garden had 'greatly prospered'[17] under Frederick Moore's management, an opinion shared by Professor Wright and other officials of the University.[18]

When Frederick Moore had been at Trinity College Botanic Garden for just over two and a half years his father died, and he set about canvassing for the curatorship of Glasnevin. There was widespread local support for him and this was voiced on 18 June 1879 by the *Daily News*:

> We presume that [Dr David Moore's] eldest son, an accomplished practical botanist, carefully trained at Glasnevin and afterwards in some of the most famous foreign schools, will be promoted from the curatorship he now holds to the more valuable post held by his lamented father.[19]

On the day after his father's death, Frederick wrote to Dr William Thiselton Dyer and Sir Joseph Hooker, asking for their support. To Sir Joseph he remarked that 'the friendship which you bore to my late father and the immense influence which your support would exert on my probable success, I take the liberty of most earnestly entreating you, if you consider that you can conscientiously do so, to recommend me as my father's successor'.[20] Other people lobbied for Frederick Moore — Mrs Alice Lindsay, of Glasnevin House, promised to write to Sir Arthur Guinness, and told Mrs Margaret Moore that she had written to Lady Cloncurry, Lady Rachel Butler and Lady Margaret Stronge, and had sent on a letter, written by Helen Moore (Frederick's sister), to Mrs Barton of Straffan House.[21] Frederick Moore wrote to the Registrar of Trinity College asking for a testimonial from the Board, and Thomas Stack replied noting that Moore had managed the College Botanic Garden to the entire satisfaction of the University during the two and a half years that he had been its curator.[22]

The vacant post was not advertised. On 23 June, exactly a fortnight after his father's death, Frederick Moore heard, in confidence, from Colonel Donnelly of the Department of Science and Art in South Kensington, that he would be appointed to Glasnevin. 'I write to inform you thus unofficially as it may be some consolation to your mother for whom I sincerely feel in her sorrow, to know that she will in all probability not have to quit the old house'.[23] By 5 July Moore's appointment was public knowledge, and was announced in such horticultural journals as the *Gardeners' Chronicle*[24] and *The Garden*, which commented that 'we believe that all who know Glasnevin will feel that the best thing has been done...'.[25] Official notification of the appointment came on 26 June from the Lords of the Committee of Council on Education:

> Though young for so responsible a post, my lords, remembering how much the Gardens owed to the great ability, untiring energy, and long service of your father, were glad to find by the excellent testimonials you forwarded, and especially by those showing the manner in which you have filled a similar post in the Botanical Gardens of Trinity College, that they would be quite justified in appointing you to it.[26]

Moore had resigned from Trinity College some time earlier, for on the same day that his appointment was officially announced in the horticultural press, Frederick Burbidge's (Fig. 92) appointment as curator of the College Botanic Garden was also announced.[27]

Frederick Moore took up his post at Glasnevin on 9 September 1879. His salary was set at £200 a year, rising by ten annual increments to £300. The Department of Science and Art carefully explained to Moore that he was appointed 'Curator', and that the possibility of appointing a 'Director who would hold with regard to the curator much the same position as is the case at Edinburgh' was under consideration. On this aspect of Moore's appointment, the *Daily News* commented:

> There is indeed a rumour, but we can hardly credit it, that the intention is to job the £100 a year [saved by appointing a curator] by creating a perfectly useless office, that of a "Scientific Director" for benefit of a gentleman, with the nature of whose claim on the Committee of Council we are entirely acquainted. The Gardens are now a branch of the "National Institution of Science and Art", and this institution has already in the person of Dr William Steele a director who, we presume, possesses all the science that is requisite for the superintendence of the Gardens and its curator. There is an adage about too many cooks, which we recommend to the attention of His Grace the Duke of Richmond, President of the Council.[29]

Frederick Moore was only just twenty-two years old when he became Curator, but his appointment, supported by the most eminent botanists and horticulturists in Ireland and Britain, was widely welcomed. And, such a young man at the head of an important international botanical garden created intense interest among experienced gardeners. Perhaps not unexpectedly, shortly after he took up the post, Moore was visited by three of his father's old friends. This self-appointed horticultural tribunal comprised John Bennett-Poë, a native of Nenagh, County Tipperary, who was an enthusiast for orchids and herbaceous perennials such as *Helleborus*, and Edward Woodall, another orchid-fancier who had gardens in Scarborough and La Selva near Nice. William Edward Gumbleton (Fig. 93), the senior member of the triumvirate, was an opinionated gentleman who had developed his passion for gardening only a decade earlier. At his house, Belgrove, near Queenstown (Cobh) in County Cork, he had assembled a choice collection of rare, tender plants and a library of exquisitely illustrated volumes on botany and horticulture. Gumbleton was an archetypal Victorian gentleman, a collector, but a discriminating one. Nearly sixty years after this visitation, Frederick Moore recalled the episode:

> It was with much trepidation that I started to take them round the garden, for three more dissimilar men could scarcely have been brought together and trouble soon began. In the Aquatic House, Mr Gumbleton took me to task severely for my pronounciation of a plant, emphasizing his remarks by banging his umbrella on the flags. Mr Woodall wanted to see the orchids; Mr Gumbleton wanted to see the florists'

92. 'Three Irish people and viper', a photograph taken about 1886 by Gertrude Jekyll. The people are Miss Owen of Gorey, County Wexford, Frederick Burbidge and Frederick Moore (kneeling). Original in Album 2, no. 376, Jekyll Papers, Department of Landscape Architecture, University of California, Berkeley (by courtesy of Dr Michael Tooley, University of Durham)

flowers out-of-doors; Mr Bennett-Poë was willing to go anywhere, and kept the peace between the other two. In front of the Curvilinear Range, Mr Gumbleton denounced a plant...as 'a Tush Plant', his term for any plant he did not like, and proceeded to beat it to bits with his umbrella...I was too timid to do more than mildly remonstrate, and bemoan the loss of a recently arrived plant.'[30]

William Gumbleton often performed such 'atrocities' in other gardens. He was prone to 'cavil at the opinions expressed by others' and he possessed 'a peculiar sensitiveness to contradiction, so far as his own views were concerned',[31] and he did not take kindly to people who did not do as he asked. While a friendship developed between him and Frederick Moore, it was put under severe strain on at least one occasion, when the Department of Science and Art refused Moore expenses to travel to Cork to visit Gumbleton. In a strongly worded letter to his superiors, dated 25 October 1898, Moore pointed out the cost of this and other refusals. Mr Gumbleton had asked Moore on several occasions to visit Belgrove, and had promised to donate some books he had purchased specially for Glasnevin. But, on being told that Moore could not travel to County Cork, Gumbleton withdrew the offer of books, and told Moore that he would never issue another invitation. Moore noted that Glasnevin had also lost an opportunity to acquire some new seedling ferns from Dr Edward Lowe of Shirenewton, Monmouthshire, when the Department refused to pay travel expenses amounting to £3 10s. 0d. The Curator told the Department that Harry Veitch, the famous Chelsea nurseryman, had allowed him personally to select expensive, rare plants from the nursery, and that Veitch had then donated these to Glasnevin. Such generous gifts would cease if Moore and his foremen were prevented from travelling to visit private gardens and nurseries.[32] This particular problem was solved, and, in the case of William Gumbleton, the threats were forgotten. Moore and Gumbleton exchanged visits regularly, and it amused Gumbleton to address Frederick Moore as 'The Regius Keeper of the Irish Kew'.

Within four months of Moore's appointment, the Department of Science and Art decided that a Scientific Superintendent should be assigned to the Botanic Gardens. This decision may have been precipitated by Professor William McNab, who, it will be remembered, was rumoured to have wished

93. William Edward Gumbleton of Belgrove, Co. Cork; bibliophile, gardener and benefactor of Glasnevin. (Photograph c. 1910, reproduced by courtesy of the late W. E. G. Bagwell, Straffan)

to become Director on the retirement of David Moore. The fact that he was not appointed, and that young Frederick Moore was made Curator, suggests that the botanical establishment did not favour McNab's promotion, quite apart from the opposition to him expressed in Dublin newspapers. Dr William Steele, the General Director of the Dublin Institutions within the Science and Art Department, wrote to McNab early in January 1880, asking him to furnish a memorandum on the duties of the holder of the proposed post. Professor McNab suggested six main responsibilities: full control over the physiology laboratory; responsibility for the educational aspects of the Gardens, including the arrangement of the family beds, and of medicinal and poisonous plants; the right to make the final decision on the naming of plants; to advise on the scientific aspects of the library and museum; to carry out experiments; and to write a guide to the Gardens.[33]

Professor McNab was appointed to the position and a modified and nicely-balanced set of duties was outlined in his letter of appointment dated 25 March 1880. The Lords of the Committee of Council on Education laid down that the 'Scientific Advisor or Superintendent' would be provided with a laboratory, under his sole control, in which he would be expected to conduct botanical research, and from time to time to carry out experiments on commercial and agricultural plants. The Superintendent was expected to report each year to the Department on the scientific work of the Gardens, and was charged with the preparation of a guide-book for the Gardens.

Professor McNab was forbidden to intervene in the administration of the Gardens, and in the technical aspects of gardening. These were within the orbit of the Curator's responsibilities, which also included supplying the Superintendent with whatever facilities he required to carry out his stated duties. The Curator was expected to seek the Superintendent's advice and assistance in the naming of plants, and the scientific and educational aspects of the Botanic Gardens.[34]

Verbena phlogiflora

Leptospermum laevigatum

Plate 1

Leptospermum laevigatum (J. Gaertner) F. Mueller
(syn. *Fabricia laevigata* J. Gaertner)
Myrtaceae

An evergreen shrub reaching 2 metres in height, with glaucous leaves about 2·5 centimetres long. The flowers are sessile, and each one has five rounded, white petals and numerous stamens grouped into five fascicles.

This tea-tree is native in the coastal regions of Tasmania and south-eastern Australia (South Australia, Victoria and New South Wales).

Originally described by Joseph Gaertner in *De fructibus et seminibus plantarum* (1788), it was probably introduced into European cultivation about that time.

In 1810, the coast tea-tree was illustrated in *Curtis's Botanical Magazine* (tab. 1304) from a specimen supplied by Glasnevin Botanic Gardens. The artist was Sydenham Edwards, and the plate was probably engraved by F. Sansom junior.

reduced to 87%

(Library, National Botanic Gardens, Glasnevin)

Plate 2

Verbena phlogiflora Chamisso
(syn. *V. tweediana* N. Niven ex W. J. Hooker)
Verbenaceae

A short-lived perennial herb, reaching perhaps 0·5 metres in height. The softly hairy leaves are arranged in opposite pairs and have coarsely toothed margins. The flowers form a terminal inflorescence about 5 centimetres in diameter; each separate flower has a five-lobed corolla, and the colour of *V. tweediana* was described as rich rosy crimson.

This plant was collected in the Argentine by John Tweedie; he sent seeds to Glasnevin Botanic Gardens where the variant shown here was raised. Ninian Niven hoped to commemorate Tweedie by designating it as a new species, but the plant was merely a colour form of a widespread and variable species that was originally described by Adelbert von Chamisso.

This illustration was published during December 1836 in *Curtis's Botanical Magazine* (tab. 3541); the plate was engraved by Joseph Swan from an original watercolour painted at Glasnevin by Ninian Niven. It seems that the same original was copied for Joseph Paxton's *Magazine of Botany* (vol. 4; tab 5).

reduced to 71%

(Library, National Botanic Gardens, Glasnevin)

Plate 3

Alopecurus pratensis Linnaeus
Poaceae

An erect, perennial grass reaching almost 1 metre in height when in flower. The leaves are slightly rough to the touch and greyish-green. The cylindrical spike is composed of numerous small flowers from which, at anthesis, the anthers dangle on very fine filaments.

Meadow foxtail is a native species in Ireland, locally abundant in the north-east but less common in other parts. It inhabits meadows and damp pastures.

This illustration was the first of two plates included in John White's *Essay on the indigenous grasses of Ireland*, which was published by the Dublin Society in 1808. The plate was engraved by Maguire, but the artist's name is not indicated. The hand-coloured plate is not as expertly finished as contemporary works published in Britain.

White's *Essay...* is one of the very few illustrated botanical monographs produced in Ireland.

reduced to 34%

(Dr E. C. Nelson, private collection)

Plate 4

Ulex europaeus Linnaeus 'Strictus'
(syn. *U. downiensis* Wade; *U. strictus* Mackay)
Papilionaceae

A fastigiate shrub, reaching over 1 metre in height, with slender, erect branches. The spine-like leaves rarely exceed 2 centimetres in length and are only about 1 millimetre broad – thus they are finer than in the common variant of gorse. The flowers are golden-yellow, but again smaller than in the principal variant.

This cultivar was discovered at Mount Stewart, County Down, by John White during 1804. It was introduced into cultivation at Glasnevin Botanic Gardens and was later sent to other gardens from Glasnevin; it was the first cultivar distributed by the National Botanic Gardens.

The original watercolour by Wendy Walsh was prepared on 19 April 1982 from a plant cultivated in Glasnevin.

reduced to 58%

(National Botanic Gardens, Glasnevin)

Ulex europaeus 'Strictus'

Alopecurus pratensis

Inula salicina

Carex buxbaumii

Plate 5

Carex buxbaumii Wahlenberg
Cyperaceae

A sedge with creeping underground stems, and grey-green leaves which are devoid of hairs. The flowering stems can be up to 0·6 metres tall and each one bears several distinct spikes of male and female flowers. The terminal spike has female flowers at the top and male flowers at the base; all the lower spikes consist of female flowers only.

The blue sedge is now extinct in Ireland; its only known habitat was on the northern shore of Lough Neagh, near Toomebridge, where it was, discovered by David Moore in 1835. It was last reported there in 1886. *C. buxbaumii* has survived in cultivation, the original stock having come from Lough Neagh.

This illustration was first published during 1837 in the second edition of the Ordnance Survey's Templemore memoir. The original watercolour by George du Noyer, prepared under Moore's supervision, is in the National Botanic Gardens, Glasnevin. The plate was engraved by G. McCoy and was hand-coloured.

reduced to 64%

(Library, National Botanic Gardens, Glasnevin)

Plate 6

Inula salicina Linnaeus
Asteraceae

This herbaceous perennial has a creeping underground stem by which means it can spread and, in gardens, forms clumps about 1 metre in diameter. The erect, leafy shoots may be 0·5 metres in height. Each stem bears one or more daisy-like flowers each of which is about 3 centimetres in diameter. The disc and ray florets are yellow.

Irish flea-bane is found only on the rocky northern and eastern shores of Lough Derg on the River Shannon. It is not recorded from Britain but is widespread in central Europe.

David Moore discovered *Inula salicina* in 1843. Because of its rarity within Ireland and Britain, the flea-bane is fully protected by law.

This original watercolour was painted by Wendy Walsh on 10 July 1982, using cultivated specimens from the National Botanic Gardens, Glasnevin; the original stock was from Lough Derg.

reduced to 60%

(Dr E. C. Nelson, private collection)

Plate 7
Abelia triflora R. Brown
Caprifoliaceae

This species forms a multi-stemmed shrub or a tree which can be up to 6 metres tall. It is deciduous, with opposite pairs of pale green, ovate leaves. The flowers are borne in clusters at the tips of side shoots, and they are very fragrant. In bud, the corolla is tinted outside with pink, but the open corolla lobes are almost white.

A native of the Himalayas, this shrub was introduced into cultivation by Edward Madden. One of his original seedlings still survives in the National Botanic Gardens, Glasnevin; it is now a small tree and from it came the specimens used by Wendy Walsh as the models for this original watercolour painted on 10 June 1982.

Abelia triflora is hardy, blooms while still a small shrub and is tolerant of lime.

reduced to 52%

(National Botanic Gardens, Glasnevin)

Plate 8
Lankesteria barteri J. D. Hooker
Acanthaceae

A subshrub reaching about 1·5 metres in height. The leaves are ovate with a pointed apex; when young they are sparsely hirsute but become glabrous. The golden or orange flowers are arranged in a terminal spike; each flower has a five-lobed corolla which is pubescent.

This elegant and showy plant was raised from seeds sent to Glasnevin by William Grant Milne who collected them in the Cameroons in West Africa. J. D. Hooker received flowering specimens from Glasnevin and used that material to describe the species and to prepare the plate reproduced here.

Lankesteria barteri is a native of Cameroun and southern Nigeria where it was collected first by Charles Barter, a Kew gardener; he was attached to the Niger Expedition and obtained this species at Abeokuta in 1857.

The illustration, from *Curtis's Botanical Magazine* (September 1865, tab. 5533) was drawn and engraved by William Fitch.

This plant is not known to be in cultivation today.

reduced to 67%

(Library, National Botanic Gardens, Glasnevin)

Lankesteria barteri

Abelia triflora

Helleborus orientalis 'Dr Moore'

Plate 9

Helleborus orientalis Linnaeus 'Dr Moore'
Ranunculaceae

An herbaceous perennial with deciduous leaves resembling in form those of the horse-chestnut. The blossoms are carried in lax clusters on tall stems. Each flower has five outer segments and numerous stamens—the flowers of 'Dr Moore' were described as rose-tinted.

Helleborus orientalis 'Dr Moore' was a seedling raised in the Glasnevin Botanic Gardens before 1878; it was named after David Moore. Messrs Barr & Sugden exhibited the cultivar at South Kensington in March 1879, and it was still in cultivation at Glasnevin in February 1921. However, it is not possible to recognize this variant among the innumerable seedlings of the Lenten rose presently in gardens.

The watercolour reproduced here is by Lydia Shackleton and was painted at Glasnevin Botanic Gardens in 1887.

reduced to 71%
(National Botanic Gardens, Glasnevin)

Publisher's note
The original painting is pencil, wash and dry brush on buff paper. The colour separators have 'dropped out' most of the background. This has led to the peculiar appearance of the reproduction. The original background can be seen as the ground colour of the main leaf and within the inflorescence.

Plate 10

Lachenalia bulbifera (Cyrillo) Ascherson & Graebner
(syn. *L. pendula* Aiton)
Liliaceae

Lachenalia is a genus of about fifty species, all native in southern Africa. They are bulbous perennials. Each bulb usually produces two leaves but as many as ten can arise. The flowers are borne in a spike; each pendulous blossom has six overlapping perianth segments.

Although not popular at present, many different forms of *Lachenalia* were cultivated during the nineteenth century; they are excellent pot-plants but are not hardy and require protection in a cool glasshouse during the winter.

An unknown number of cultivars of this genus, mostly of hybrid origin, were raised at the Botanic Gardens, Glasnevin during the late 1800s. Lydia Shackleton painted twenty-five of the plants in the Glasnevin collection; the original watercolour portraits often have herbarium specimens mounted alongside.

The plant in the painting reproduced here was named *Lachenalia pendula* which is a synonym of *L. bulbifera*. The varietal name (*gigantea*) was used by Frederick Moore for a form that increased 'very slowly' and flowered after the type.

None of Miss Shackleton's *Lachenalia* portraits is dated, but the paper is watermarked 1890.

reduced to 56%

(National Botanic Gardens, Glasnevin)

Plate 11

Sarracenia rubra Walter
var. *acuminata* de Candolle
Sarraceniaceae

This insectivorous plant is an herbaceous perennial. The flask-shaped, hooded leaves arise from a stout rhizome; the tallest leaves may be 0·7 metres long. The leaves are green or variously tinted with red. The nodding flowers have maroon petals and purple sepals.

This species is very variable in colouring of the leaves and flowers, and in gross morphology. The variety depicted here is not maintained in a recent monograph on the genus.

The red pitcher-plant occurs in the south-eastern USA, from southern Mississippi eastwards into Florida and the Carolinas.

The plant painted by Lydia Shackleton in this original watercolour was just one of many insectivorous plants cultivated in Glasnevin Botanic Gardens during the late 1800s. The painting was executed in May 1885.

reduced to 46%

(National Botanic Gardens, Glasnevin)

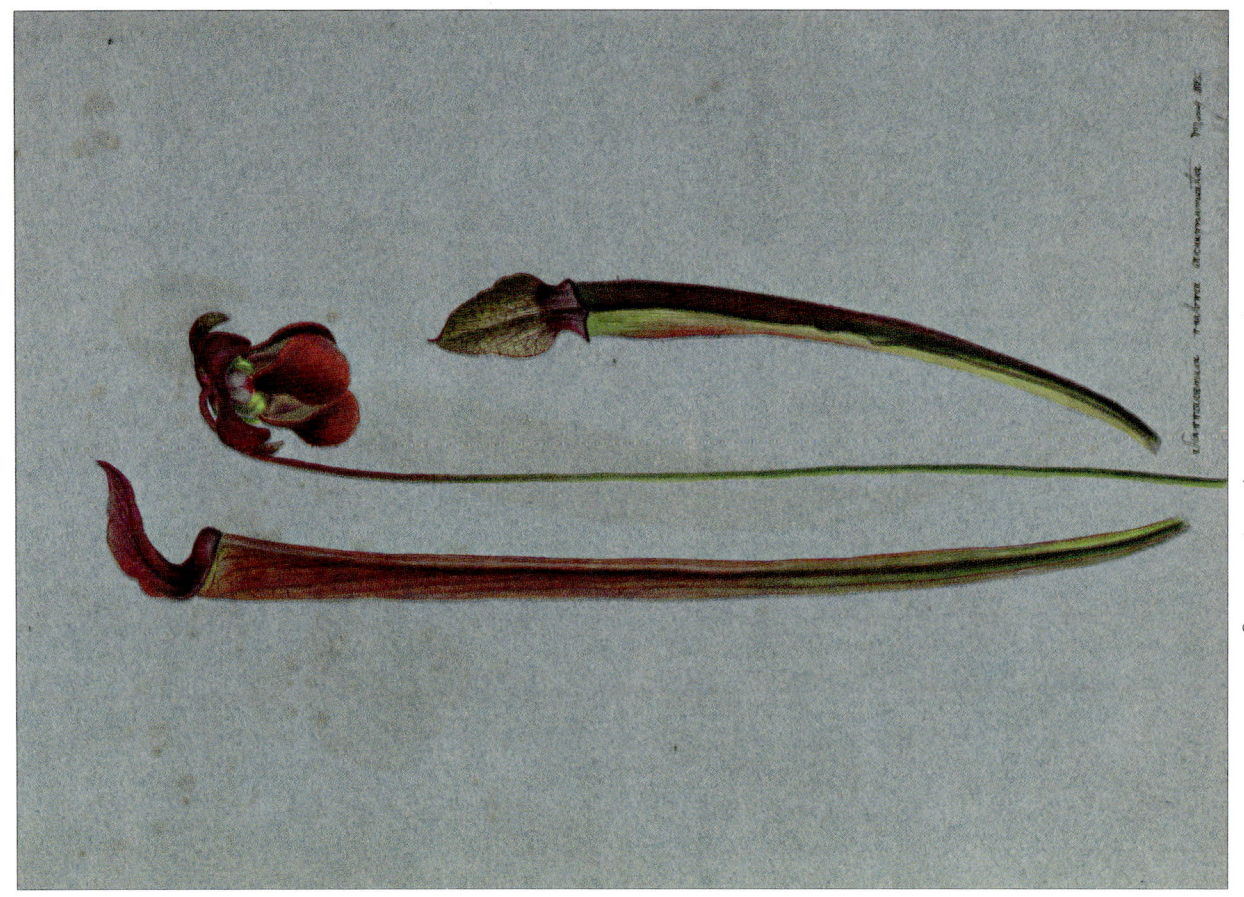

Sarracenia rubra var. *acuminata*

Lachenalia bulbifera

Anguloa virginalis

Paphiopedilum dayanum

Plate 12

Paphiopedilum dayanum (Lindley) Pfitzer
(syn. *Cypripedium burbidgei* H. G. Reichenbach)
Orchidaceae

The genus *Paphiopedilum* contains all the tropical species of terrestrial slipper-orchids; originally they were included in *Cypripedium* which is now restricted to plants from temperate areas.

Like so many orchid genera, *Paphiopedilum* is extremely variable; the species often have numerous forms differing especially in the colour and morphology of the flowers. Because of some minor differences in the shape of the petals and sepals, and the markings on them, this plant was originally placed in a discrete species named after Frederick W. Burbidge—the epithet 'burbidgeanum' is erroneous. However, modern opinion places it within *P. dayanum* which is native in northern Borneo.

This terrestrial orchid does not have a pseudobulb; the leaves arise from a very short stem and are characteristically mottled.

This is one of over one thousand orchid portraits painted in watercolours by Lydia Shackleton at the behest of Frederick Moore and based on plants in cultivation at Glasnevin. It is dated January 1896.

reduced to 85%

(National Botanic Gardens, Glasnevin)

Plate 13

Anguloa virginalis Schlechter
Orchidaceae

A terrestrial or epiphytic perennial herb with clustered pseudobulbs each one about 15 centimetres long. The leaves are usually produced in pairs at the apex of each pseudobulbil. The fleshy flowers are solitary; they are usually white, but pink-flowered variants (like this one) are occasionally found.

This orchid is native in Colombia, where it inhabits moist tropical forests. It is sometimes regarded as a form of *Anguloa uniflora* Ruiz & Pavon which was first collected in the late eighteenth century and was introduced into cultivation in Europe in the 1830s.

The original watercolour reproduced here was executed by Alice Jacob in December 1909.

The source of this plant is not known; the inscription suggests that it came from Linden's Nursery in Belgium.

reduced to 48%

(Library, National Botanic Gardens, Glasnevin)

Plate 14
Escallonia 'C. F. Ball'
Escalloniaceae

An evergreen shrub with slender, somewhat pendulous branches. The leaves are glossy, dark green above, with shallow marginal teeth. Each flower has fine dark, blood-red petals and the anthers are yellow. The shrub blooms in June and July; the blossoms are not perfumed.

This plant is a cultivar raised at the Botanic Gardens before 1915; it was later named after Charles Frederick Ball who was killed at Gallipoli. The parentage of the plant is not recorded but it is often considered to be a variant of *Escallonia rubra* Ruiz & Pavon. It is one of a suite of three, apparently unrelated, *Escallonia* cultivars selected at Glasnevin–'Alice' (named by Ball after his wife) and 'Glasnevin Hybrid' are the other two.

'C. F. Ball' is hardy, tolerant of lime-rich soil, and still relatively common in cultivation.

The original watercolour by Wendy Walsh was painted on 15 June 1982 using shoots from a long-established shrub in the National Botanic Gardens, Glasnevin; this particular plant, growing in front of the Orchid House, may be the original.

reduced to 55%

(Dr E. C. Nelson, private collection)

Plate 15
Tulipa sylvestris Linnaeus
(syn. *T. australis* Link)
Liliaceae

A dwarf, bulbous herb, with from two to four deciduous leaves. Each bulb produces a solitary flower–the buds nod–composed of six yellow or cream perianth segments of which the three outer ones frequently are red-tinted on the back.

It is not clear where this species originated; it is now widely naturalized in Europe, north Africa, western Asia and Siberia. It is also very variable, but the plants presently cultivated can be segregated into three distinct groups. One of these was named *Tulipa australis* Link; it has more slender stems and smaller flowers (not more than 3·5 centimetres long) than the type.

This plate, originally published during December 1925 in *Curtis's Botanical Magazine* (tab. 9078), was drawn and lithographed by William Trevithick. The tulip came from the Marquis of Headfort's garden at Kells, County Meath; Lord Headfort had collected the bulbs in Algeria in 1923.

reduced to 69%

(Library, National Botanic Gardens, Glasnevin)

Tulipa sylvestris

Escallonia 'C. F. Ball'

Dr McNab was not granted the right to live in the residence then occupied by the Moore family. He was told to buy a house near the Botanic Gardens, but finding a suitable house took a long time. The McNabs eventually went to live in Cabra, a suburb adjacent to the Gardens.

Whatever Frederick Moore and William McNab thought of the situation, they seem to have succeeded in presenting to the world a tolerably cordial relationship. At the end of his first year, Professor McNab was able to report that plants were supplied for all his practical classes; some had been specially grown and others forced to have them available at the appropriate time. However, there were disagreements and at times the association was seriously strained — for example, there was a prolonged row over the production of an annual list of seeds for exchange in 1886.[35]

Dr McNab's most significant contribution to the Botanic Gardens was the writing of the guide-book, as required by the Department. This was published in 1885, and was essentially a revised and greatly expanded edition of David Moore's earlier guide which had been out of print for many years. The new guide-book was illustrated with steel engravings of the glasshouses, including the Great Palm House which had been completed the previous autumn.[36] Several woodcuts which had been used in earlier booklets were also included, and there was a finely executed coloured map of the Botanic Gardens (Fig. 94) which shows that the layout of the Gardens today is almost unchanged from what it was a century ago.

While preparing the guide-book, Dr McNab sought the opinion of the Department about the proper title for the Glasnevin Gardens. The Director of the Dublin Institutions, Valentine Ball, replied that 'official authority, popular usage, and the early ideas of the promoter' justified the use of the plural *Gardens*, and pointed out that the prefix *Royal* had been adopted as the outcome of correspondence between the Department and the Lord Lieutenant. Professor McNab was not satisfied and asked Mr Ball to look into the matter again; McNab felt that the plural was a recent corruption of the title, and he wanted Ball to find the document in which *Royal* was used for the first time. McNab received the following reply to his second enquiry:

> Sir
> The authorised and official title of the Gardens is "Royal Botanic Gardens, Glasnevin"; when informed by the Department of the decision on the subject arrived at by them and the late Lord Lieutenant, I was further given to understand that My Lords did not desire the question to be reopened..[37]

Professor McNab made good use of the plant collections during the two decades he was connected with Glasnevin. In 1885 he made a study of the production of cones by *Macrozamia denisonii* (= *Lepidozamia peroffskyana*) a cycad growing in the Great Palm House, and published his observations in the *Proceedings of the Royal Dublin Society*.[38] Later, he investigated phyllotaxis in the same cycad, which is a native of Australia and was among a consignment of plants sent to Glasnevin from Sydney before 1865 by Frederick Moore's uncle, Charles Moore.

William Ramsay McNab died on 2 December 1889 leaving a widow in such straitened circumstances that she had to take a lodger into her home. To raise money she was forced to sell her husband's extensive private herbarium, his library and his scientific instruments. Frederick Moore, Frederick Burbidge and Alexander More[39] were asked to examine the herbarium and assess its value. They found that it was in good condition and some of the specimens (mainly those collected in the British Isles) were of interest, but they valued it at less than one hundred pounds. The herbarium was offered to the British Museum (Natural History) in London, and to the Royal Botanic Gardens, Kew; while Dr Thiselton Dyer was keen to purchase for Kew some specimens of South Africa *Erica* species which McNab had inherited, the two London institutions were not impressed by the collection. Mrs McNab then offered her late husband's herbarium, library and scientific instruments to the University of Toronto, but the University declined to purchase them. Eventually, the Department of Science and Art agreed to pay sixty pounds for the herbarium, and it was deposited in the Natural History section of the National Museum in Kildare Street, Dublin. The books and scientific instruments were sold separately, and one of Dr McNab's microscopes was acquired for the Museum.[40] The McNab herbarium and the microscope are now in the National Botanic Gardens, having been transferred there from the National Museum in 1970.[41]

INDEX.

No. 1. Octagon House.
2. Victoria regia House.
3. Australian or New Holland House.
4. Great Central House.
5. Great Tropical Stove.
6. Orchid and Fern House.
7. Great Palm House.
8. Camellia House.

A. Herbarium, Library, Office, and Botanical Laboratory.
B. Curator's Office, attached to the Curator's House.
C. Botanical Museum.
D. General Arrangement of Hardy Herbaceous Plants.
E. Arrangement of British Plants.
F. Hardy Medicinal Plants of British Pharmacopœia.
G. Rock Garden.
H. Addison's Walk.
J. Lake.
K. River Tolka.
L. Salicetum or Willow Garden.
M. Experimental or Horticultural and Agricultural Divisions.
N. Arboretum and Fruticetum.
O. Reserve Specimen Ground.
P. Rose Garden.
Q. Propagating Department.
R. Chain Tent.

94. Botanic Gardens in 1885; map from William Ramsay McNab's *Guide to the Royal Botanic Gardens, Glasnevin*

Some bundles of loose herbarium specimens from Professor McNab's collection were sent to Dr Thiselton Dyer in Kew, who paid ten pounds for the material which included, it seems, the Cape heathers collected by James Niven. Those specimens not wanted by Kew went to Dr Isaac Bayley Balfour for the Royal Botanic Garden, Edinburgh.

In April 1890, within a few months of Professor McNab's death, the position of Scientific Superintendent was abolished. Frederick Moore told William Thiselton Dyer that he was greatly relieved by this action.[42] In a confidential memorandum from Dublin to London it was pointed out that the situation in Glasnevin had changed substantially since Moore and the late Professor McNab had been appointed; then Moore had been young and unknown, but

> time has shown that he is quite capable of carrying on the management of the Gardens and collections without the assistance of any other officer. In the kind of scientific knowledge required for his work he had special training in Leiden. The nomenclature of cultivated plants is not a subject with which scientific botanists are generally conversant.[43]

The memorandum added that to put a new, young professor in the position of Scientific Superintendent would cause friction. The document discloses some aspects of William McNab's character, showing him to have been very ambitious and demanding. He had sought to have the exclusive use of one of the gardeners, but this was not sanctioned, and, had it been approved, the authority of the Curator would have been seriously challenged. Professor McNab had tried to stop all teaching in the Gardens except that undertaken by himself. And, he had demanded that the position of Scientific Superintendent should be independent even of the authority of the Director of Science and Art Institutions. These demands were seen by officials in the Department of Science and Art as attempts by McNab 'to obtain privileges for his post which were not compatible with the good order of the Gardens'. In one respect, that of naming plants, it was acknowledged that McNab had stopped interfering and Frederick Moore was undertaking this work himself.[43]

In 1891, when Thomas Johnson was appointed as McNab's successor to the chair of botany in the Royal College of Science, he was not given the post of Scientific Superintendent in the Royal Botanic Gardens. He was made Keeper of the herbarium in the National Museum and, later, set up a seed testing service for the Department of Agriculture and Technical Instruction.

Professor Johnson had access to the laboratory at Glasnevin and was given the use of an office, but he had no function with respect to the management of the Royal Botanic Gardens, and was not consulted about plant naming or arrangement. He was asked to lecture to the Gardens' apprentices and shared with Moore the job of teaching basic botany to those young men. Johnson availed of the facilities at Glasnevin for a number of years, but by 1907 he was only using the laboratory when teaching, and as he rarely occupied the office allotted to him, this was taken over for the new Assistant Keeper, C. F. Ball. Professor Thomas Johnson had inherited the set of keys to the conservatories which William McNab had been so insistent on obtaining, and in the fulness of time, handed them on to Dr Joseph Doyle, his successor as Professor of Botany in University College, Dublin. Dr Doyle returned the keys to Glasnevin, thereby effectively relinquishing the use of the glasshouses.

When the Department of Science and Art abolished the post of Scientific Superintendent in 1890 it was decided to upgrade Frederick Moore's title from Curator to Keeper, and to raise his salary to £400 a year. When informed of this decision in December 1890, Moore objected to the change of title, saying that Curator was adequate. The Department sent a memorandum to Valentine Ball, Director of the Dublin Institutions, pointing out that in the Imperial Calendar the Curator of the Royal Botanic Gardens, Kew, was listed as plain 'Mr', while the Keeper of the Herbarium and Library was dignified by the title 'Esq'. Moore, finding that he was overruled, acquiesced.[44]

With the entire responsibility for horticultural matters falling on him, it is not surprising that Frederick Moore gained a reputation as a horticulturist and gardener, but not as a botanist. Indeed, he was 'temperamentally...the antithesis of the patient, painstaking scientist'.[45] In his formative years he had received basic training in nurseries and gardens, as had his father, but in contrast to his father, Frederick Moore showed little inclination to tackle pure botany, and only rarely carried out field work in Ireland. He did travel to Kerry with David McArdle in July and August 1881, to collect

95. Orchids in bloom in Glasnevin about 1910 (National Botanic Gardens, Dublin)

native plants, including the Killarney fern, and specimens of native mosses and liverworts for the Gardens' herbarium, and also took the opportunity to visit William Gumbleton's garden at Belgrove, and Lakelands, the garden owned by William Crawford. Moore wished particularly to obtain some of the new hybrids of the tropical genus *Brownea* that had been raised at Lakelands, and in this he was not disappointed.[46]

Thus, from the time of Frederick Moore's appointment a change of emphasis is apparent in the work of the Glasnevin Botanic Gardens. For the next ninety years, no substantial research on native plants was carried out by the staff of the Botanic Gardens, with the exception of that undertaken by David McArdle, who collected mosses and liverworts while employed as a clerk. In 1890, a botanical section with herbarium was formed in the National Museum in Kildare Street, Dublin, and scientific work on native plants was concentrated there, leaving the Royal Botanic Gardens as a purely horticultural institute.

There were other remarkable changes at Glasnevin after 1879, especially in the acquisition of plants. Much more material was acquired by purchases from major nurseries, and the bill for seeds and plants jumped from about one hundred pounds a year to three times that sum. Nevertheless, as in his father's time, large numbers of plants and seeds were obtained by exchange, or as gifts from leading gardeners and botanical gardens, as well as from casual donors. Moore actively sought to exchange plants. His personal contacts and his friendship with other accomplished gardeners paid off handsomely. In July 1880, for example, he travelled to Belgium and Holland to visit his former workplaces and returned home with many valuable new plants.

The change in the management was also signalled in another way. In March 1880, a printed catalogue listing seeds available for exchange, was published for the first time; the seeds had been collected in the previous autumn and about 1,200 different species and varieties were offered. One remarkable characteristic of this first list was the large number of different grasses—about 180 species. A seed list, or *Index Seminum*, has been published every year since 1880, and the complete series provides excellent evidence of the richness of the plant collections. In 1880 one hundred copies of the list were printed, but by 1910 the number had increased to 250. Today, 650 copies are mailed to botanical gardens throughout the world. Another indication of policy changes came in November 1880, when official sanction was given for Frederick Moore to attend a plant auction in London to bid for orchids. In succeeding years the orchid collection at Glasnevin was frequently augmented by purchases at auctions, as well as by direct purchases from nurseries and specialist orchid dealers, and by exchanges and gifts.

Frederick Moore had his favourite plants: 'to possess a fondness for certain classes of plants, and to have means, facilities, and sufficient independence to gratify this love constitutes one of the pleasing episodes in life that gardeners reflect on with pleasure'.[47] He had, of course, all the facilities of a major botanic garden and indulged himself fully. In some cases Moore preferred to collect and grow pure species. His policy in collecting orchids followed this pattern, and the orchid collection (Fig. 95) at Glasnevin became world-renowned for the high proportion of species rather than artificial hybrids. However, he collected the garden varieties (cultivars) of other genera, and assembled at Glasnevin important collections of *Paeonia*, *Helleborus* (Christmas and Lenten roses) including one cultivar named for his father (Plate 9), *Trollius* and other hardy herbaceous perennials as well as more tender plants like *Nerine*, *Lachenalia* (Plate 10) and *Sarracenia* (Plate 11). In a strictly scientific garden, such garden varieties would have been considered inappropriate, but Frederick Moore did not eschew cultivars, and worked on the principle that the role of a botanical garden was to 'collect good plants and distribute them',[48] both species and cultivars.

Frederick Moore's overriding passion was tropical orchids. From his first months in Glasnevin until he retired, the orchid collection was his greatest concern. Unlike his father, who pioneered the cultivation of orchids from seed, Frederick Moore does not appear to have attempted to raise his own orchid plants. No hybrid orchids were ever produced in the Botanic Gardens, and those labelled "Glasnevin Variety" are simply forms picked out by Moore from batches of new importations. Frederick Moore had 'a wonderful memory and a quick eye for a new plant'[49] so that a number of the orchids he obtained from the tables of the auction rooms were later designated as new species.

Although orchids were Moore's main interest he wrote little about them, but he did give lectures which were sometimes published. In an address on orchids to the Irish Gardeners' Association in 1903, he began by correcting the impression which many people had that orchids were 'air plants': 'air and water are the pabulum on which the orchid could no more support its framework than any other plant'.[50] Like other plants, orchids required nitrogen and potash as well as calcium, and Moore noted that Frederick Burbidge was successful in growing orchids on bones at Trinity College Botanic Garden because the bones provided 'a good supply of lime and other mineral matters'. Moore showed slides to illustrate the development of orchids from seed and discussed their fertilization, stressing the interrelationship between some orchids and insects. In 1907, he lectured to the Royal Horticultural Society in London on lesser known orchids. His 'special pets', he declared, were species of *Masdevallia*. Moore commented that many orchid growers neglected those species with small flowers, and as a result he was often allowed to pick over new importations, removing such unprofitable 'pygmies' for Glasnevin.[50] In this way Frederick Moore was able to build up a comprehensive collection of *Masdevallia* species, and plants from Glasnevin were used to illustrate a monograph on this genus published in 1896 by Miss Florence Woolward.[51]

It is not easy to calculate the number of orchids in the collections at Glasnevin during Frederick Moore's years. Records survive, but as there were always many unnamed and duplicate plants in Glasnevin, these records only provide approximate figures. A list prepared in November 1879 includes 500 named species, but a total of 1,650 individual plants. The collections increased rapidly. In March

1880, for example, one hundred new orchids were sent by Mr Howard from London. In 1883, nearly 360 species of orchids flowered in the Botanic Gardens, including over sixty blooming for the first time. In 1903, the *Orchid Review* noted that 'it is doubtful whether there is another collection in existence' that can equal that in the Royal Botanic Gardens, Glasnevin, 'more especially in what are called botanical rarities'.[52] When Frederick Moore retired in 1922, there were at least 1,330 named species in the Gardens and about half that number of orchid hybrids. In 1960, there were about 1,100 species and 375 hybrids.[53]

The high standards demanded by Moore placed considerable responsibilities on the gardeners who tended the orchid collection. William Pope cared for the plants until his retirement on 18 November 1899, when he was succeeded as indoor foreman by his son, Paddy, who had been propagator since 6 October 1887. Paddy Pope's successor as propagator was Michael Conway, one of Moore's protégés. Conway had come to Glasnevin as an apprentice in March 1893, and on Frederick Moore's recommendation went to Trentham House, the Duke of Sutherland's garden in Staffordshire, in November 1896. He got on well there and was eventually promoted to foreman. At Moore's request, Conway moved to Veitch's nurseries in Chelsea in April 1899, to receive training for the propagator's post in Glasnevin. When Paddy Pope retired as indoor foreman in December 1934, Michael Conway was promoted to that position. Conway was often entrusted with the task of conducting foreign visitors round Glasnevin, as he could read the Latin names on the plant labels — even at the beginning of this century many of the Gardens' staff were illiterate. Michael Conway retired at the end of January 1942, but his health was poor, and within three months he died.[54]

Given his interest in species rather than hybrids, Frederick Moore often acquired unnamed orchids. As he was not a botanist, he relied on those botanists in Kew and the British Museum (Natural History) whose special concern was orchids to identify the unnamed plants; Moore's main contacts were Robert Rolfe at Kew, and Henry Ridley at the Natural History Museum. Quite a few of the orchids sent from Glasnevin to London represented undescribed species, and, not surprisingly, some of these new orchids were named after Frederick Moore. A beautiful white-flowered one was named *Coelogyne mooreana* (Fig. 96) in 1903 — it still grows in the National Botanic Gardens. The original plant, purchased from Messrs Sanders of St Albans, had been collected by Wilhelm Micholitz in Annam (now part of Vietnam). Other species named for Moore include *Cymbidium mooreanum*, *Epidendrum mooreanum* and *Physosiphon moorei*.

Among orchids sent from Glasnevin for naming in 1890 was one which Robert Rolfe could not identify. He wrote to Moore saying, 'I arrive at the conclusion that it is a new genus — cannot fit it in any existing one I can find'. A few weeks later, after seeing a watercolour of the plant painted in Glasnevin by Lydia Shackleton, Rolfe told Moore that he proposed to call the new species *Moorea irrorata*, 'unless you object'.[55] Unfortunately, this was contrary to the rules governing plant names. The pampas grass introduced to Glasnevin from South America by John Tweedie had already been given the name *Moorea*, so Rolfe had to invent another generic term, and he decided upon *Neomoorea*. Ralph Arnold has described *Neomoorea wallisii* (formerly *Neomoorea irrorata*) as 'a fine orchid to keep green the memory of a great and honoured figure in the realms of botany and horticulture'.[56] This orchid (Fig. 97), from the Andes in Colombia and Peru, is very rare in cultivation — it is not in the Glasnevin collection at the present time, but was illustrated in the *Botanical Magazine* in 1891, one of the many orchids from Glasnevin which Moore supplied to this journal.

Frederick Moore was fortunate to find in Lydia Shackleton (Fig. 100) an artist capable of painting the Glasnevin orchid collection. She was a Quaker from Ballitore in County Kildare, and was over fifty when she started painting for Frederick Moore. The earliest of her paintings in the Glasnevin collection is dated September 1884. It is not known how she was commissioned, for no records of official payments to her survive.[57] She is known to have requested ten shillings for one orchid painting, and as she did 1,500 or more paintings at Glasnevin, if that was the rate paid, the cost would have averaged about thirty-five pounds a year.

Lydia Shackleton was trained in the School of Art, a sister institution of the Botanic Gardens, established by the Royal Dublin Society. She attended classes there in 1850, and afterwards acted

96. *Coelogyne mooreana* photographed in Glasnevin about 1920

97. *Neomoorea wallisii* photographed in Glasnevin about 1920; perhaps the original plant

98. A group of hardy orchids in the rock garden at Glasnevin about 1910, including *Dactylorhiza elata* 'Glasnevin' (left back), *Dactylorhiza maculata* var. *immaculata* (centre) and *Dactylorhiza maculata* var. *rosea* (right) — see also Fig. 99

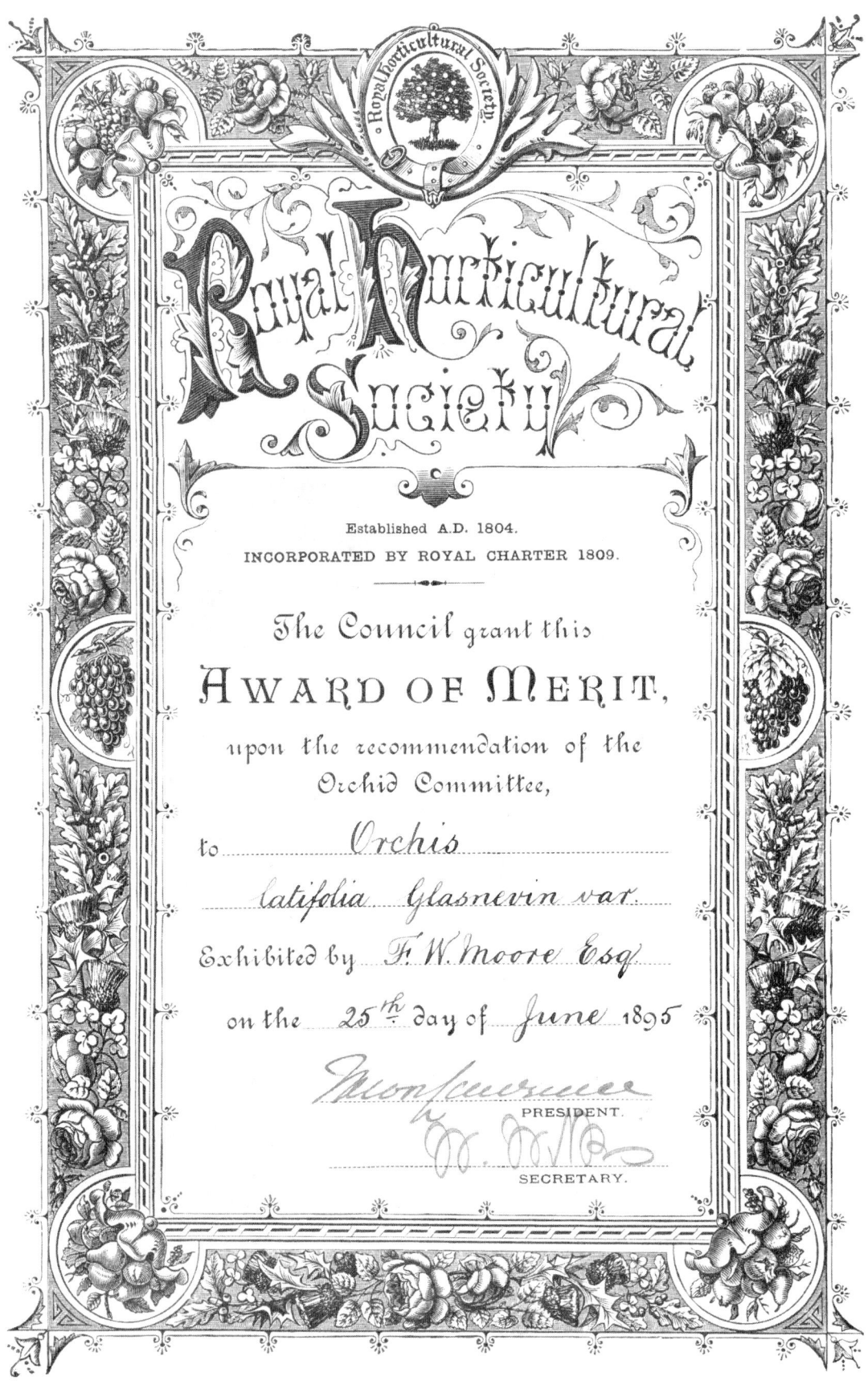

99. One of many certificates for the Award of Merit given to Frederick Moore for orchids—this certificate is for the Glasnevin cultivar of *Dactylorhiza elata* (= *Orchis latifolia*)—the plant gained a second A.M. in 1939

100. Lydia Shackleton: a photograph taken about 1885 at Lucan (by courtesy of the Shackleton family)

as housekeeper for her brother in the family flour-mill at Straffan in County Kildare.[58] As her brother, Joseph Fisher Shackleton, was a friend of David Moore—he accompanied Moore and Isaac Carroll on a visit to Scandinavia in 1863—she probably knew Frederick Moore when he was a boy.

During the twenty-three years she painted flowers at the Botanic Gardens, Lydia Shackleton executed fine watercolour portraits of over 1,000 orchids and several hundred other paintings of species and cultivars of *Helleborus* (Plate 9), *Lachenalia* (Plate 10), *Paeonia* and *Sarracenia* (Plate 11).[59] Few of the orchid portraits show the habit, leaves and stems, but they all depict individual life-sized blossoms (see Plate 12). *Paeonia* was one of Moore's favourite genera, and Lydia Shackleton painted thirty-three of the cultivars grown in Glasnevin. These portraits are probably her finest watercolours. One peony raised at Glasnevin was named *Paeonia* 'Emodoff', because Moore believed it was a hybrid between *Paeonia officinalis* 'Alba' and *Paeonia emodi*, a species for which he gained an Award of Merit from the Royal Horticultural Society of London.[60]

Miss Shackleton's paintings provide a unique record of Frederick Moore's achievements and the skill of Glasnevin's gardeners. The orchid paintings especially testify to the outstanding importance of the Royal Botanic Gardens, Glasnevin, as a centre for horticulture and botany in the late 1800s and early 1900s. Moore was anxious to record his plants, and at a time when colour photography was unknown, engaged artists to paint the flowers. During the 1890s another woman, Josephine Humphries,[61] was employed by Moore—she often portrayed the same individual flower as did Lydia Shackleton, so they must have worked side by side. Miss Humphries' work is not as skilful as that of Lydia Shackleton, and there are only ninety-three paintings by her in the collection. Miss Shackleton herself painted until 1907, when her rapidly failing eyesight made it impossible for her to continue the work. In November 1908, her place was taken by Alice Jacob, who was also of Quaker stock and taught design and ornament in the Metropolitan School of Art. Miss Jacob was well-known for her lace-work and graphic designs, and she had painted a set of Belleek china for Professor Wright of Trinity College. The illustrations for the Reverend F. C. Hayes' *A handy book of horticulture*, published in 1900, were sketched by Alice Jacob.[62] Miss Jacob's engagement was officially sanctioned and

Moore was permitted to spend not more than three pounds each year commissioning her work, but she had to do the paintings in her own time.[63] Alice Jacob painted over one hundred orchids. Her work is quite unlike that of Lydia Shackleton—she often included enlarged dissections and portrayed the flowers from the side as well as in front view (Plate 13).

The South African genera *Nerine* and *Lachenalia* attracted Moore's special attention, but, unlike the orchids, these genera were the subjects of intensive breeding programmes at Glasnevin.

The raising of *Nerine* cultivars commenced at Glasnevin in the early 1900s. It seems likely that Moore collaborated with William Gumbleton and John Bennett-Poë, both of whom donated plants to the Botanic Gardens. About ninety named cultivars were produced, and numerous unnamed seedlings also existed. The new *Nerine* cultivars were named after friends of Frederick Moore, or after some of the lady gardeners who were trained at Glasnevin.[64] 'May Crosbie', 'Oonagh McCarthy' and 'Elsie Miller' are three examples—these girls enrolled as pupils in Glasnevin in 1905, 1911 and 1912 respectively. Severe frosts in the early 1930s wiped out the entire *Nerine* collection, and sadly no watercolours were made of these exquisite flowering bulbs, and no colour photographs were taken.

By the 1920s the *Nerine* collection at Glasnevin also contained plants raised by William Gumbleton, John Bennett-Poë, Henry Elwes and Miss Wilmott. Ellen Ann Wilmott was a brilliant, extravagant and eccentric lady who owned three gardens, the most famous being Warley Place, in Essex. She was a friend of the irascible William Gumbleton and visited Ireland at least once, in 1905, when she spent several days examining the plants in Glasnevin. Writing to Moore after her visit she told him of a widespread rumour that he was to succeed Sir William Thiselton Dyer as Director of the Royal Botanic Gardens, Kew.[65] In later years she exchanged material with Moore and donated sets of *Primula* hybrids which had been raised at Warley Place. In return Moore sent her some rare plants, including *Anemone obtusiloba* f. *patula*.

One of Moore's early favourites was *Lachenalia*. He read a paper on the genus to the Royal Horticultural Society of London in April 1891.[66] By that time a number of seedlings had been raised at Glasnevin, some of which were of 'considerable merit, but...[lacked] the brilliant colours' of other hybrids (Fig. 101). Moore continued his efforts to produce interesting varieties, and in 1895 he was gratified to receive an Award of Merit from the Royal Horticultural Society for *Lachenalia quadricolor*. In March 1901 he showed two hybrids of his own raising, 'Phylis Paul' and 'Kathleen Paul' and both received Awards of Merit—Phylis Paul was his fiancée[67] and Kathleen her sister. In following years Awards of Merit were given to Moore for *Lachenalia* 'W. E. Gumbleton' (1902), 'Ruth Lane' (1903), 'Brilliant' and 'Jean Rogers' (1905). In 1906, *Lachenalia* 'Brightness' was the last to receive this coveted prize. Over forty cultivars were raised by Moore and many of these were exhibited at Royal Horticultural Shows without receiving awards. Lydia Shackleton painted a series of *Lachenalia* species and hybrids (Plate 10), and nearly every painting is accompanied by a dried herbarium specimen of the plant.

Frederick Moore maintained the important collection of carnivorous plants started by his father. *Nepenthes distillatoria* scrambled like a vine in the rafters of the Orchid House, and the famous *Nepenthes rajah*, discovered in Borneo and introduced by Frederick Burbidge, flourished in the Botanic Gardens. The Glasnevin plant was one of the few of the species in cultivation, and a pitcher from it was sent to London to be modelled for the Museum at Kew (Fig. 102), as there was then no plant of *N. rajah* in London gardens.[68] David Moore had raised the first artificial hybrids of *Sarracenia* (Plate 11), and Frederick Moore produced more hybrids; he even cross-pollinated the hybrids among themselves. His father had only effected three crosses, *S. leucophylla* × *S. flava* (*S.* × *moorei*, twice, in 1868 and 1878), *S. rubra* × *S. flava* (*S.* × *popei* in 1872), and *S. flava* × *S. purpurea* (*S.* × *catesbaei* in 1879). In 1880 *S.* × *popei* was crossed with an unnamed species and also with *S. purpurea*. *S.* × *moorei* was pollinated with *S. leucophylla* var. *alba* and several other combinations were attempted. Twelve crosses were tried and seven different plants successfully raised. In 1880, fourteen crosses were made, including *S.* × *popei* against *S.* × *chelsonii*, the first interhybrid cross. This breeding programme continued for many years. Some of the seedlings with complicated parentage were selected and given names. For example, *Sarracenia* 'Rosamund Pollock' was called after one of the lady gardeners who later became

101. *Lachenalia* hybrid (*L. rubra* × *L. aureliana*) raised at Glasnevin in July 1903

102. The model in wax (now in the Museum, Royal Botanic Gardens, Kew) by Miss Eccles of one leaf of *Nepenthes rajah* grown in Glasnevin (By courtesy of Royal Botanic Gardens, Kew)

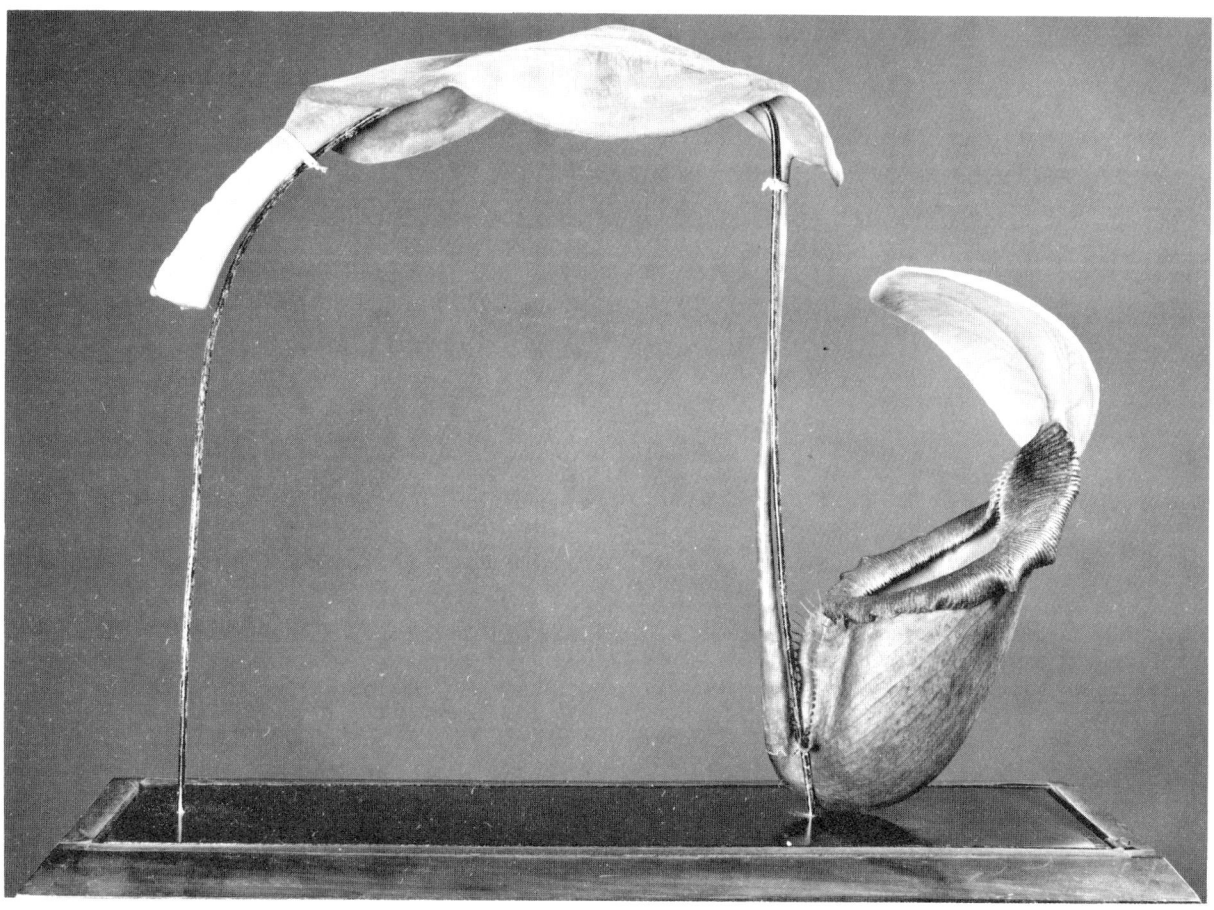

technical assistant in Glasnevin; it was raised by cross-pollinating *S.* × *moorei* and *S.* × *patersonii* (= *S.* × *catesbaei*).[68] Just as his father had won recognition for the hybrids he had raised, Frederick Moore received a first class certificate for a collection of pitcher plants shown at the Biological Exhibition in September 1898, during the visit to Bristol of the British Association for the Advancement of Science. Nine superb paintings of pitcher plants by Lydia Shackleton are at Glasnevin (Plate 11), and it is probable that the illustration published in *The Garden* on 16 October 1886 was prepared from her original watercolours.[69]

Glasnevin's prominent position in the horticultural world was enhanced by a steady stream of notices in various horticultural journals giving details of the new plants raised there. Moore sent many of the more interesting species to Kew for illustration in the *Botanical Magazine*. As William Gumbleton also supplied material for the *Botanical Magazine*, the pages of that journal were liberally sprinkled with illustrations of Irish garden plants by the end of the century. And, in 1918, Moore confessed that it 'bucks me up when we are able to get a plant each year in the *Botanical Magazine*'.[70]

Frederick Moore was interested in the establishment of difficult and tender plants in sheltered Irish gardens—an interest he shared with his father. Frederick Burbidge remarked in articles in the *Gardeners' Chronicle* during 1884[71] on the half-hardy plants and bulbs which throve so well at Glasnevin and 'seem so well nowhere else'. Californian species such as *Romneya coulteri* and *Dicentra chrysantha* flourished out-of-doors, alongside *Hedychium* from the Himalayas and such South African species as *Kniphofia caulescens* and *Crinum moorei*. Moore attributed the successful culture of bulbous plants, including *Hippeastrum* and *Crinum*, which were usually grown under glass, to the practice of deep-planting in well-drained soil against a wall and placing a handful of coconut fibre on the crowns of the bulbs in winter.

There were many other developments and improvements initiated by Frederick Moore within Glasnevin. The arboretum was extended and a new rock-garden established. Weirs were constructed on the River Tolka. Some of the existing buildings were repaired and new ones erected. The paths in the original part of the Gardens, many of which had no foundations, were remade with stones and gravel, because the number of visitors had risen from 224,000 in the 1860s to 394,000 a year by the end of the century. Moore also had the library reorganized.

The area of the Gardens was increased in 1878 by nine statute acres. The additional ground lay on the west side, and had been offered to the Society as early as 1860, when the Committee of Botany of the Royal Dublin Society had recommended its acquisition. But it was not until 1879 that the government sanctioned its addition to the Gardens. The following year Frederick Moore was granted access to it. The area was levelled, graded and laid out in walks. At the same time clumps of laurels and other common shrubs were removed from the older part of the arboretum and the elms felled and replaced by conifers to form a pinetum on what is now called Pine Hill (Fig. 103, 104). Six extra labourers were employed for the work. In all, the cost of the extension and the changes in the arboretum came to £275, including twenty-five pounds for plants. Young trees, already raised in the nursery by David Moore in anticipation of the extension, and duplicates from Kew sent by Sir Joseph Hooker, were planted out in the new arboretum.

The Far Grounds, as the new area was called, was divided by paths into four sections. The section next to the Pinetum was planted with *Fraxinus* (ash), and the adjacent quadrant with *Populus* (poplar), *Ilex* (holly) and *Ulmus* (elm). *Aesculus* (horse chestnut), *Alnus* (alder), *Betula* (birch), *Fagus* (beech) and *Carpinus* (hornbeam) occupied the south-western section, while *Acer* (maple) filled the east quadrant. An oak grove was made on the site of the old shrubbery (Fruticetum) on the south side of the Pinetum.

The final land acquisition was made in 1899, when the South Field was bought for use as a nursery plot for trees and to accommodate the collection of economic plants and vegetables. This increased the area of the Gardens to forty-seven statute acres. Thus, in a century, the Botanic Gardens had almost doubled in size. But the new nursery was remote from the rest of the Gardens and suffered at the hands of trespassers. In 1905, Lord Ardilaun donated thirty holm oaks to be used as a screen to hide some unsightly terrace houses that had been built beside the new field. The acquisition of the South Field allowed Moore to open a separate goods entrance. Formerly everything had to be

103. *Cedrus atlantica* f. *pendula* planted before 1879 in the Botanic Gardens—note the sapling of *Cedrus atlantica* 'Glauca' behind and compare with Fig. 104. This photograph was taken about 1910. (National Botanic Gardens, Dublin)

104. *Cedrus atlantica* f. *pendula* in the National Botanic Gardens in November 1981 (photographed by D. Davison), overshadowed by *C. atlantica* 'Glauca' (see Fig. 103 which shows the same trees about seventy years ago)

105. Photograph of the rock garden at Glasnevin in the early twentieth century

106. The rock garden at Glasnevin photographed in 1920 by William Edward Trevithick, before the acquisition and planting of the dwarf conifer collection (see Fig. 131) (National Botanic Gardens, Dublin)

107. The Great Palm House photographed by David Davison in November 1981

brought into the Gardens through the main entrance gates, and apart from the inconvenience to visitors when loads of manure, coal and other supplies were being brought in, the haulage carts caused serious damage to the surface of the paths. A second goods entrance was opened in 1928.

The rock-garden in front of the Great Palm House was dismantled and a new rockery built at the western end of the ancient Yew Walk. This rockery was made of stone quarried nearby in Finglas; it was believed that Howth stone was superior, but unfortunately Howth stone was more expensive. Reginald Farrer, a well-to-do plant-hunter, opinionated author and enthusiast about rock-gardens, described the Glasnevin rockery in *My rock-garden*, published in 1908. He repeated the same unflattering description in *The English rock-garden*, a best-seller published for the first time in 1919. To Farrer, the Glasnevin rockery was one of the finest examples of the "Devil's lapful" category of rock-garden. 'The plan is simplicity itself. You take a hundred or a thousand cartloads of bare square-faced boulders. You next drop them all about absolutely anyhow; and you then plant things amongst them. The chaotic hideousness of the result is something to be remembered with shudders ever after'.[72] This "Devil's lapful" rock-garden is still a feature of the National Botanic Gardens (Figs. 105, 106).

During Frederick Moore's keepership four new glasshouses were built. The Great Palm House, which had been caustically criticized by William Robinson in 1864, was so badly designed it swayed visibly in high winds. In the 1870s it had been strengthened internally with extra supports. On 25 September 1883, a gale from the north-north-west blew in four sashes and broke eight huge panes of glass. The wood was so rotten that the sashes were beyond repair. The whole glasshouse was deemed unsafe, and was boarded up for the rest of the winter. On 20 February 1884, another gale blew in two large sashes and broke more glass. However, by that time, it had already been decided that this ugly and unstable house must be demolished. Frederick Moore was asked to state what was required as a replacement, and James Boyd of Paisley, Scotland, was asked to build a new dome according to Moore's specifications. The plans were ready by January 1884. Work began on 19 March and was completed by 1 October 1884 at a cost of just under £800. The Palm House (Figs. 107,

108) was sixty-five feet high, eighty feet wide, and one hundred feet from front to rear. The sides were constructed of teak, bound with wrought iron, and the foundations were of concrete. The curved roof had iron glazing bars. Inside, a walk-way encircled the dome twenty-five feet above the ground and there were galleries outside. A spiral staircase led to the inner gallery, from which visitors could look down on the plants in what was claimed as the best lighted palm house in Europe. It was heated by water-filled pipes, a system supplied by Weeks of London. To control condensation on the roof a three-inch pipe fed from a separate boiler circled the spring of the upper roof.

The two side wings of the Palm House Range were retained, one for orchids and the other for camellias. Minor structural alterations were carried out in 1881 on the wing which housed the orchids. The entrance at the middle of the wing was closed and a new entrance made through a small porch at the east end. The original roof of grooved glass was replaced by clear glass in 1887, and in 1892 the brick staging was replaced by open ironwork. The west wing, the Camellia House, was reglazed by Boyd and Company in 1891 with larger sheets of glass, and the sides were extended by nine inches. Repairs were also carried out to the Curvilinear Range, then the oldest conservatory in the Gardens. James Boyd's firm installed new boilers in 1882. In 1884, much of the curved glass was replaced and the inside painted dull red instead of white. Iron and slate shelving was erected inside when the original wooden shelving was removed in 1899.

Up to this time the Botanic Gardens had lacked a purpose-built fern house to which the public could be admitted. The hardier ferns were grown outside and those demanding protection were housed in the Long Range and in the Orchid Wing of the Palm House Range. In 1860 a small fernery was planted against the north wall of the Curvilinear Range, and this was rebuilt in 1870 after the Curvilinear Range was enlarged. There is still a small lean-to fern house behind the Range. The original little 'secret-looking lean-to' was watered by a stream that was contrived to trickle from the roof and flow down a built-up wall of turf matted with native species of *Hymenophyllum* (filmy fern) interspersed with luxuriant masses of the Killarney fern. It was a sight that, in 1878, a writer to the *Gardeners' Chronicle* asserted almost merited travelling from London to Dublin to see.[73]

The old Octagon House was demolished in 1886 when it was sixty-seven years old, a great age for a wooden conservatory. On its site the Gardens' first fern house was erected in 1887. This had a square foundation—the sides were thirty-two and one half feet long—and the twenty-three feet high curved roof had glazing bars of iron. The new Fern House (Fig. 109) was connected to the Aquatic House by a short passage lined with glazed cases in which the Killarney fern and various species of *Todea* from New Zealand were grown. The house was divided into two compartments: the main one was for ferns from temperate regions, and the heated porch housed tropical ferns. In 1898, a twelve-foot lean-to glasshouse was built against the east wall of the Fern House for species from cooler habitats.

There was no special accommodation for cacti and other succulent plants at Glasnevin until the latter years of the last century. Cacti were introduced into Europe in large numbers after the second quarter of the nineteenth century. During the 1830s the most extensive English collection belonged to the Duke of Bedford, and in Ireland the Trinity College Botanic Garden had a fine collection including cacti sent from Mexico by Dr Thomas Coulter. Towards the end of that decade Glasnevin received cacti from John Tweedie, and the collection was further enriched in 1856 by the gift of eighty-five species from Edouard Ortgies of Zurich. So the last of the glasshouses built in the 1880s was devoted to cacti and succulent plants. The Succulent House (Fig. 109), which is still standing, was erected on the western side of the Victoria House and linked to it by a short passage. It has a square floor-plan, and a double span-roof. Originally the house was divided into two compartments and contained not only cacti but representative species from other families and genera that inhabit arid regions, for example, *Lithops, Aloe, Agave, Euphorbia* (spurge), *Mesembryanthemum* and *Gasteria*.

During 1888 an iron greenhouse was transferred from the Chief Secretary's Lodge in Phoenix Park to the upper yard behind the Curvilinear Range at Glasnevin. It was used for housing plants that did not need heat, and it pleased Moore to have this additional room. The iron greenhouse still stands today, and in 1985 was restored and refurbished. It is a unique glasshouse for it bears the

108. The interior of the Great Palm House about 1900 showing the luxuriant growth of palms, ferns and aroids

maker's plaque 'Turner, Dublin'; no other Turner greenhouse bears such a plaque. Its history prior to 1888 is not known.

During the last few years of the nineteenth century plans were prepared for a series of large greenhouses, including, in December 1898, some for a new palm house by James Boyd; architect's drawings dated January 1896 also survive.[74] The most dramatic proposal was for a two-chambered fern house, surmounted by twin domes (Fig. 111).[75] None of these buildings was erected and nothing is recorded about them in the archives of the Botanic Gardens. They may just have been 'flights of fancy' sent by Boyd in the hope that Frederick Moore would press them on the Board of Works. Moore's annual reports for this period do contain requests for a tropical fern house, and plans survive for such a house prepared by James Boyd in October 1899, but again this was not built.

Several non-horticultural buildings were put up during the late 1800s. New houses were built for those members of the Gardens' staff who were entitled to a residence. The two gate-lodges, which had long been in disrepair, were remodelled in 1891, an upper storey being added to each lodge (Figs. 32, 33) and the internal arrangements reorganized.

When the lodges were ready for occupation, the library was moved back into the right-hand lodge. A few non-botanical volumes were sent to the National Library in Kildare Street, and some sets of botanical journals were completed by purchasing missing numbers. The library was greatly enriched in 1911 when William Edward Gumbleton left his library to the Royal Botanic Gardens. Frederick Moore travelled to Belgrove to select the finest of the superb illustrated botanical books for Glasnevin; the remainder were donated to the Lindley Library of the Royal Horticultural Society in London.[76]

109. Glasshouses in the Botanic Gardens about 1910 — a photograph from the Lawrence collection, reproduced by courtesy of the National Library of Ireland, Dublin. In the foreground is the Cactus House (erected 1890), in the background the high Fern House (erected 1887) with the Victoria House in the middle (note the elaborate facade of the Victoria House)

110. The same range as Fig. 109 in November 1981 photographed by David Davison. Note that the original Fern House has been replaced by a stark aluminium house (open 1966) and that the elaborate decoration has been removed from the Victoria House

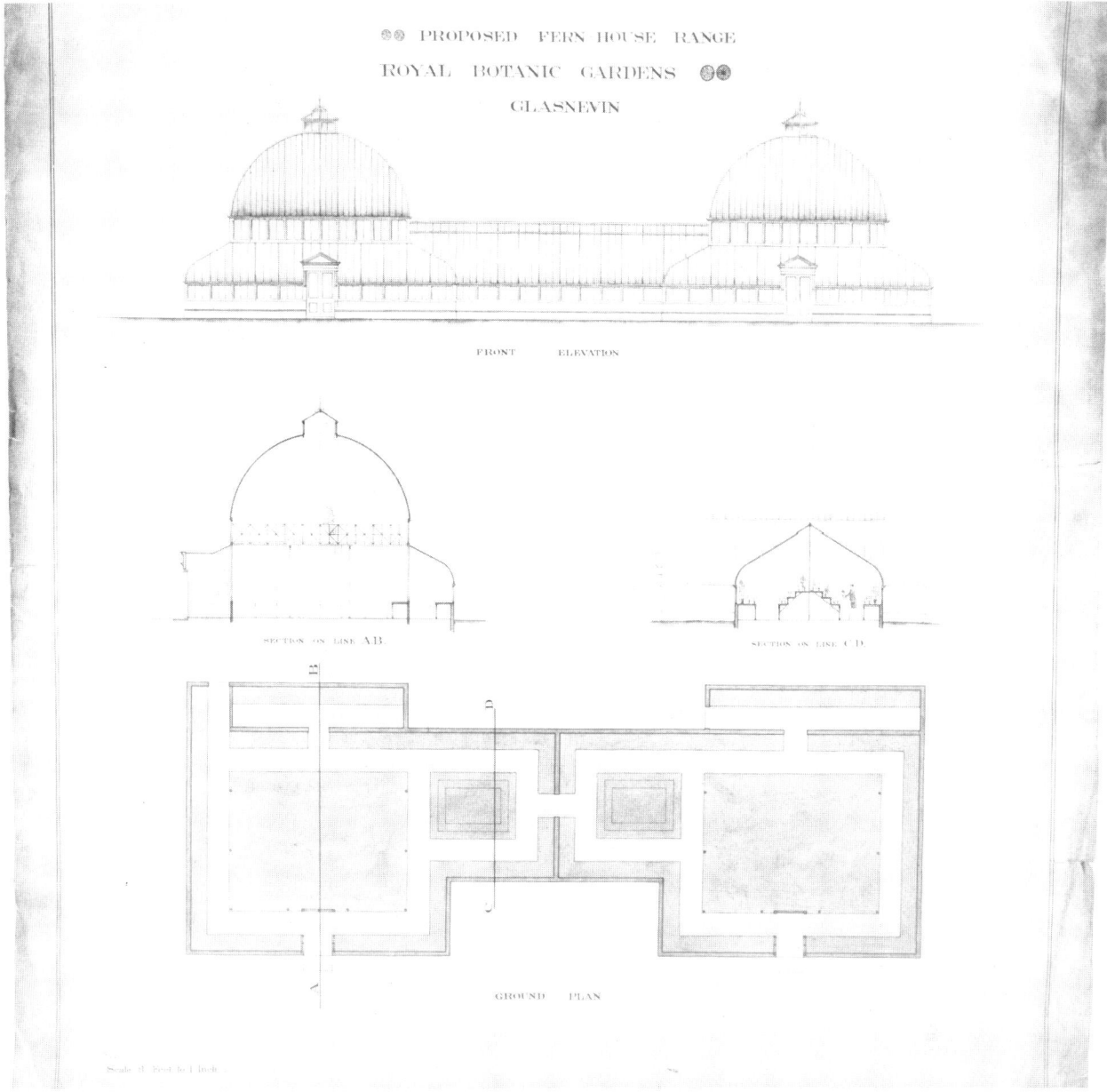

111. Plan for a fern house that was not built (Reproduced by courtesy of the Commissioners of Public Works)

Frederick Moore complained repeatedly to his superiors about the facilities provided for the Gardens' staff. He campaigned for better accommodation for the eight resident apprentices, who were housed in a lodge that contained two dormitories, a living room and a reading room. There was no bathroom, and even as late as 1898 there were no sanitary arrangements in the gate-lodges or in the bothy. In 1907, bathrooms were added to the residences, and seven years later the entire sewage system in the Gardens was overhauled and improved. In the apprentices' bothy no provision existed for a sick-bay—when anyone was ill he had to remain in the dormitory with three other apprentices. An additional room was added in 1914.

As for the non-resident staff, they had nowhere to eat their mid-day meal except in the furnace rooms, where they sat on piles of coal. Apart from the hygienic undesirability of this, the men often built up the furnace fires so that the conservatories became overheated. A mess-room was provided in 1907, and zinc-lined presses were supplied for the men's provisions.

In keeping with contemporary ideas, more concern was shown during the first half of the nineteenth century about securing a reliable water supply for the conservatories than about installing modern sanitation. Indeed, originally, the sewage from the Director's residence flowed down an open furrow which was periodically cleaned out and the deposits used as fertilizer.

The Royal Dublin Society had not felt any obligation to provide toilets for visitors, though in 1859 it drew the attention of the Board of Works to 'the want experienced in the Botanic Gardens for water-closets and urinals for the accommodation of male and female visitors'. Not until 1898 were these installed in the waiting rooms, and as late as 1910 'very satisfactory earth closets were provided within the grounds for the use of the public'. Modern water-closets were built in the Far Grounds in 1944.

The first hydraulic ram in Ireland had been constructed on the Mill Race at Glasnevin in 1832. It was used to pump water up to the conservatories, but was inefficient and expensive to service. Repairs cost approximately seven guineas each year, and eventually a new, more efficient pump was installed in the old ram-house in 1903. But the River Tolka could not be relied upon for a continuous supply of water, and during exceptionally dry summers, when there was no water flowing in the Mill Race, holes had to be dug in the exposed bed of the river to obtain water for the Gardens. In the 1860s the Royal Dublin Society unsuccessfully negotiated with both the Royal Canal Company and the Midland Great Western Railway Company for a supply of piped water. Eventually, in 1880, pipes were installed in the Botanic Gardens, and water was supplied from the Vartry Reservoir. Two small drinking fountains, one near the gate and one in the centre of the Gardens, were fed from this new clean and reliable water supply.

The amenities provided in the Royal Botanic Gardens for visitors were inadequate throughout the nineteenth century, especially when the distance from Dublin and the large number of visitors are considered. Even getting to Glasnevin from Dublin raised problems. The road was in very poor condition until the end of the century. Residents in Glasnevin seemed not to mind the mud, but strangers complained bitterly about it. Another annoyance, the turnpike on the Dublin side of the Gardens, was not removed until 1853 in spite of efforts by the Royal Dublin Society to have it resited north of the village. As no public transport was available until 1877, when the North Dublin Tramway was opened, people without private carriages had either to hire a cab or walk.

During the 1830s, when 20,000 visitors came each year, there was only a couple of dozen seats on which one could rest after the three-mile walk from Dublin, and even by the time the Royal Dublin Society relinquished control of the Gardens in 1877 only a further three dozen had been provided. For many years no shelter against the rain was available, for until the 1860s the conservatories were kept locked. The open shelter at the entrance, whose walls are decorated with photographs of trees, and the shelter at the western end of the Gardens in the centre of the Far Grounds, were erected in 1894.

Bicycles, prams and children's chaises became a problem at the beginning of this century. Sometimes there were nearly one hundred in the Gardens at a time. To control the numbers, prams and chaises were only admitted on the production of a permit—this system was abandoned in 1908 (Fig. 112). An open shed behind the tea rooms served as a bicycle park until 1912, when a proper bicycle shed was built.

It is curious that in the long history of the Gardens, it was only for a very short period, 1908 to 1911, that any effort was made to provide refreshments. During these four years, refreshments were sold in a cottage outside the gates on the left hand side of the entrance (Fig. 113).

The end of the nineteenth century saw Frederick Moore half way through his career and he could have contemplated his personal and professional achievements with some satisfaction. He must have been delighted to read the editorial in the *Gardeners' Chronicle* on 22 April 1897 in which Glasnevin was lauded as 'one of the most interesting and beautiful gardens in the northern world'. On the personal side, he married Phylis Paul, daughter of Robert Paul, J.P., of Dublin, in Rutland Square Presbyterian Church on 19 November 1901. Frederick Moore was a noted athlete: he had played as a rugby international several times, he was a great oarsman, a good horseman and a crack shot. And, professionally, membership of various bodies was thrown open to him in recognition of his considerable competence as a horticulturist and gardener. During his second year at Glasnevin he was made an

DEPARTMENT OF AGRICULTURE
AND
TECHNICAL INSTRUCTION

ROYAL BOTANIC GARDENS, GLASNEVIN.
PERAMBULATOR PASSES.

It has been the custom for some years to allow persons to bring perambulators and children's carts into these gardens, although the practice is strictly forbidden at other Botanic Gardens. So much inconvenience, however, has been caused by the perambulators and carts that it is necessary to stop this, but in order to lessen the unavoidable inconvenience as much as possible, those persons who have already been allowed to bring in a perambulator may have for the present a special pass on which it will be admitted.

Every applicant must fill in and sign the regular form of application, and if the applicant proves that he or she has been regularly sending a perambulator into the gardens in previous years a Pass will be issued.

The Passes printed on White Paper are no longer of use, only those printed on Pink Paper are available for the year 1903.

DEPARTMENT OF AGRICULTURE AND TECHNICAL INSTRUCTION.

ROYAL BOTANIC GARDENS,
GLASNEVIN, DUBLIN.

REFRESHMENT ROOM.

LIST OF PRICES.

Tea, with Milk and Sugar, Small Cup	2d	GingerBeer	2d. & 3d.
Tea in Teapot (2 or three persons) with Milk and Sugar	3d each	Lemon Soda	2d. & 3d.
		Seltzer Water	2d. & 3d.
Coffee, Large Cup, with Milk and Sugar and Cream	3d.	Ginger Ale	2d. & 3d.
		Bun, Large	1d.
Bread (Brown), or Large Roll or Scone	1d.	Bath Bun	1d.
Butter	1d.	Genoa, Tennis, Seed, or other Cakes	2d. per slice
Cocoa, with Milk and Sugar (Breakfast Cup)	4d.	Bread and Butter (Two Slices)	2d.
Do., Smaller	3d	Jam or Marmalade, Small Pot	1d.
Glass of Milk, Hot or Cold	1d.	Boiled Eggs	2d. each
Cream, Tablespoonful	1d.	Eggs, Poached, or Scrambled on Toast	4d. each
Extract of Beef, or Bovril	3d.		
Soda and Milk	2d. & 3d. Per Glass	Banana, Orange, Apple, or other Fruit in Season	½d. & 1d.
Lemon Squash	2d. "	Strawberries & Cream, in Season	4d. & 6d.
Lemonade, Home-made	2d & 3d "	Ices	4d. & 6d.

BICYCLES AND CHILDREN'S CARRIAGES STORED FREE.

Gratuities to attendants are Prohibited.

Contractors.— **ANNE LYNCH & Co.,**
"CAFE CAIRO,"
59 Grafton Street, Dublin.

113. Notice concerning refreshments

honorary member of the Royal Horticultural Society of Ireland. In 1887 he was elected to membership of the Royal Irish Academy, and he served as a Council member in 1898 and 1899. He was appointed consulting botanist to the Royal Dublin Society in 1890, with an annual honorarium of ten pounds, and in 1891 he was elected a member of the Society. Frederick Moore became lecturer in botany to the Albert National Agricultural Training Institute, Glasnevin, in 1894, and he gave thirty lectures a year for a fee of fifty pounds. He was one of the sixty original recipients of the Victoria Medal of Honour awarded in 1897 by the Royal Horticultural Society of London to distinguished horticulturists in commemoration of the sixtieth anniversary of Queen Victoria's accession to the throne.

Other honours were bestowed on Frederick Moore during the early 1900s. The Royal University of Ireland conferred on him the degree of Master of Arts *honoris causa* in 1908, in recognition of his services to its Department of Botany. State recognition of his ability came in 1911 when King George V conferred a knighthood on him. Moore (Fig. 114) was told of the proposed honour only the day before he was knighted. On board R.M.S. FRANCONIA *en route* for Canada and the United States, where among other places he visited the Arnold Arboretum and certain fruit growing districts, he wrote to Sir Arthur Hill at Kew:

> The affair came as a great surprise to me and to everyone else. I only heard of it at 7 p.m. the evening before the ceremony and had grave doubts about accepting. However, I was persuaded that for the sake of horticulture I should and I gave in.[77]

In 1901, the control of the Royal Botanic Gardens was transferred from the Department of Science and Art to the Department of Agriculture and Technical Instruction. One development which followed the transfer was a scheme to encourage fruit-growing by smallholders and farmers. This was entrusted to Frederick Moore, who made visits to fruit-growing areas in England and Scotland and to different parts of Ireland to assess the potential of the local soils and weather conditions. To implement the scheme, Moore selected several young head gardeners as instructors. They already had practical experience of fruit-growing, and after undergoing instruction in the Albert Agricultural College they were sent to selected Irish counties. Moore remained in charge of these activities until 1919.

Frederick Moore was ever keen to support schemes for the education of young gardeners, and at a time when women had relatively few opportunities to pursue independent careers, he started a training course for women in 1898. Earlier, in October 1890, Moore was persuaded to assist in setting up a course at Alexandra College for girls who wished to receive horticultural training. In the Gardens' archives there is a letter written by Moore, dated October 1890, seeking permission to give lessons on horticulture at Alexandra College.[78] The story of Frederick Moore's participation in the College scheme was told, many years later, by Lady Moore.[79] According to her account, Miss Henrietta White, principal of Alexandra College, and Miss Fanny Currey, a noted gardener from Lismore, arrived at Glasnevin by jaunting car. These two formidable ladies demanded to see Moore, who was 'still a shy bachelor'. They were granted an interview but 'could not be induced to leave [Moore's] office' until he promised to pass on their request to the Department of Science and Art.

The initiation of the training course for women at Glasnevin appears to have been entirely unofficial. In January 1898, a Miss Wilkinson, who was connected with the 'Ladies Employment Association', approached Moore and asked him if two young women, who had just completed two years at Swanley Horticultural College, Kent, could come to Glasnevin for practical training.[80] On 31 January Moore wrote that there was 'no chance of their coming...if they expected to receive payment'. However, he qualified this seemingly unsympathetic statement by adding that the girls could come to 'work under conditions given, without payment, and if [he] found them intelligent and anxious to learn [he] would endeavour to teach them all [he] could, and allow them to attend...lectures'. The conditions laid down were that they would work from 9 o'clock in the morning until 5 o'clock in the evening during week-days (9.45 a.m. until dark in the winter) and until 1 o'clock on Saturdays. They would receive three weeks holidays during the year, but were not to be absent from duty without permission. These conditions were accepted, and so on 20 June 1898, Gertrude Webb and Mary Graves arrived from Swanley to begin work at Glasnevin as the first lady pupils. Moore kept the

114. Three generations of the Moore family, photographed about 1904 at the front door of the Keeper's residence. Mrs Margaret Moore, mother of Frederick Moore is on the right. Mrs Phylis Moore and their son Frederick Moore are in the centre. (Reproduced by courtesy of Major-General F. D. Moore)

matter as quiet as possible—indeed, he did not inform his superiors until a week after the two girls had arrived and begun work.[80] He anticipated questions about the 'lady gardeners', and numerous visitors did ask about them, but Moore refused all requests to meet the girls, saying categorically that 'they are not here "on show"'.[80] He was more than satisfied with Miss Webb and Miss Graves, and told the Department that 'if they are only left alone, I am sure the experiment will be a success'.

Frederick Moore's unofficial scheme seems to have been quietly approved, and it was a success. During the next thirty years, sixty young women received a thorough training at Glasnevin.[81] Between 1905 and 1927 only four failed to complete the one year course. Many more girls applied for training than could be accepted, yet about sixty who were offered places did not avail of the opportunity—most of the non-starters were applicants who lived in Britain.

The women pupils carried out the same work as the male apprentices, but they were not allowed to mow grass as this was still mainly done by using scythes (Fig. 115). In the evenings, from 7 o'clock until 8.30, they attended lectures given by Frederick Moore. All the pupils were given the opportunity, during their course, to work in each section of the Gardens so that they gained experience of all types of plants and conditions of cultivation. The girls were given a few special tasks, including the cleaning of the seeds that were sent to other botanical gardens. They also helped with the hand-pollination of Glasnevin's famous blue-flowered *Cineraria*, seed of which had to be saved every year. The same task was performed with the red and yellow polyanthus and a blue-flowered primrose that were grown in the Gardens.[82]

May Crosbie, who entered Glasnevin in May 1905, later wrote an article[83] on the training of lady gardeners, in which she stressed that thorough practical experience was essential for any girl intending to make a career in horticulture. She gave it as her opinion that work of a practical nature was preferable to the entirely theoretical training given in some horticultural colleges, and noted that Frederick Moore spared no trouble in providing such a course for girls. Miss Crosbie wrote that prospective lady pupils would require 'health, strength and a liking for the work...as the work is hard and hours as

115. M. Dowling and N. Lynch with the gang-mower in the Botanic Gardens, 1912 (photograph by Elsie Miller, reproduced by courtesy of Miss Miller)

116. Miss Rosamund Pollock—a photograph by Elsie Miller, 1912. (Reproduced by courtesy of Miss Miller)

a rule long; but there is no doubt that for a strong girl the life is both healthy and exceedingly interesting'.

Having started this type of training scheme for lady gardeners, Moore widened its scope and 'gentlemen gardeners' were also allowed to gain practical experience at Glasnevin. Like the women, these men were not paid. One of these 'gentlemen gardeners' was G. O. Sherrard, who entered Glasnevin about 1903. Sherrard is best known as a plant-breeder, whose particular interest was Brussels sprouts, and in 1938 he was Professor of Horticulture in University College, Dublin. His distinguished contribution to Irish horticulture culminated in 1955, when he was elected President of the Royal Horticultural Society of Ireland. Professor Sherrard died in 1959.

Few of the women who trained as gardeners in the Royal Botanic Gardens under Frederick Moore became well-known horticulturists, but several did pursue careers in gardening and associated occupations.

May Crosbie managed a bulb farm at Rush, County Dublin, and often went to London with Lionel Richardson, the famous daffodil breeder from Waterford, to stage daffodils at shows of the Royal Horticultural Society. Rosamund Pollock (Fig. 116), who entered Glasnevin as a lady pupil in 1902, was appointed technical assistant and secretary to Frederick Moore in 1904. She contributed numerous notes and articles about the Botanic Gardens to *Irish Gardening*. Miss Pollock left in September 1918 to work for Captain Hunter. Later she married Harry Richardson, the uncle of Lionel Richardson. In 1904, Emmeline Crocker, aged forty-six, entered Glasnevin as a pupil. She was an experienced traveller, having undertaken a trip round the world during which she visited botanical gardens in Singapore, Hong Kong and Canada. While in Japan she met Reginald Farrer, that somewhat eccentric Yorkshireman who was later to become a well-known writer, explorer and plant-collector. For a while they travelled about Japan together. After returning to Europe, Miss Crocker came to Glasnevin to become 'conversant with practical garden work'. On leaving she rejoined Reginald Farrer who had, by then, established a nursery in Yorkshire, but, according to Lady Moore, Miss Crocker found that although a woman of fifty and a youth of twenty-five might travel together amicably, it was quite another matter to work in a small garden without quarrelling'.[84]

117. Peter Daly and Miss Oonagh McCarthy—a photograph by Elsie Miller, 1912. (Reproduced by courtesy of Miss Miller)

118. Elsie Miller. (Photograph reproduced by courtesy of Miss Miller)

Many of the women who trained at Glasnevin in the early years of this century married and left horticulture. Others continued to work in the general field of botany. Elsie Miller, a native of County Limerick, enrolled as a pupil in January 1912 and completed the one-year course. She was employed in the seed testing unit in the Royal College of Science for three years, and returned to Glasnevin in November 1918 on the resignation of Miss Pollock. Miss Miller (Fig. 118) took over as secretary to Sir Frederick Moore and worked for him until he retired in 1922. She was also the Gardens' unofficial photographer (see Figs. 117, 118, 119); her photographs were used to illustrate *Irish Gardening*, and she also provided photographs for a book on the garden at Mount Usher, and for research papers written by Professor Augustine Henry. Another of Elsie Miller's duties was the cataloguing of the plants in the Gardens. She left Glasnevin in 1927 and returned to the Seed Testing Service of the Department of Agriculture.[85]

Augustine Henry, for whom Elsie Miller provided illustrations, was closely connected with Glasnevin after his appointment in 1913 to the chair of forestry in the Royal College of Science.[86] Dr Henry's best-known work is a seven volume monograph *The Trees of Great Britain and Ireland*, which he wrote in collaboration with Henry J. Elwes, a friend of Sir Frederick Moore and author of *Monograph of the genus Lilium*. Augustine Henry played a large part in the development of the Irish forestry industry. He suggested the planting of fast-growing coniferous trees, especially species like Douglas fir from the western United States of America. He also tried producing artificial hybrids of some well-known deciduous trees, hoping to obtain vigorous progeny which would be useful as timber trees. The facilities

119. Seedling of *Populus* × *generosa* in the Botanic Gardens, Glasnevin, 1914. The gardener is Peter Daly (see also Fig. 117). This photograph was taken for Professor Augustine Henry (National Botanic Gardens, Dublin)

of the Botanic Gardens were at his disposal, and in 1914 there was a twenty-seven month old plant of Henry's artificial hybrid poplar (*Populus* × *generosa*) standing 10 feet 1 inch high—it was measured every day and that year did not stop growing until 15 September (Fig. 119). Pollen from a poplar (*Populus* 'Robusta') growing at Glasnevin was sent to Kew for pollinating *Populus candicans* in 1914, and at Glasnevin Dr Henry attempted to cross-pollinate *Populus angulata* and a species of *Fraxinus*, but these experiments failed. Professor Henry often brought his students to the Gardens to study. He was sufficiently intimate with the Moores to stay in their house, and on at least one occasion Henry, Moore and Elwes toured Irish gardens and demesnes together.

Major changes in the staff of the Royal Botanic Gardens at Glasnevin were precipitated in 1906 by the illness and death of William Parnell, who had been foreman since 1868. For several months before Parnell's death from cancer on 28 November, Frederick Moore had been sounding out his friends in Britain about suitable men to fill a vacancy when it arose, and he had two names by the time Parnell died. Towards the end of September 1906, Moore's close friend, William Watson, Curator of the Royal Botanic Gardens, Kew, had suggested Charles Frederick Ball, described by Watson as 'an excellent fellow in every way, gentlemanly, quiet, good-looking...rings true however tried and loyal'.[87] Ball was then twenty-seven years old, and a sub-foreman at Kew. Watson offered to allow Ball to travel to Dublin to meet Moore 'if you would like to sample him'.

C. F. Ball—his close friends called him Fred—was a native of Leicestershire, and he had started his career as an apprentice to William Barron, a nurseryman at Barrowash in Derbyshire. He joined Barron on 7 April 1896, and after working through every department in the nursery, Ball left on 6 May 1899 'only to better himself'. He worked with Barr and Sons of Long Ditton for one year before entering the Royal Botanic Gardens, Kew. At Kew he progressed rapidly, becoming sub-foreman in the herbaceous and alpine department. In 1903, C. F. Ball gained the Hooker Prize from the Mutual Improvement Society at Kew and obtained a first class certificate in advanced botany at South

120. The staff in the Botanic Gardens 1913. No key to the men photographed exists, but the following may be in the photograph: P. Pope, D. McArdle, M. Conway, M. Doherty, H. McDonnell, A. Doyle, P. Daly, J. McArdle, R. Goff, T. McCormack, E. Gartland, R. Gainford, S. Reilly, F. Parnell, J. Kenny, T. Fagan, M. McGovern, T. Goggins, J. Woods, T. Brady, M. Dowling, P. McCormack, L. Smith, J. Harris, W. Kinsley, J. McDonnell, S. Glass, N. Lynch, J. McCann, R. Usher, J. Morgan, R. Drumgoole, J. Day

Kensington. He received honours certificates in drawing and French. Ball left Kew to join his brother who had established a nursery and market garden near Nottingham, but after some months they quarrelled and Fred left. William Watson took him back at Kew, and reappointed him as a sub-foreman. But Ball was only biding his time until he could get a better permanent job either in England or abroad.

The day after Moore received the commendation of Ball from William Watson, a letter arrived from Arthur Bulley of Ness in Cheshire. Bulley suggested another young Kew-trained gardener, John Besant, whom Moore had met during a visit to Bulley's nursery—Bees Limited—at Chester. Besant had left Bulley's employ some weeks earlier following a difference of opinion on nursery practice, but Bulley told Moore that it was not necessary to discuss this disagreement 'save to say that each side remained unconvinced'.[89] Arthur Bulley praised Besant as a 'first class man'. Two days later, another letter arrived from Bulley urging Moore to choose Besant. William Watson was consulted about both men, but was unable to say which was the better one. He suggested that Besant would not humour Bulley, and that that was the reason for their parting.

Frederick Moore tried to persuade the Department of Agriculture and Technical Instruction to appoint an assistant to the ailing Parnell, but while the negotiations were still in train William Parnell died. On the day following his death, Moore sent a minute urging the immediate appointment of either Ball or Besant, stating that none of the existing staff was sufficiently well-educated for the job—several were illiterate.[90] The Department questioned the wisdom of a non-Irish appointment,[91] but Moore responded that he wanted a man with 'a thorough technical training and knowledge of botanic garden work...well up in British [sic] flora, and who has an all round knowledge of outdoor

121. 'Preparing a beech stump for blasting', Winter 1912-13. When originally published in December 1913 (*Irish Gardening*, vol. 8, p. 185) John Besant provided the following commentary: At Glasnevin, where a good deal of blasting is done most winters, Nobel's dynamite is found very safe and satisfactory. Here it is preferred to gunpowder, being more powerful and perfectly safe when reasonable precautions are observed. Only a careful man accustomed to the work and who can be relied upon to take no risks, should be put in charge of the operations...All workmen, except the man in charge who fires the fuse, must...get out of danger and place themselves round the scene of operations so as to prevent any straggler from penetrating the danger zone. When all are safe the workman left at the fuse now proceeds to light it with a fusée match, and immediately the powder is properly burning he too must get to safe quarters...no one should be near the stump a second after the fuse is lit.
The accompanying illustration depicts the stump of a very large beech tree, felled here last winter with a workman preparing the charge.

plants, herbaceous plants, alpine plants, trees and shrubs".[92] The man should have trained in a botanical garden. Frederick Moore reiterated that there was no suitable candidate in Glasnevin because the best trainees inevitably left Ireland for further training in Kew, whence they went overseas — he noted that several Glasnevin men were in government service, as managers of tea plantations in the colonies. The Department accepted Moore's arguments, and on 10 December 1906 C. F. Ball arrived for a trial period. After one month's probation his appointment as outdoor foreman was confirmed.

On 11 January 1907, Moore was granted permission to appoint another outdoor foreman but, before that post could be filled, Moore sought the appointment of an Assistant Keeper. He pointed out that he had to do all the scientific work, plant labelling, herbarium work and correspondence, and suggested that a man with botanical and practical horticultural experience would be an asset. In early April, Frederick Moore wrote to John Besant asking if he was still looking for a job, saying that C. F. Ball would be likely to get promotion to Assistant Keeper. Besant was still available, so Moore recommended his appointment, and on 23 May he was offered the job of foreman. John Besant arrived in Glasnevin on 1 July 1907. Meanwhile Ball sat an examination for the post of Assistant Keeper, and was duly promoted.

122. Charles Frederick Ball – photograph about 1910 (National Botanic Gardens, Dublin)

Charles Frederick Ball (Fig. 122) was interested in plant breeding and carried out some important work at Glasnevin. He may have assisted in the breeding programmes for *Nerine* and *Lachenalia*, but he was concerned chiefly with the production of new garden varieties of *Calceolaria*, *Escallonia*, *Berberis* and *Mahonia*, as well as *Ribes* and *Campanula*. In 1916 Moore recorded that there were unflowered hybrid seedlings of these genera, full of promise, in Glasnevin.[93] *Calceolaria* × *ballii*, now lost to cultivation in its original form, was produced when Ball cross-pollinated two Peruvian species (*C. deflexa* × *C. forgeti*); it was a very floriferous greenhouse plant with lemon-yellow blossoms. No records of the *Berberis*, *Mahonia* or *Ribes* hybrids remain in the Glasnevin archives, but Ball's work on *Escallonia* is perpetuated in three excellent cultivars. *Escallonia* 'Glasnevin Hybrid' has large, deep red flowers. *Escallonia* 'Alice' was named by Ball after his wife (they were married in December 1914); it still grows at Glasnevin and, although not now seen in nurserymen's catalogues, is cultivated as far afield as Seattle. The plant, however, by which Charles Frederick Ball is best remembered is the eponymous *Escallonia rubra* 'C. F. Ball' with blood red petals and bright gold stamens (Plate 14).[94]

Before the outbreak of World War I, Ball took part in several plant-hunting expeditions to central and southern Europe. In June 1911 he travelled to Bulgaria, where he met The O'Mahony of Kerry who had set up an orphanage in the capital, Sofia. On Saturday, 17 June, the day after his arrival, Ball was received by King Ferdinand I, 'The Fox of the Balkans', who was a keen ornithologist, gardener and amateur botanist. 'After finding that I had not come to strip the country ruthlessly of rare plants', Ball noted, 'His Majesty gave every help and facility in the way of guides, ponies and even a railway carriage, to our party for some of the excursions, and also kindly said he would like to show us the children of his garden, as he fondly termed his alpines in the palace garden'.[95]

Fred Ball and Herbert Cowley, a fellow Kewite, made their first excursion to Vitosh Mountain near Sofia; it was once the main habitat of *Lilium jankae* but the lily was virtually extinct on the mountain by 1911 due to the depredations of collectors. Ball and Cowley found such plants as honey balm (*Melissa melissophyllum*), *Saxifraga rotundifolia* and *Geranium macrorhizum* (Fig. 123), as well as dwarf willows and brooms. In boggy places *Salix laponnicum* and *Geum coccineum* abounded.

Then, with Pierce O'Mahony, they travelled eastwards to Stara Zagora, where they visited another of the royal gardens, situated on a nearby limestone hill. On the slopes near the garden Spanish broom (*Spartium junceum*) was 'a blaze of yellow'. From Stara Zagora the botanists and Herr Kellerer, the royal gardener, travelled north to the Shipka Pass, the site of a battle in 1878 between the Bulgarian and Russian armies and the Turks. Here also brooms abounded, and on shaded cliffs *Haberlea rhodopensis* was in full flower. Some of the plants formed tufts up to one metre across, and Ball noted that the colours and size of the flowers varied. 'Suddenly', he recalled, 'Herr Kellerer shouted out "Weiss, weiss" in an excited tone and came along bearing a lovely tuft with white flowers...The sight of the *Haberlea* is one which is stamped on the memory'.[95] After driving through Shipka village, they entered the famed Valley of Roses, and stopped at Kazanluk. This district is the centre for the production of attar of roses; Ball described it as 'the biggest rose garden known, stretching 80 miles with nearly 170 villages devoted to the culture' of *Rosa damascena* from which the attar is distilled.

After returning to Stara Zagora, Ball and his companions travelled by train through Plovdiv (Phillipoplis) to Kostenetz, where they stayed in The O'Mahony's mountain home. In the garden there was a plant of the very rare thorn *Crataegus altaica*, and Ball collected specimens of it. From this base they visited the Bellmachan mountains, collecting *Geum bulgaricum*, a large-leaved species with small, pale yellow flowers, that had interested William Gumbleton. They also found *Rhododendron myrtifolium*. After taking his leave of The O'Mahony and Cowley, Ball travelled on to Tchanhouri, where he again met Kellerer, and visited several other gardens owned by the Bulgarian royal family. At the beginning of July 1911, Ball and Kellerer climbed the highest mountain in Bulgaria, Musala (2,926 m). Primroses, including 'a great stretch of *P. deorum* at its best...in the full sun the thousands and thousands of half-nodding rich purple violet flowers, were a feast for the gods'. As they neared the summit, they crossed snow patches and walked past masses of the fairy primrose (*Primula minima*) in flower. On the summit grew saxifrages, *Anemone vernalis*, and an alpine pink (*Dianthus microlepis*) that was flowering so freely its tufted leaves were completely hidden.

123. Wild flowers in Bulgaria, photographed by C. F. Ball—*Melittis melisophyllum* (lower left), *Saxifraga rotundifolia*, and *Geranium macrorhizum* (right) growing on Mount Vitosha (originally published in *The Gardeners' Chronicle*, 20 April 1912)

In the following year C. F. Ball visited the Maritime Alps.[96] He particularly wanted to collect two species, *Viola valderia* and *Saxifraga florulenta*. He found *Viola valderia*, a pansy having downy, grey-green leaves and reddish-lilac flowers with bright yellow centres, at Valdieri, but after searching many hundreds of plants, he was rewarded with only a few seeds. Later, with his companion, H. M'Clanaghan, he walked over the Ciriega Pass (Fig. 124) to the valley of the Boreas River, and found another violet, *V. nummularieafolia*, but because it was only just coming into flower he was unable to collect seed. His main quarry was *Saxifraga florulenta*, a species with shining green rosettes up to fifteen centimetres across, and he discovered it growing in crevices among the granite boulders on the mountain sides.

C. F. Ball took over the editorship of the monthly journal *Irish Gardening* in 1912. Since its launch in 1906 Ball had contributed many articles and some photographs—he also wrote for other periodicals including the *Gardeners' Chronicle*. Following the publication of a note in *Irish Gardening*, William Gumbleton wrote one of his typically blunt memoranda to Ball, addressing him as 'The Assistant Keeper of the Irish Kew'. In it Gumbleton expressed disappointment at Ball's condemnation of *Geum bulgaricum* 'which I have just bought from the Minor Prophet Amos[97] at Enfield on the strength of his statement in his catalogue that it had handsome large golden yellow flowers'.[98] A few days later, Gumbleton sent another missive complimenting Ball on his proposal for an article on *Olearia*, a genus 'so exceptionally well represented at Glasnevin, and [I] hope you will send it when written to The Times of Horticulture, the Gardeners' Chronicle'.[99] The article was published in January 1911. Although he greatly admired *Olearia*, William Gumbleton commented that a variety of *Olearia virgata* 'got from the ignorant most presumptuous Gauntlett..is utter Tush'.[100]

In November 1914, three months after the outbreak of the Great War, Fred Ball volunteered and was enrolled in the 7th Royal Dublin Fusiliers; he was a member of D Company, nicknamed 'The Pals' or 'Football Corps'. The regiment was sent to Gallipoli, and although he was involved in the heavy fighting, Ball found time to write long letters to his friends in Glasnevin, extracts from which were published in *Irish Gardening*. These letters were often sprinkled with gentle humour and remarks about the local flora. On 24 August 1915, Ball quipped

> If, like birds, we required a certain amount of grit for digestion, we might be more comfortable, for helped by a breeze, the sand seems to pervade everywhere—food, eyes, mouth.[101]

Five days later, he wrote to Sir Frederick Moore noting that he had volunteered with some others to fetch the mail. They had walked four miles over sand, passing a salt-lake, to the sea.

124. Ciriega Pass, August 1913, photographed by C. F. Ball. On the left is a local guide and on the right H. McClanaghan. (Originally published in *Irish Gardening* 1914)

When going for the parcels I had a dip in the sea, and was very glad to get it, for it was the first wash I had for a week! I came across a pretty lot of Maidenhair fern this morning growing near a spring. A tiny Love in the Mist also grows wild here, Olives, Brooms, Sea Lavender, Sea Holly, and many other interesting plants. A knowledge of plants and botany always makes a walk interesting, and conveys much useful information. Plants have their tastes as well as we have.[102]

Ball described dog-fights between aeroplanes and reported that there were many casualties when the regiment landed—he himself had been bowled over by an exploding shell but suffered only bruises. On 13 September 1915, after heavy fighting, Ball's detachment was sent to a rest camp for a respite. Whilst here, a fragment of shell struck Ball, fatally wounding him.

The death of Charles Frederick Ball was deeply felt in Glasnevin. His quiet good humour had been much appreciated. He was generous, helpful, yet shy and studious. Many tributes were paid to him. One writer commented that 'it was a particularly fine thing for a man of his peaceful habits to join, and only those who knew him well will ever thoroughly appreciate how much [Frederick Ball] gave up and what a wrench it was for him to throw up the work he loved so much'.[103] Many of Ball's duties and responsibilities, including the editorship of *Irish Gardening* were taken over by John Besant, who was formally appointed Assistant Keeper in 1920.

As well as Ball, eleven other members of the Gardens' staff joined up during the Great War. Of these, ten came safely through the war. The other casualty was Stephen Rose, who had been an office worker and photographer in Glasnevin since 1907. He drowned when his ship was sunk in the Mediterranean Sea on its way to Egypt.[104]

The Easter Rising of 1916 had as little effect on the Botanic Gardens as it had on other outer suburbs of Dublin. There were many visitors that Easter Monday, and the next day all the staff turned up except the Gardens' constable. At no time were the gates shut. No damage was done to the Gardens despite the large attendance during the week, and the only disturbance to the normal tenor of life was that passes had to be obtained for the staff who lived outside the military cordon.[105] English botanists, anxious about their friends in Ireland, appealed to Sir Frederick Moore for information. In a letter to Sir David Prain of Kew, dated 1 May, Moore was able to report that the Botanic Gardens was safe, and that only a few bullets had fallen in it.[106] Writing to Arthur Hill, the Assistant Director at Kew, Moore said that both Dr Robert Praeger and Professor Henry Dixon of Trinity College were unharmed. 'Not one of my men', he continued, 'was absent, they are all accounted for, those who had the holiday on Monday were back on Tuesday. I am very glad of this. I know that some would have liked to go, but the fact remains, they did not'.[107]

The remaining years of Sir Frederick Moore's keepership at Glasnevin were devoted to the routine maintenance of the Royal Botanic Gardens and to the consolidation of its international reputation. There had been some disruption in the supply of plants during the war, but once the conflict had ended and freedom of movement returned, there was an upsurge in Glasnevin's acquisitions. Sir Frederick Moore had long since given up attempts to get novel plants through the usual nursery catalogues. But there were other sources, particularly the specially commissioned seed-collectors who were working in places like Burma and central China. Through the expedition sponsors or friends, Glasnevin acquired hundreds of packets of seeds from these collectors of whom George Forrest, William Purdom and Frank Kingdon Ward were the most important. Many of the seeds germinated, and when the seedlings were robust, they were planted in the Gardens. From these collections a number of new species were introduced into cultivation through Glasnevin.

Sir Frederick Moore took advantage of an early retirement scheme offered to civil servants on the transfer of power from London to Dublin and retired on 30 April 1922, five months before his sixty-fifth birthday.

On the occasion of his retirement many tributes were paid to Sir Frederick Moore, and of these the one by William Watson, Curator of the Royal Botanic Garden, Kew, who also retired in 1922, seems to convey best the essence of the man. Watson recalled that he had known Frederick Moore since 1879.

> Few years have passed without our meeting either for a plant foray together or for some celebration. Moore was great at both. He took charge of me when we went plant hunting in Continental countries, for he speaks the French, German, Dutch and Flemish languages like a native, and his personality is of the kind which wins through where the average man often gets stuck. Hundreds of men have felt the magnetism of Moore and have been all the better for his company, either when there was work to be done or when pleasure was the object. The gods were good to him in giving him a big, generous soul and a strong body (he hasn't had a day's illness since he became Keeper of Glasnevin), both of which he has made great use of. If Moore has an enemy or traducer, I have never heard of him. For many years he has been the doyen of our calling. I have become acquainted with many of the best men in it, and he stands first, an easy first, in my opinion. Glasnevin has long been the Mecca of gardeners, amateur and professional. Rich collections of all kinds of plants were there, and good cultivation. There was also the Keeper, the life and soul of the whole. If it were worth while, I could expiate on the great work Sir Frederick Moore has done in Ireland, on his wonderful knowledge of plants of all kinds, on his achievements as a collector and cultivator, and on his brilliant success as the head of an important teaching, scientific establishment. But these things are known to the cognoscenti, and the others may go and see for themselves...Acquainted as I am with most of the leading botanical gardens in Europe, [I] say deliberately that, taken all round, Glasnevin is unsurpassed [*sic*] by none and equalled by very few. All this has been accomplished by Moore, single-handed. Ever since I have known him he managed everything in such a way that all his men adored him. Moore's attitude towards scientific horticulture has always been correct...He has a gardener's feeling for plants, he loves them, and is unhappy when they are not happy. We have looked upon things together which have drawn from him strong words of condemnation of the wasted effort, the hopeless mess that ignorance and stupidity have resulted in. I, like many others am sorry that the limitations of age have compelled Sir Frederick to put down his tools, for he has been a real master of the craft, a splendid example of devotion to duty, and is "a worthy knight", if ever there was one.[108]

Sir Frederick and Lady Moore moved to Willbrook House in Rathfarnham, in the foothills of the Dublin Mountains, and both continued to take great interest in gardening and horticulture in Ireland. They made numerous trips to Britain, where they visited their friends, including such redoubtable figures as William Robinson.[109] In 1935 the Moores went to the United States and they assisted in the founding of the American Rock Garden Society.[110]

Sir Frederick devoted much of his retirement to working for voluntary organizations, especially the Royal Horticultural Society of Ireland—he was appointed its honorary secretary in 1906 and retained that post until 1945, when he was elected President following the death of the Marquis of Headfort.[111] In 1939, Moore and Headfort were among the first recipients of the Society's Medal of Honour, and in the same year the University of Dublin conferred the honorary degree of Doctor of Science on Sir Frederick Moore.[112]

While Sir Frederick Moore wrote several short papers and articles on a variety of gardening topics—wall plants, orchids, water-plants and *Lachenalia*—his written work does not bear comparison with

125. Lady Moore, photograph taken in the USA in 1935.
(Reproduced by courtesy of Major-General F. D. Moore)

126. Sir Frederick Moore, pencil portrait drawn c. 1922.
(Reproduced by courtesy of Major-General F. D. Moore)

that of his father either in quantity or quality. He himself said that he had no time for writing, but he did contribute to the small book on Mount Usher, in which he recalled the garden's development from 1885. In addition to a description there are asides which are delightful:

> To spend a few days as a guest was a treat to look forward to...One was shown how squirrels and birds tore away the soft bark of the redwoods to make their nests. How the top branches of the weeping Sequoia changed its direction from weeping to upright to replace the leader which had been broken off in a storm, how the golden-crested wren built its nest at the end of a slender branch where it hung safe from intruders. One watched while the dipper flew through the water coming over the fall to her nest on the ledge behind, and saw where a salmon had made a depression in the soft gravel to spawn![113]

At Willbrook, Sir Frederick and Lady Moore created a new garden, which was as frequently visited by good gardeners as was Glasnevin. By 1948 Sir Frederick was in declining health and his memory was fading. He died, aged ninety-two, on 23 August 1949. Between father and son, David and Frederick Moore, Glasnevin had been raised to the peak of its international standing. They had worked in the Botanic Gardens for a total of eighty-four years, and had had contact with Glasnevin for a period spanning more than 120 years. Sir Frederick Moore was described as 'a doughty champion' by his friend and contemporary, Robert Lloyd Praeger, who noted that Moore 'independent of mind...[had] fought many a bureacratic battle for the independence of the Gardens and its importance as a scientific institution'.[114] But most people best remember Sir Frederick as a generous man, distributing rare plants to other gardeners, thereby ensuring that species which were not suited for Glasnevin's soil and climate were placed in fine gardens that were more favourably situated. He was, beyond doubt, 'the Grand Old Man of Irish Horticulture'.[115]

National Botanic Gardens 1922-1968

SIR Frederick Moore's retirement coincided with a political revolution in Ireland. The Government of Ireland Act of 1920 had partitioned the island and granted the six Unionist-dominated counties of Ulster a government and parliament of their own. Following the Anglo-Irish Treaty of 1921, the Irish Free State was established, incorporating the remaining twenty-six counties, and power was transferred to a provisional government. The Royal Botanic Gardens at Glasnevin passed to the control of the Irish Department of Agriculture and its name changed to the National Botanic Gardens, Garraithe Náisiúnta na Lus.

On 1 May 1922, John William Besant was given charge of the Gardens but only with the title of Acting Keeper. Besant, the last Scotsman to become director of Ireland's premier botanical garden, was born on 15 August 1878 at Longforgan in Perthshire, the son of James and Elizabeth Besant. James Besant worked as head gardener at Castle Huntly, and John and his two younger brothers followed their father's example and chose careers in horticulture. After attending Harris Academy in Dundee, John Besant served his gardening apprenticeship in Rossie Priory, Alloa Park and Callander, three of Scotland's most important private gardens. For almost two years, beginning in 1899, he was a student gardener in the Royal Botanic Garden, Glasgow, before moving, in February 1901, to the Royal Botanic Gardens in Kew. Here he was soon promoted to the position of sub-foreman with responsibility for the flower-garden department. While at Kew, Besant studied surveying, levelling and systematic botany at Paddington Technical College. He was an excellent student, and a keen member of the Gardens' Mutual Improvement Society. He read papers at several meetings of the Society and, like Ball, he received a Hooker Prize in 1903. But, John Besant left plenty of time for social life and he proved to be a 'very useful steady bat and change bowler on the cricket field'.[1]

In June 1905, John Besant left Kew and went to Bees' Nursery at Ness in Cheshire as assistant manager. This nursery had been established by Arthur Kilpin Bulley, an enthusiastic plantsman and avid collector of exotic species, who had explained to his friend, Dr Augustine Henry, that his aim in life was to introduce beautiful flowers to the gardeners of England. It was at Bulley's nursery that John Besant first met Frederick Moore, who later recalled that meeting—'I was very much impressed by [Besant's] knowledge of plants. It was on a Saturday morning, and when close to one o'clock I said I knew it was closing time, and the Saturday half-holiday, and that he should go. With the fine Scottish independence which he never lost [Besant] replied "I'll not go, I will stay with you and show you more..."'.[2] Moore and Besant then spent the rest of the afternoon looking at plants in the nursery.

Besant resigned from Bees' Nursery after he had been there for less than two years, because he and Bulley disagreed about nursery practice and neither man would give way. But, when Frederick Moore showed an interest in securing Besant for Glasnevin, Arthur Bulley stated his opinion of Besant fairly. He told Moore that John Besant was 'a first class man...[whose] conduct here has been devotion beyond words and [whose] knowledge of hardy plants is extensive'. Bulley believed that if Besant went to Glasnevin, to replace William Parnell, he would be more than suitable because he '...adds to knowledge and enthusiasm an excellent tact for government'.[3]

William Watson, curator of Kew Gardens, advised Moore that Besant was a dour Scot who 'stands no damned nonsense from anyone, holds his own like a man, is just the sort I like, but would you like him?' As to whether Charles Frederick Ball or John Besant was the better man to replace Parnell, Watson declined to offer his opinion. Indeed, he told Moore that there was yet another man at Kew, William Purdom, who was in some ways better than Besant. 'Purdom [is] a jewel', Watson wrote,

127. J. W. Besant, photograph about 1930
(Reproduced by courtesy of Mrs Margaret Mansfield)

'but, he is a socialist, atheist, everything that's bad from the parson's wife's point of view'.[4] Accepting Watson's judgement that C. F. Ball was 'more gentlemanly', Moore selected Ball, who was appointed outdoor foreman. But, Moore continued to press the Department of Agriculture and Technical Instruction for more staff in Glasnevin. He pointed out that not only had he to direct the running of the Gardens, he also had to carry out all the work of a scientific nature, including the labelling of plants, the herbarium work and all correspondence. An experienced man of about thirty with a sound education and a good background in botany and practical horticulture was needed on the Gardens' staff. Moore really required an assistant keeper. Eventually the government sanctioned such an appointment; C. F. Ball was promoted to the position of Assistant Keeper, and John Besant was invited to accept the job of outdoor foreman vacated by Ball.

John Besant (Fig. 127) arrived in Glasnevin on 1 July 1907. Four weeks later he was given leave to go home to Scotland, and on 7 August, in Huntley, Aberdeenshire, he married Helen Watt. The couple reached Dublin the following day, and for the next fifteen years lived in the foreman's cottage beside the River Tolka.[5] Although the assistant keepership fell vacant in 1915 on the death of C. F. Ball, it was not until 1920 that Besant was promoted to that position. When Sir Frederick and Lady Moore moved out of the Keeper's residence in 1922, the Besant family moved from their cottage into the bigger house.

John and Helen Besant had three children, one son and two daughters. Recalling her childhood days in Glasnevin, Mrs Margaret Mansfield (née Besant) remembers the tranquility of the Gardens, the kingfishers darting along the river, which was not polluted in those days, the red squirrels scampering across the lawns, the old mulberry tree outside the front door of the Keeper's house, and, in the glasshouses, such fascinating plants as the giant waterlilies, the orchids and the carnivorous pitcher plants. And, on warm, still, summer evenings, the three children were excited by their father's trick with the burning bush (*Dictamnus fraxinifolius*).[6] He would apply a burning match to the base of the stem, the flame would run up the shoot without harming it and flicker out, leaving a scent like incense in the air.

As John Besant had already worked in Glasnevin for fifteen years before becoming Acting Keeper in 1922, he had the advantage of being thoroughly familiar with the place, its needs and problems.

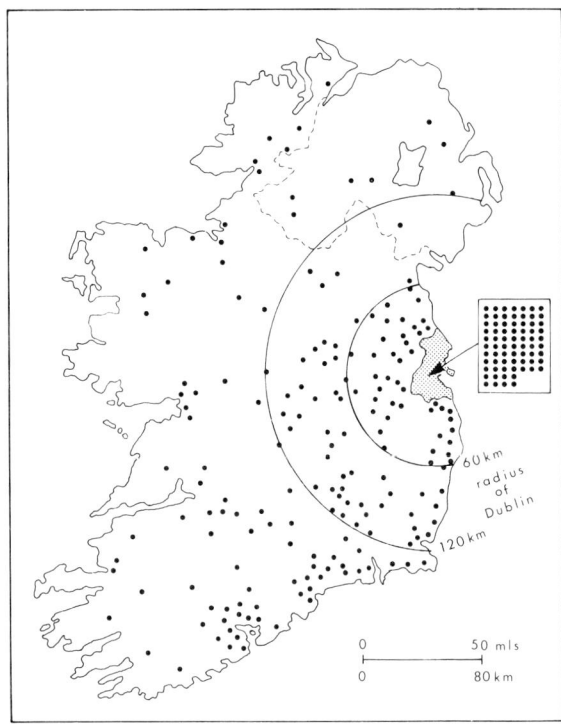

128. Map showing the home towns of apprentices who trained at the Botanic Gardens before 1932

The Botanic Gardens was seriously undermanned at this period, but Besant was able to maintain the high standards his predecessor had set, and he successfully took the Gardens through the 1920s and 1930s, when the political and economic dispute between the Irish Free State and the United Kingdom greatly hindered the commercial development of the country, and restricted the amount of money available for institutions like the National Botanic Gardens.

Besant continued as best he could to increase and diversify the plant collections and was closely involved with the introduction into Ireland of many new plants from the Far East. When he took over as Acting Keeper, he began listing the plants already growing in Glasnevin, and this work was still proceeding slowly in 1925. In his report for the year ending in March 1925, Besant noted that one of the problems in cataloguing the collections was that many of the newly introduced species were still only labelled with their collection number. Without the skilled assistance of a taxonomist, the naming of these new plants was proving difficult. He returned to this problem in a lengthy memorandum[7] to the Department of Agriculture in October 1927. After pointing out the differences between a botanical garden and a public park, and underlining the educational and research roles of the Glasnevin Gardens as he saw them, Besant stressed that the senior staff must be skilled in plant identification as well as in practical horticulture. He reiterated his opinion that the Botanic Gardens was understaffed; work often had to be postponed due to the illness of gardeners or foremen. One of the tasks that sometimes had to be left incomplete was the keeping of plant records. In Moore's time, the Assistant Keeper—Besant himself—had supervised this work, which was carried out by the plant collector and clerk. But the assistant keepership had been vacant since Besant's promotion, and in September 1923, David McArdle, the clerk, had retired. While William Rose was made temporary plant collector and clerk on 8 November 1923, he does not seem to have been able to carry out the complicated task of keeping the records, which involved checking cross-references and looking up synonyms. In his annual report for 1926-1927, John Besant noted that record keeping would have been at a standstill for the past five years but for the help of Elsie Miller, who had done most of the work in addition to her normal duties as secretary. Rose's appointment as plant collector was confirmed in April 1927, but thirteen months later he was dismissed. No replacement was appointed until September 1929, when Daniel Egan joined the Gardens—he remained in the post until 27 June 1945.

129. William E. Trevithick. (Reproduced by courtesy of Royal Botanic Gardens, Kew)

The training of young men and women for careers in horticulture continued to be a vital part of the day-by-day work in the National Botanic Gardens after the establishment of the Irish Free State. Most of the apprentices entered Glasnevin at eighteen to twenty years of age, after they had gained some practical experience in private gardens or nurseries. About two-thirds of the young men who served apprenticeships in the Gardens before 1932 came from provincial towns and villages;[8] the other third were from Dublin (Fig. 128). When they had completed their course, the majority left Ireland—this had also been the pattern in Moore's time—and in his annual report for 1932-1933 John Besant had pleasure in recording some of the achievements of former students. Brendan Mansfield, soon to become his son-in-law, was Superintendent of Parks and Reserves in Invercargill, New Zealand; he had worked as garden boy and apprentice at Glasnevin and later became Director of Christchurch Botanic Gardens. George Farley, who had enrolled as a student gardener in January 1924, was Superintendent of the Government Garden in Madras, India; Denis Grogan, a pupil between January 1925 and March 1927, was then studying in New York Botanic Garden.

Another former pupil, William Edward Trevithick (Fig. 129) was working in the Royal Botanic Gardens, Kew. His father had been the head gardener at Headfort, near Kells in County Meath, and the young Trevithick received practical training there and at Belvoir Castle, home of the Duke of Rutland, in Leicestershire. William Trevithick returned to Ireland in April 1916 to become a student gardener in the Botanic Gardens. When his apprenticeship was completed on 31 March 1920, he moved to Kew, where he was promoted to sub-foreman before entering the herbarium as a technical assistant. Trevithick was a competent photographer and a skilful artist; he drew about seventy plates for the *Botanical Magazine*, including a number illustrating plants supplied by the Marquis of

130. Andy Doyle at work in the Cactus and Succulent House in 1930. (Reproduced by courtesy of B. O. Mulligan)

Headfort—*Lilium formosanum*, *Berberis vernae*, two conifers (*Tsuga chinensis* and *Abies forrestii*) and a tulip (Plate 15) being a few examples.⁹

One of the lady gardeners during Besant's keepership was Charlotte (*alias* Charles) Lane-Poole, whose father held the post of Commonwealth Inspector-General of Forests in Canberra. She was related to the Yeats family, and Miss Lane-Poole lived with Lily and Lolly Yeats, sisters of the poet William Butler Yeats. Charlotte Lane-Poole worked at Glasnevin between March 1935 and March 1936 and, having a great interest in trees, she occasionally visited the larger Irish gardens with Mrs Alice Henry, Elsie Miller and other lady gardeners.

Two men, David McArdle and William Pope, retired during John Besant's keepership, bringing to an end their families' links with the Gardens which had been unbroken since the early nineteenth century. David McArdle was the first to depart; he retired on 29 September 1923 after fifty-four years in the Botanic Gardens. His father, Patrick McArdle, had come to work as a labourer in Glasnevin in 1829, when he was twenty-one years old; he was promoted rapidly, and after only twelve years became outdoor foreman. One of his duties was to keep the meteorological records at the Gardens. Patrick McArdle retired on pension in 1869, at the unusually early age of sixty-one, but for the next eight years he continued to do work for the Gardens, writing labels for the plants—one of the entries on the last page of the minute book of the Royal Dublin Society's Committee of Botany, that for 25 June 1877, records the payment of £3 18s. 6d to Patrick McArdle for lettering.

David McArdle was born to Patrick and May McArdle on 28 November 1849, and started work at Glasnevin when he was twenty, the year his father retired. Within two years David had been appointed plant collector and clerk, and was put in charge of the seed exchange and the library. He became a competent cryptogamic botanist and travelled throughout Ireland collecting mosses and liverworts. In later years his research was partly supported by grants from the Royal Irish Academy, in whose *Proceedings* he published a check-list of Irish Hepaticae in 1904, still an important source of information on native liverworts, and in 1917 an account of the mosses and liverworts of the Glen of the Downs, County Wicklow. Patrick McArdle, who died on 10 November 1883 and was buried in the graveyard of Saint Mobhi's Parish Church, Glasnevin, and David McArdle, who died at Ilford in Essex on 2 June 1934, between them served in the Botanic Gardens for a total of ninety-six years.

According to one source,[10] when William Pope joined the staff on 19 May 1853, his father and grandfather were already working in the Gardens. But, there were no Popes included on the Gardens' payroll in 1836 and we can trace no surviving record of the original Pope; only Patrick Pope was listed as a gardener in 1841. His son, William, who was born in January 1834, is said to have been a shrewd judge of the weather and could forecast it with uncanny accuracy. For several years he worked as Gardens constable on Sundays, and, on 1 January 1869, was appointed indoor foreman. William Pope's particular skill was the cultivation of orchids, and it is claimed[10] that plants he raised were sent to Charles Darwin, but there is no evidence to support this claim in the lists of plants despatched from Glasnevin. Pope was a tireless worker and often ignored the 6 o'clock bell that marked the end of each working day. He left Glasnevin on 29 November 1899, and enjoyed sixteen years of retirement before dying on 22 December 1915.

William Pope's son, Patrick, born on 8 April 1869, joined the Gardens' staff on 15 January 1883, and, in October 1887, was promoted to the post of propagator. Twelve years later, Paddy Pope succeeded his father as indoor foreman, and he retained that job until he retired on 8 December 1934. As Paddy Pope worked in Glasnevin for over fifty years, the four generations of Popes must have spanned a full century, if not more.

Although John Besant did not have the Moores' passion for orchids, the tropical collections were certainly not neglected during his term (Fig. 130). Besant's interests were in hardy plants, shrubs, herbaceous perennials and alpines, and his special pride was the Alpine House. The first Alpine House had been built in 1907, but was replaced in 1930 by a new wooden greenhouse, measuring thirteen feet by thirty-three feet, with a ridge height of ten feet. This was kept full of choice alpine plants, each one grown in a pot (Fig. 131). The rock-garden was replanted and also partly rebuilt in the 1930s under Besant's guidance (Fig. 132).

One of John Besant's first tasks as Acting Keeper was to accept an outstanding collection of dwarf and slow-growing conifers presented by Murray Hornibrook, a resident magistrate, who lived at Knapton near Abbeyleix. Hornibrook decided to leave Ireland on the foundation of the Irish Free State, but he left most of his plants with the National Botanic Gardens. Parting with this collection was a great wrench for him, as he had painstakingly built it up over many years.[11] However, Hornibrook was content that the best plants should remain in Ireland, and his gift meant that Glasnevin then possessed the finest collection of dwarf conifers in the world—many of Hornibrook's plants still grow in the National Botanic Gardens but are no longer dwarf trees (see Figs. 132, 133). Some of the specimens were unique—the only plants of certain cultivars—and had not been propagated, so Hornibrook asked Besant to arrange for this to be done at Glasnevin, and to send him some of the young conifers for his new garden. Before Murray Hornibrook left Ireland he invited Besant to come to Knapton and look at the other plants in the garden, so that he could select those required for the Botanic Gardens. Besant obtained numerous alpine plants and some unidentified shrubs that Hornibrook had raised from Chinese seed.[12]

Other gifts of lesser significance came to the Botanic Gardens in the 1920s, thereby enhancing the collections and indicating that the National Botanic Gardens under Besant retained its high international standing. Baron Rothschild of Tring, in Hertfordshire, bequeathed part of his collection of *Iris* cultivars. These were planted along the Vine Border in a series of display beds together with other herbaceous perennials. Then, as today, *Helleborus, Aster, Paeonia* and *Iris* cultivars were grown in groups so that visitors could see the range of fine plants available. The Honourable Vicary Gibbs, whose garden at Aldenham House near Elstree in Hertfordshire was famous for Michaelmas daisies and shrubs, donated many fine young trees and shrubs to Glasnevin in the 1920s. A fourth benefactor was Lieutenant-Colonel Sir George Holford, who gave a collection of *Hippeastrum* cultivars and Javanese species of *Rhododendron* from his garden at Westonbirt in Gloucestershire.

The first four decades of this century were particularly exciting for gardeners in Britain and Ireland because large quantities of seeds reached them from collectors working in remote areas of the Far

131. The display of alpine and rock plants in the Alpine House about 1925. (National Botanic Gardens, Dublin)

East. The triangle of Tibet, Burma and western China, a region of high mountains and deep valleys, was the most productive hunting-ground and there the collectors Frank Kingdon Ward, George Forrest, William Purdom, Roland Cooper and Reginald Farrer worked. Glasnevin received many packets of seeds from these men, and the seedlings raised were treated with great care. Thousands of plants were introduced at this period, ranging from tiny gentians, through herbs including many fine poppies, to roses, rhododendrons, maples, pines and junipers. So much new material was available at Glasnevin that new shrubberies had to be created, and these were filled almost entirely with species from the Orient. Many of these plants had no names, but when they flowered for the first time, herbarium specimens were collected and sent to botanical gardens in Edinburgh and Kew for identification. If an exciting new species flowered at Glasnevin, Besant sent fresh specimens to Kew for painting, and these illustrations were published in the *Botanical Magazine*. Like Frederick Moore, John Besant was keen to see Glasnevin represented in this prestigious periodical.

Glasnevin's other link with the Far East was through that great plant collector, Augustine Henry, who had been appointed Professor of Forestry at University College, Dublin, in 1913. Dr Henry continued to use the facilities at the Botanic Gardens for his research, and he regularly brought students to Glasnevin to study the living trees. After an association with the Gardens lasting nearly a quarter of a century, Augustine Henry died on 23 March 1930.[13] Dr Henry's widow, Alice, donated his extensive herbarium of tree specimens to the National Botanic Gardens, and this was housed in a room specially set aside in one of the gate-lodges. The herbarium comprises specimens obtained by Augustine Henry and his patron, Henry Elwes, from gardens throughout Ireland and Britain between 1903 and 1913, when they were working on the monograph *Trees of Great Britain and Ireland*, and many other specimens that Dr Henry had collected in later years. Mrs Henry spent several years putting the herbarium specimens into botanical order. She also gave fifty volumes from her late husband's library to the Gardens, including some of the books he had used while in China between 1881 and 1900, his personal copy of *Trees of Great Britain and Ireland* which contains copious annotations, and his Chinese diaries and note books.[14]

These books, manuscripts and herbarium specimens form a remarkable archive of the work of a man whose influence on botany and horticulture was profound. In the year before he died, Chinese

132. The rock garden and (on left of path) the Hornibrook collection of dwarf and slow-growing conifers, photographed about 1925

133. *Cedrus libani* 'Comte de Dijon', one of the dwarf conifers donated by Murray Hornibrook, photographed about 1965. This plant appears in Fig. 132, centre front, on the low mound of the rockery. Thomas Crawford, Curator of the National Botanic Gardens from 1965 to 1985, is seen on the left.
(National Botanic Gardens, Dublin)

134. Garden staff in 1928: 1 P. Doyle; 2 A. Doyle; 3 T. Fagan; 4 G. Greene; 5 B. Bacon; 6 P. Lowry; 7 M. Jordan; 8 F. Barry; 9 L. Smyth; 10 M. Doherty; 11 J. Woods; 12 P. Daly; 13 B. Usher; 14 P. Emmet, 15 D. Lyons; 16 E. Byrne; 17 T. Goggins; 18 J. Harris; 19 A. McCormack; 20 J. Byrne; 21 J. Kenny; 22 J. McCann; 23 J. Butler; 25 F. Parnell; 26 T. Darby; 27 S. McCormack; 28 P. Maher; 29 T. Brady; 30 N. Lynch; 31 B. Lacey; 32 J. McDonald; 33 T. Hearne; 34 T. Melia; 36 J. Parkes; 37 P. Burke; 38 O'Brien; 42 Mick Reilly; 44 J. Friel; 45 J. Bean; 46 J. Greene Jnr; 47 Bob Marks

botanists had honoured Dr Henry (Fig. 135) by dedicating the second volume of *Icones Plantarum Sinicarum* in these terms:

> To Augustine Henry through whose assiduous botanical exploration of central and south-western China the knowledge of our flora has been greatly extended.

John Besant also paid tribute to Henry, remarking that if he had 'done no more than make known the marvellous riches of China he would have achieved more than most men, and happily so many plants bear his name that while trees and shrubs are cultivated, his memory will remain in every garden and arboretum for long years to come'.[15] Five years after Dr Henry's death, Besant wrote a short article in which he recorded some of the plants named for Henry or discovered by him—*Plantae Henryanae*—that were then growing in Ireland, and in particular in the garden of Mrs Henry's home in Ranelagh.[16]

The flow of seeds from the Far East did not abate during the 1930s, and the National Botanic Gardens continued to benefit from innumerable novel introductions. The Gardens had not the financial resources to subscribe towards the expenses of plant-hunting expeditions, but several keen gardeners, most notably Lord Headfort and Lord Rosse, did. The many new plants continued to present Besant with problems—there was very little space available in existing shrubberies, so another new shrub border was created for Chinese plants. *Rhododendron* seeds from the collections of Joseph Rock, George Forrest and Frank Kingdon Ward were raised in abundance not only at Glasnevin but also by Lord Headfort at Kells and Richard Grove Annesley at Castletownroche in County Cork. Just one example of the wealth of material need be given; in 1934, over three hundred packets of *Rhododendron* seeds from Joseph Rock alone were sown in the nursery at the Botanic Gardens. Lord Headfort and his

135. Augustine Henry, a portrait in oils by Celia Harrison now in the National Botanic Gardens, Glasnevin

fellow enthusiasts regularly gave surplus seedlings to Glasnevin. The lime-rich soil in the Botanic Gardens is generally unsuitable for such plants as *Rhododendron*, but several beds of rhododendrons were planted in the Mill Field and were reasonably successful. However, Besant preferred to follow Moore's example and sent duplicate seedlings to more suitable gardens; in this way rhododendrons went to Muckross near Killarney, Annes Grove at Castletownroche, Greenfields in County Tipperary, and Headfort. Many smaller gardens also benefited through this exchange and distribution arrangement.[17]

In 1938 the first consignment of seeds collected by Yu Dejung, a native Chinese botanist, arrived in Ireland and proved to be most interesting. Yu had been engaged by Dr Hu Hsien-Hsu, Professor of Botany in the Botanical Institute, Beijing, to gather seeds for a group of Irish and British horticulturists. Such an expedition had been suggested by the Earl of Rosse following his honeymoon in China in 1936, when he had met Professor Hu.[18] Yu Dejung visited areas of western China, some of which had not been searched before, and obtained seeds of many new plants of horticultural merit, and, most importantly, plants that were to prove hardy in Irish gardens. Yu's seeds were germinated at Glasnevin and Birr Castle, and the fine shrubberies in both gardens now contain mature plants bearing Yu Dejung's collecting numbers.

Although he was always delighted to obtain seeds and plants from such expeditions, John Besant remained as cautious as ever. In 1938 he wrote:

> Practically every year expeditions are ransacking the less known parts of the world [and] at the present time many thousands of seeds are germinating...[But] it must be understood that years often elapse before the plants grown from these seeds are big enough to put out into the public collections.[19]

By the autumn of 1942, some young shrubs raised from Yu's seeds were ready for planting out, although most of the plants had not been named. There were no fewer than six species of *Berberis* and four different roses, as well as species of *Philadelphus, Prunus, Lonicera* and at least one rare maple (*Acer pentaphyllum*). In the following autumn more barberries and roses, and unidentified species of *Jasminum, Astilbe* and *Sophora* joined the plants already growing in the Chinese shrubbery.

John Besant was a meticulous Keeper. Every autumn he prepared detailed lists of plants to be removed, or pruned, or planted. Some of these planting lists survive in the archives at Glasnevin, and they provide an insight into the work carried out each year during the latter half of Besant's term.[20] Useless plants were burnt, surplus ones were given away, and only those required to enhance the collections were despatched to the shrubberies for final planting. Besant carefully noted those plants that had to be identified, and any labels that were missing or incorrect. In 1936, a surplus *Buddleja davidii* was marked 'Burn', whereas a plant labelled '*Ilex fragilis*' was recorded as 'Wrong, keep for naming'. In 1937, he decided that the bog-beds in the rockery had to be remade, and that new scree beds should be constructed nearby. Besant's notes indicate that in the *Clematis* collection there were several plants without labels, some of the herbaceous *Clematis* were not named, and Besant wrote, 'Where is *Clematis orientalis, C. tangutica* and *C. obtusiuscula*?' His instructions about sickly plants were that they should be fed or lifted to the nursery, and, as in previous years, plants of no merit were to be removed and destroyed.

His attention to detail is also seen in the paper that he published in the *Journal of the Royal Horticultural Society* in 1942 on the use of standardized composts.[21] John Besant suggested that while such mixtures as John Innes composts were all very well for the many nurseries and public gardens which grew only large quantities of a few different plants, the use of such composts in botanical gardens was most inappropriate. As each species needed careful individual attention, Besant argued that it would require courage to change over to these standardized mixtures. So many plants deserved special treatment, especially those from southern Australia and South Africa, that these composts should not be used in botanical gardens.

As an author, John Besant was prolific. He contributed brief notes and longer articles to many of the leading horticultural periodicals in Ireland and Britain, including a regular series 'Notes from Glasnevin' in the *Gardeners' Chronicle*. While still outdoor foreman at the Gardens, Besant took over the editorship of *Irish Gardening* following the tragic death of C. F. Ball in 1915, and he edited the magazine until it ceased publication in April 1922. One of John Besant's more substantial essays was written for a book, published in 1928, about Edward Walpole's magnificent garden at Mount Usher in County Wicklow. This elegant volume only contains twenty-nine pages of text and Besant's contribution occupies more than a third of those pages. The other contributors were Sir Frederick Moore, who recorded his recollections of Mount Usher,[22] and Frederick William Millard, an alpine enthusiast from Sussex, who wrote five pages on his impressions of the garden. Besant's lucid chapter described the plants then growing in Mount Usher. *Mount Usher 1868-1928* was edited by the owner, E. H. Walpole, and included some photographs by Elsie Miller, as well as maps and reproductions of paintings by Letitia Hamilton and J. E. Hodgkin.[22]

It is proper that we should acknowledge John Besant's article on the origin, history and development of the National Botanic Gardens, which he published in the *Journal of the Department of Agriculture for Ireland*.[23] It was reprinted and made available as a guide to the Gardens. This short article is not a scholarly history of Glasnevin, for it relies on the brief historical précis that prefaced the guidebook prepared by William McNab in 1885. Besant therefore made no useful contribution to the historical bibliography of the Botanic Gardens. He recounted the difficulties that the site presented; the poor soil underlain by gravel, the exposure to wind, and the low rainfall. But what comes across clearly in this article is the pride John Besant had in the Gardens his predecessors had created, and in the marvellous collection of trees, shrubs and herbs which was renowned throughout the world.

The highest possible standards of horticulture were maintained by Besant despite the undermanning and restricted budgets, and the excellence of Glasnevin continued to impress its many visitors. In July 1931, Sir Arthur Hill, Director of the Royal Botanic Gardens, Kew, paid this tribute to Glasnevin:

> ...I have had the opportunity of paying two long visits to the Gardens, under the able guidance of Mr J. W. Besant and Sir Frederick Moore, to whom the botanical and horticultural world owes so deep a debt of gratitude...I have been [greatly impressed] by the admirable condition of the Gardens and by the great interest and value of the collections at the present time. In renewing my acquaintance with this beautiful domain I am particularly pleased to see that your Government are fully aware of the importance of the Gardens, and I know I am expressing the opinion of botanists and horticulturists in the British Isles, and throughout the Empire, in saying how grateful we are to your Government for maintaining the Glasnevin Botanic Gardens in such beautiful condition and in so high a state of efficiency. Long may this recognition of the scientific value of Glasnevin be continued.[24]

John Besant believed that a botanical garden was not simply a place in which pretty flowers were displayed for visitors to admire. In common with the Moores, he held the view that scientific research was an important function of such a garden, and although Glasnevin did not have scientists on its staff while he was Keeper, botanists from other institutes regularly used the living collections for their research. Dr Robert Lloyd Praeger, the foremost botanist in Ireland during the first half of this century, is best known for his evocative books about Ireland and its native plants and animals. What is not so widely known is that he studied *Sedum* and *Sempervivum*.[25] Plants that Praeger collected from wild habitats in southern Europe and on the Atlantic islands were grown for his work at Glasnevin. Using these living plants, as well as pressed specimens preserved in many herbaria, Praeger classified these two genera, describing and naming many new species. Sir Frederick Stern, an international authority on *Paeonia* and creator of the famous garden at Highdown in Sussex, was sent peonies from Glasnevin for his monograph.[26] William Stearn, at the time librarian in the Royal Horticultural Society of London, obtained *Epimedium* specimens for his research.[27] And, as has been mentioned, Professor Augustine Henry often used the facilities provided in the National Botanic Gardens for his research on poplars and larches in particular.

The plants growing in the Gardens were constantly in demand for teaching purpose. Specimens were cut for use by the Veterinary College, the Pharmaceutical Society and the departments of botany in the two university colleges. Although Trinity College still had its own botanical garden at Ballsbridge, there were some plants only available in Glasnevin. University College, Dublin, the successor to the Royal College of Science, had no botanical garden. A supply of flowers was sent regularly to the School of Art for drawing classes. Decorative plants were also supplied for state functions in Dublin Castle, and for special events such as the bicentenary celebrations of the Royal Dublin Society in 1931, and the Eucharistic Congress in 1932.

John Besant's contribution to the progress of horticulture in Ireland was acknowledged in 1939 by the Royal Horticultural Society of Ireland. Along with Sir Frederick Moore, the Marquis of Headfort, E. H. Walpole of Mount Usher, and that great gardener from Rowallane, Hugh Armytage Moore, Besant was one of the first recipients of the Society's Gold Medal of Honour. He was chairman of the Society's Council in 1940 and 1941, and also served for many years as its honorary secretary. The Royal Horticultural Society of London made him an Associate of Honour in 1931, and he was President of the Dublin Naturalists' Field Club and of the Kew Guild.

On 18 September 1944 John Besant died at Glasnevin. Sir Frederick Moore paid tribute to his successor, describing him as 'a man of modest, reserved nature, noted for his sincerity and uprightness of character, his shrewd and kindly advice was freely given, especially to his pupils in whose work and careers he took great interest'.[28] His son-in-law and former pupil, Brendan Mansfield, recorded that John Besant had shouldered the full burden of reponsibility for the Gardens alone for almost twenty-two years, as no assistant keeper was appointed until six months before his death. Yet the National Botanic Gardens prospered under his care and was his 'living monument'.[29]

John William Besant played an important part in the Gardens' development, and had maintained its exemplary standards during a time of change and uncertainty. It was no small achievement to have followed Sir Frederick Moore and to have received the unreserved praise of that doyen of Irish gardeners.

By 1944 nearly a century and a half had elapsed since the Glasnevin Botanic Gardens was established, and in that time traditions had evolved. One such tradition was maintained by John Besant throughout his years as Keeper. Every year, at the first stroke of midnight on New Year's Eve, the bell on the wall of the Great Palm House was rung to herald the New Year. This little ceremony was attended by the Keeper and his family, and by the stoker who was on duty that night keeping the furnaces alight.[30] It is sad that this pleasant tradition has been abandoned. There is now no need to stoke the furnaces as they burn natural gas, and the old bell is only rung to signal the end of the working day and closing time for visitors.

In April 1944, Thomas James Walsh (Fig. 135) had been appointed Assistant Keeper, and he took over as Acting Keeper on the death of John Besant. It fell to Tom Walsh to cope with the legacy of problems caused by war and weather. Ireland had remained neutral during the 'Emergency'—a euphemism for the Second World War. The National Botanic Gardens continued to function, but some of its activities had to be curtailed. The only source of living plants was Britain, and it is a remarkable thing that cuttings and plants were frequently exchanged between Glasnevin and gardens throughout Great Britain during the 'Emergency'. All botanical gardens try to maintain exchange programmes, particularly by sending seeds collected from their own plants to sister gardens throughout the world, and by receiving other seeds in return. In 1938 and 1939 over 5,200 packets of seeds were sent from Glasnevin, and about 1,700 packets were received—the imbalance is of no importance but the international contacts are vital. After the outbreak of hostilities, the number of packets despatched from Glasnevin dropped to about 2,200, and approximately 1,000 were received in return. The figures continued to decline. In 1944 only 1,500 packets were sent and only 500 received, mainly from gardens in Britain and the United States of America. As the conflict drew to a close, Glasnevin made increased collections of seeds to help the continental botanical gardens and, by March 1946, over 7,000 packets had been despatched.

The number of visitors also fell from approximately 320,000 in 1939 to 245,000 in 1945. This decline was due to transport difficulties within Ireland, and to the impossibility of foreign travel. But, as the Botanic Gardens did not depend on visitors for income—no entrance fee was charged—the drop in numbers had no detrimental effects.

There was a fuel shortage during the 'Emergency'. The furnaces that provided heat for the glasshouses were used as little as possible. In November 1944, just after Tom Walsh became Acting Keeper, the shortage of coke, made from imported coal, became acute and turf from Irish bogs was substituted. This meant that the temperature in the houses could not be kept at the required levels, and in the severe weather of January 1945 many tender tropical plants suffered. Another problem that had to be overcome was a lack of *Osmunda* fibre for growing orchids—an unsatisfactory substitute was tried but despite this it was possible to maintain the orchid collection in a tolerable state without the fibre.

The frosts in January 1945 were as severe as those of January 1940, which had killed many plants that were growing outdoors. Australasian species and those from the western coast of North America were particularly badly affected, but fortunately duplicates were available. During the following months many dead and redundant trees and shrubs were removed to make way for new plantings in the spring of 1945.

Thomas Walsh was born on 8 October 1902 in Clogher, County Tyrone. His father was the head-teacher in the local national school and owned a small farm outside the town at Slevin. Tom Walsh went to a local primary school and later attended St Malachi's College in Belfast. In 1922, following the partition of Ireland, the Walsh family moved to Dublin, to a farm at Cypress Grove in Templeogue. For ten years Tom Walsh helped to run this farm and he also gave riding lessons.[31]

In 1932, when he was thirty, Thomas Walsh entered University College, Dublin, to study horticulture. During the vacations he went to various nurseries to gain practical experience—among these were McGredy's at Portadown in Northern Ireland, Frederick Secrett's at Walton-on-Thames, Seabrook's at Chelmsford, and, like his predecessor Sir Frederick Moore, Louis van Houtte's nursery in Ghent. Walsh was an excellent debator and won four gold medals for oratory during his student days.[31] He graduated in 1939, and after a short spell at Kew, was appointed an inspector in the Irish Department of Agriculture.

Soon after his arrival in Glasnevin, Thomas Walsh began cataloguing the specimens in the Henry Forestry Herbarium (Fig. 137). Mrs Henry had already sorted the specimens into genera and species, and she gave Walsh valuable help in his work. In 1950, the University of Dublin awarded Thomas Walsh the degree of doctor of philosophy for the catalogue, and Mrs Henry attended the graduation ceremony (Fig. 136). Seven years later, the catalogue of Dr Henry's specimens was published,[32] and to mark Alice Henry's work and her gifts of the herbarium and books, a plaque was placed in the room housing the Henry Forestry Herbarium.

The post of Assistant Keeper, which again fell vacant on Walsh's appointment as Keeper, was not filled until 1949, when John Fanning returned to the Botanic Gardens. Fanning, a native of Ahullen,

136. Dr Thomas Walsh and Mrs Alice Henry on the occasion of the conferring of the degree of doctor of philosophy by the University of Dublin in 1950. (Photograph by courtesy of Dr M. Walsh)

County Wexford, had enrolled as a student gardener in Glasnevin on 8 October 1940, at the age of eighteen. Later, he moved to University College, Dublin, where he studied horticulture, and after graduating, joined the Department of Agriculture in 1945 as a lecturer. For four years he was based in Johnstown Castle, near Wexford, but went back to Glasnevin in 1949 when he was engaged as assistant to Thomas Walsh. During his time in the Gardens he took a special interest in daffodils.

John Fanning (Fig. 138) became well-known throughout Ireland as a broadcaster on gardening and horticulture, and he wrote many articles for magazines and newspapers. In 1963, Fanning was promoted within the Department of Agriculture, becoming a senior inspector, and he left the Botanic Gardens. But he retained his concern for and interest in the Gardens, and remained its very good friend and helper until his early death at the age of forty-nine on Christmas Eve 1971.[33]

In the same year that John Fanning was appointed Assistant Keeper, 1949, Frederick Parnell (Fig. 139) retired from the job of outdoor foreman. Frederick Parnell's father, William, had also been outdoor foreman from 1869 until 1906, and father and son together had served in Glasnevin for one hundred and one years. Freddie Parnell's departure thus severed the last human link with the Glasnevin of the nineteenth century. William Parnell had joined as an apprentice in 1848, and Frederick, born on 11 August 1882, began his four-year apprenticeship half a century later, in May 1898. When his time as an apprentice ended in June 1902, Freddie Parnell went as a student gardener to the Royal Botanic Garden in Edinburgh, but is said to have been so homesick that he came back to Dublin after just three weeks![34] He tried working for Lord Headfort at Kells, but did not stay long in that post, and, with his sister, he emigrated to the United States. Once again Freddie Parnell was driven back to Dublin by homesickness. In July 1904 he was re-employed at Glasnevin, and eventually became gardener in charge of alpine plants. On 23 September 1920, Parnell succeeded John Besant as outdoor foreman and held that job until his retirement on 29 June 1949. Freddie Parnell was a great story-teller, and he amused the other gardeners with what must have been purely fictional tales of his adventures. He told of his infatuation with a Russian princess, of running a beer garden in Austria, of captaining an emigrant ship, and of his wasted years studying for the ministry in the Anglican Church![35]

Frederick Parnell was succeeded as outdoor foreman by Brendan O'Reilly, who only remained in that post from November 1949 until March 1951. In turn O'Reilly was succeeded by Thomas Crawford on 1 May 1952. The foreman in charge of the glasshouses then was Robert Ernley ('Bob') James, who had come to Glasnevin as propagator on 31 August 1939. Following the retirement of Michael Conway, James was promoted to indoor foreman in February 1942. The post of propagator

137. The Henry Forestry Herbarium as housed in the National Botanic Gardens, Glasnevin (photographed about 1955). The portrait of Dr Henry is reproduced in Fig. 135. (National Botanic Gardens, Dublin)

138. John Fanning. (National Botanic Gardens, Dublin)

139. Grand Old Men in the National Botanic Gardens, 1947, (left to right): Mick Doherty, Thomas Goggins, Frederick Parnell, Andy Doyle and Tommy Brady. These men had served in total more than 220 years when the photograph was published in *Feature Magazine*. (By courtesy of the National Library of Ireland)

was given to James McCann, but in December 1949 the title was changed from propagator to nursery foreman.

In 1950 several innovations were made in the Gardens. Horses were replaced by a tractor with a three-way tipping trailer. The South Field, part of which was formerly used for grazing the Gardens' horses, was wholly converted into a nursery. Selective weed-killers were first used as a matter of routine in 1950, while steam sterilization of soil was introduced in the following year, 1951.

Replenishing the collections was in full progress by the early 1950s. In April 1950 the Thomas Moore Society presented the 'Last Rose of Summer' to the Gardens. The particular bush had been raised from a cutting taken off the rose growing at Jenkinstown House, County Kilkenny, that was said to have inspired Thomas Moore's famous ballad. The 'Last Rose of Summer' is one of the cultivars of the China roses (*Rosa* 'Old Blush') introduced into Europe at the beginning of the nineteenth century. The rose was planted near the main entrance to the Gardens, and was surrounded by a fence — it is still there and is a major attraction for visitors.[36]

In 1951, over forty new orchids were received from Gothenburg Botanic Gardens, and during the next twenty years the orchid collection was improved by purchases and by gifts from private growers, including Maurice Mason of King's Lynn and Ambrose Congreve of Waterford. The botanical gardens in Montreal, St Louis and Ghent also donated plants. Dr Walsh did much to improve the quality of the orchid collection by visiting nurseries and gardens in Britain and through his personal contacts with enthusiastic amateur growers.

By 1953 Walsh could report that the plant losses sustained during the frosts and the 'Emergency' had been made good, and throughout the remaining fifteen years of his keepership, Walsh maintained the plant collections at a highly creditable level. The Gardens were well-stocked and there was little room left to accommodate new plants by the 1960s. The trees planted in the Far Grounds when Frederick Moore was Keeper were mature, and the shrubberies that John Besant had created with

140. Model of a proposed fern house for the National Botanic Gardens, dated June 1960—this was not built.
(National Botanic Gardens, Dublin)

Chinese plants were coming into their prime. No substantial new collections were begun by Dr Walsh, but he did make several much-needed changes in the Gardens.

The area surrounding the Alpine House was remodelled between 1947 and 1949. The decorative stone arch at the entrance to the Alpine Yard and the facing of the Victoria (Aquatic) House were created by Robert Bacon, a skilled stonemason, under the supervision of the architect to the Office of Public Works, Christopher Pemberton (1884-1953).

In 1954 an old cottage, once occupied by the Parnells, on the north bank of the River Tolka, was demolished and work started on forming a rose garden on the site. The new Rose Garden was opened to the public in 1960, and now contains a small but representative collection of modern roses, including many raised by famous Irish rose-breeders. Sam McGredy of Portadown gave a valuable collection which formed the original planting in the garden and which included a cultivar that was named 'Glasnevin'—this rose, alas, has not survived in cultivation. Today, Dicksons of Hawlmark and Sean McCann of Dublin are represented in the rose garden, as well as foreign rose breeders. The old Rose Garden, with its semi-circular pergola and beds, was converted into a display area for half-hardy annuals.

The Fern House (Fig. 141), erected in 1887 when the Octagon House was demolished, had become unsafe by 1949 and was closed to the public. Although plans for a replacement were prepared early in 1952 and revised in 1953, a new fern house was not built until the mid-1960s. A photograph exists, dated June 1960 (Fig. 140), of a model of the proposed replacement—it was a striking glass pyramid that, had it been built, would have been remarkably innovative. To provide temporary accommodation for the tree-ferns displaced from the old crumbling Fern House, a fine teak house was erected near the entrance to the nursery. In 1966, the present incongruous aluminium and glass Fern House was completed and opened to the public (Fig. 142). This tall, gaunt glasshouse is uncompromising and dull, quite out of character with the lower Aquatic House beside which it stands. The new Fern House is divided into three compartments; the largest is for tree-ferns, a smaller one for tropical species, and a locked section in which it was hoped the Killarney fern might flourish.

141. A view looking towards the first Fern House, taken by W. Lawrence about 1900. (Reproduced by courtesy of the National Library of Ireland, Dublin)

142. The same view as Fig. 141 photographed by David Davison in November 1981 showing the modern Fern House completed in 1966

143. Students in the Botanic Gardens 1948-1949 (centre back row Aidan Brady, present Director of the National Botanic Gardens. (National Botanic Gardens, Dublin)

The need for a catalogue of the plants growing in the Gardens was still not satisfied when Thomas Walsh became Keeper. John Besant had tried to prepare one, but only the trees and orchids had been completely catalogued in his time. Under Dr Walsh a card index of the entire collection was prepared, and some of the plant names were brought up to date. The small herbarium, representing plants cultivated in the Gardens during Besant's keepership, was rehoused in 1958; few new specimens were added while Dr Walsh was Keeper. In the 1950s the library was also catalogued. It contained the beautiful illustrated books presented by William Gumbleton, sets of important horticultural journals and periodicals, including a complete set of the *Botanical Magazine*, and a fairly comprehensive collection of books on horticulture, gardening and botany. Some minor gifts of books were made during Walsh's term; for example, J. J. O'Connor of Merrion Square, Dublin, donated over sixty horticultural books.

In June 1962 a plan was put forward[37] for transferring the Botanical Section of the National Museum, then accommodated in the museum complex in Kildare Street, to the National Botanic Gardens. The amalgamation of the Gardens' own library and herbarium with the collections from the Museum would create a botanical research institute with a comprehensive botanical library, a world-wide herbarium, and a rich collection of living plants. These possibilities were discussed for many years, and eventually were accomplished in 1970 after Dr Walsh's retirement.

In the 1950s and 1960s the Botanic Gardens continued to train young gardeners (Figs. 143, 144, 145). Indeed, the educational facilities at Glasnevin were enlarged in the early sixties when the Department of Agriculture decided to transfer its training course from Johnstown Castle, Wexford, to the Botanic Gardens. The Gardens' apprenticeship scheme was abolished, and an integrated two-year course in horticulture was established in its place. As there was no lecturer in horticulture based in the Botanic Gardens, Aidan Brady took up duty in Glasnevin on 1 June 1960 and he organized the new instruction course. When John Fanning was promoted and left Glasnevin in 1963, Aidan Brady became the new Assistant Director — at the same time Dr Walsh's title was altered from Keeper to

144. First year students in 1959-1960 (back row from left, C. Crosbie, W. Hannon, T. O'Reilly, E. Paterson, W. Lacy, F. Hardiman, F. Nea; front row from left, J. McNally, J. Kilgallon, J. Niland)

145. Second year students in 1959-1960 (back row from left, J. Shiels, T. Bergin, M. Clancy, D. Dixon, S. Murphy, P. Delany, M. McNamara; seated from left, P. Brown, M. J. Loughnane, G. Nevin)

146. *Todea barbara*, the original plant, in the Fern House at the National Botanic Gardens, Glasnevin, 1970. (Photographed by Dr B. Morley, National Botanic Gardens)

Director. Thomas Walsh remained in charge at Glasnevin until his retirement in 1968, and Aidan Brady, the present Director, then succeeded him.

By 1968 the National Botanic Gardens had been in existence for 173 years, and the Trinity College Botanic Garden at Ballsbridge was 152 years old. But that year the College closed its garden because the lease was soon to expire, and began to move some of the plants to a much smaller experimental garden near Palmerston Park, and then gave up the land at Ballsbridge. The site of that memorable old garden is now occupied by two hotels. An Australian tree-fern (*Todea barbara*) which had grown in the College Botanic Garden since 1892, was donated to the National Botanic Gardens in October 1969, and was planted in the recently completed Fern House. That particular fern had been sent to the Ballsbridge Botanic Garden by Baron Ferdinand von Mueller, Government Botanist and Director of the Royal Botanic Gardens in Melbourne, to mark the tercentenary of the University of Dublin (Trinity College). Mueller estimated that the fern was as old as the College, and thus the plant (Fig. 146), which is still alive, is now perhaps four centuries old.[38] In 1995, when the National Botanic Gardens at Glasnevin will have been 200 years a-growing, it will only be half the age of the humble tree-fern that was already a century old when the Board of Trinity College decided, on 25 June 1687, to establish in Dublin Ireland's first botanical garden.[39]

Postscript

THE National Botanic Gardens at Glasnevin is now 190 years old. The present boundaries of the Gardens were fixed about a century ago when the Far Grounds and South Field were added, and the layout of the paths can be traced back to the 1850s and earlier. The glasshouses span more than 140 years. Of the plants now growing in the Gardens, most were planted within the past one hundred years—very few of the trees and shrubs are older than that. Today, the Botanic Gardens functions as a quiet oasis in the expanding conurbation of Dublin, a place for relaxation and recreation, but it also follows the ancient traditions of the physic garden in that it is a research centre and a teaching institution.

147. Aidan Brady, Director of National Botanic Gardens, Glasnevin
(photographed in the Victoria House, 1984)

Since the retirement of Dr Thomas Walsh in 1968, Aidan Brady, a native of Elphin in County Roscommon, has been Director. He was a scholarship student in Glasnevin from 3 September 1948 (Figs. 143, 147) and on completing the course in September 1949, entered University College, Dublin, whence he graduated in 1954! In 1965, the new position of Curator was created, and the outdoor foreman, Thomas Crawford, was promoted to fill it. Tom Crawford received his training in the garden at Howth Castle, and came to Glasnevin in 1952. After twenty years as Curator, he retired on 6 June 1985. The Curator is responsible for the day-to-day work in the Gardens and the three foremen,

148. *Columnea gloriosa* in the Botanic Gardens 1911 — a colour version, one of the earliest colour photographs taken in the National Botanic Gardens, is extant (a Lumière 'Autochrome' plate). This was probably photographed by C. F. Ball

who separately oversee the outdoor collections, the glasshouses and the nursery, are responsible to the Curator.[2] There are also eighteen fully-trained gardeners on the staff, six of them have the rank of assistant foreman.

The naming of the plants in the collections is a skilled task, requiring expert knowledge of botanical classification and the rules governing plant nomenclature. Until the late 1960s, the Director (or Keeper) and his assistant identified and named plants as best they could, and they relied on the botanists of the Royal Botanic Gardens in Edinburgh and Kew for help with the difficult groups, such as orchids and rhododendrons. It is important for a botanical garden to have all plants correctly and fully labelled, and the need for a taxonomist, based in Glasnevin, had been keenly felt for many years. There was considerable support for this view from John Fanning, who believed that a taxonomist attached to the National Botanic Gardens would not only provide a badly needed technical service, but could also carry out botanical research of international value and importance.[3] When the assistant directorship was abolished in 1968, the post of taxonomist was created and Dr Brian Morley was appointed. Brian Morley, a graduate of the University of Wales, had spent three years in Jamaica carrying out research on New World species of *Columnea*, a genus of tropical epiphytes which, incidentally, had interested John Besant (Fig. 148). Dr Morley continued working on the genus after coming to Glasnevin in February 1969, and raised several artificial hybrids.[4] Some of these were of horticultural potential, so he named them and introduced them into cultivation. Brian Morley moved to the Botanic Garden in Adelaide, South Australia, in 1975, and was replaced in August 1976 by Dr Charles Nelson.

The National Botanic Gardens is the only institution of its kind in Ireland and it provides plant identification facilities for the general public and other institutes. In 1969, agreement was reached between the Department of Education and the Department of Agriculture for the transfer of the herbarium and botanical books from the Natural History Section of the National Museum to Glasnevin;

149. Valerie Waters, one of the A.C.O.T. students, working in the rock garden; Pat Leonard, 1985

as has been noted, this had been suggested as early as 1962 and the idea was vigorously supported by John Fanning. The two botanists in charge of the Museum's collections, Miss Maura J. P. Scannell and Donal Synnott, moved to the Botanic Gardens, and, with the transfer complete in 1970,[5] Glasnevin regained its role as the principal centre for the study of Ireland's native flora.

The herbarium assembled in the National Museum was founded, in part, on collections originally housed in the Botanic Gardens, including those of Samuel Litton and David Moore[6] Some of Moore's specimens had remained in the Gardens, and these were incorporated into the National Herbarium when it was received at Glasnevin. This unified herbarium now contains the most important assemblage in Ireland of specimens of native flowering plants, gymnosperms and cryptogams, as well as extensive collections of specimens from Great Britain and from all other parts of the world.

After an interval of ninety years, research on the native flora is once more being carried out at Glasnevin. In 1973, Maura Scannell and Donal Synnott published a census catalogue of all native flowering plants, gymnosperms and ferns. It provides details of the distribution of these plants, bringing up to date the data that David Moore and Alexander More had included in their *Cybele Hibernica* in 1866[7]

Scientific research is not completed until it is published for the use of fellow scholars. In order to facilitate publication of the work of its own botanists, a journal, titled *Glasra*, was established by the National Botanic Gardens and published for the first time in 1976. This has appeared about once a year since then, and has included research papers on botany and the history of botany and gardening in Ireland. *Glasra* (an Irish word meaning verdure, vegetation, greens or vegetables) is distributed to botanical gardens and universities in many countries. One benefit which the National Botanic Gardens has gained by publishing *Glasra* is that it now receives about 200 botanical journals in exchange[8]

Every year a list of fruits and seeds gathered from the plants growing in Glasnevin is prepared and published. This *Index Seminum* (seed list) is sent to over 600 botanical gardens and research institutes worldwide. The exchange of seeds has been going on for many centuries between botanical gardens — an *Index Seminum* published by Halle Botanic Garden in 1815 is in the archives at Glasnevin — and without

150. *Zelkova carpinifolia*, one of the finest trees in the National Botanic Gardens, Glasnevin, photographed in November 1981 by D. Davison

the free international exchange, it would be very difficult for botanical gardens to maintain comprehensive plant collections.

Every botanical garden aims to display within its grounds and glasshouses as large a number of different species as possible. The plants are not all pretty: some are of interest only to scientists; some may be grown so that students can see living examples of particular plant families or of important economic plants. Thus a botanical garden functions as a plant 'zoo', where students may learn about the diversity of the plant kingdom, and where gardeners can get new ideas for their own gardens. Among the many plants in Glasnevin are a few that are now extremely rare in the wild, and these are given special care and attention. Indeed, the National Botanic Gardens contains a handful of plants that are not known to be in cultivation in any other botanical garden.[9]

Ireland's National Botanic Gardens continues to help enrich private gardens by introducing new cultivars and species, and by distributing these to gardeners and nurserymen. *Ulex europaeus* 'Strictus' was the first cultivar to be introduced from Glasnevin, and in recent decades other plants have been named and distributed. *Erica cinerea* 'Glasnevin Red', a chance seedling found in the heather collection, received an Award of Merit from the Royal Horticultural Society of London in 1968. About a dozen plants have been named after the Gardens, but only two are common in cultivation—*Solanum crispum* 'Glasnevin' and *Escallonia* 'Glasnevin Hybrid'. A new tassel bush with deep red markings on its catkins is shortly to be introduced into general cultivation from the Botanic Gardens, and it will be called *Garrya* × *issaquahensis* 'Glasnevin Wine'.[10] The National Botanic Gardens was also the birthplace of *Escallonia* 'C. F. Ball' and *Escallonia* 'Alice', and, as has been recorded already, the first European home of many exotic species, including the pampas grass (*Cortaderia selloana*), *Crinum moorei* from Natal, and the beautiful Chatham Island daisy-bush (*Olearia semidentata*).

In fulfilling its educational role, the training of young men and women as gardeners continues. Each year, about twenty-five students complete the three-year course of lectures and practical training which is organized by ACOT (An Chomhairle Oiliúna Talmháiochta-Council for Development in Agriculture). They depart to take up jobs in the larger private and public gardens, or to go on to university. And every year thousands of school children visit the Botanic Gardens to look and learn.

A casual visitor to the Gardens perhaps does not realise that these diverse activities are important aspects of the function of an institute of this kind. For such a visitor, the 'Botanics' is just another public garden, but one with marvellous carpet-bedding, a vegetable garden, fine glasshouses, a colourful rose-garden, and a monster of a water-lily. Yet, it is a traditional botanical garden, a place 'for Knowledge as for Pleasure made',[11] maintained by the Irish state for the benefit and enjoyment of all people, for the pursuit of knowledge, and as a living museum of plants. More than that, Glasnevin holds an historical collection of dried specimens, a library of great scientific and artistic importance, and an archive of Irish botany and horticulture.

Alle of the herbys o Ierlonde
Here thow shalt ham knowe eueri onde.[12]

The National Botanic Gardens, Glasnevin, was created by two men whose patriotism and personal ambitions drove them to form this bright jewel. It is an integral part of the cultural heritage of this island. Today, as in the days of Dean Swift and Dr Delany, the name of Glasnevin is recognized throughout the world. In that bygone era it was merely the home of a handful of savants, yet now it is the site of one of the world's greatest gardens.

Appendix 1

Sketches of the Medical Profession in Ireland
Dr Litton
Professor of Botany to the Royal Society of Dublin

THE GLASNEVIN BOTANIC GARDEN

UNTIL a recent period botany was scarcely considered one of the sciences in Ireland, and the low state into which its cultivation had fallen, might have well justified the estimate formed of its rank. Except by a few individuals whose natural energies of mind freed them from the bondage of vulgar prejudice, it was either neglected altogether, or pursued with views but little more elevated than those of the Rhizotomists or herb-gatherers of Athens. A revolution, as rapid as it is general, has however taken place in the public mind on this subject; and it may now be said to be the most popular, as it was once the most despised of sciences in the Irish capital. Like the aloe, whose bloom is prized in proportion to the slowness of its advent, it has at length burst its plural envelopes, and captivated society by displaying its beauty and its utility. Not only are its claims to attention beginning to be recognised by that profession which so long treated it as an alien or an outcast of their *curriculum*; but by every department of the community aspiring to the possession of refined and useful information. We have now many young ladies amongst us who are better acquainted with its principles and application, than some of our remaining professors of materia medica;* while the crowded audiences in the Glasnevin Garden, amounting often to three and four hundred persons, demonstrate the favour in which the science is at present held.

The slow progress which the science has made in Ireland, admits of an easy solution; and may be instructively cited as one of the mischievous errors committed by the directors of medical education. The College of Surgeons, at the time of its establishment, and for many years subsequently, succeeded, by a laudable attention to a single branch of knowledge, in obtaining the almost exclusive management of education in the country, and in giving tone to the public mind on that subject. The University, owing to the perverse construction of its statutes, had neither pupils nor influence to advance the study of the science, though possessed of a professor and a garden for that purpose. The Corporation of Apothecaries, which might be expected, from the immediate connexion of the objects of its institution with this science, to take an active interest in its cultivation, did not until very lately even appoint a professor for its tuition, and is still without a perch of ground to grow specimens for its pupils' instruction, notwithstanding the increasing prosperity of that body. The trading revenues of this establishment, which must be considerable, instead of originating a 'Chelsea' and a 'Miller' on the banks of the Liffey, were directed, to the last farthing, into the channels of private embezzlement. By that department, therefore, of the profession which alone by its influence could have promoted the diffusion of the science, it was not only neglected, but even publicly ridiculed. So far from being required as an essential in the education of the licentiates and members of the College of Surgeons, they were taught to look on its study as so much useless labour; and its practical application as an employment fitting only for old women. The incomparable ingenuity of the system and nomenclature of Linnaeus, and the profound and beautiful philosophy of the method and writings of Jussieu, were looked up as the dreams of idle speculatists, unprofitable as the pursuits of the alchemists in search of the 'philosopher's stone'. Concentrating their views on the anatomy and diseases of a single being — on

---

* We allude more particularly to the first attempt of the kind made in Ireland by Miss B....y [Baily] to compile an Irish Flora — a small unpretending work which we may, without compromise between truth and gallantry, recommend as a useful companion to the botanical student in his excursions through that country. As an illustration of our text, we may further remark that the professor of materia medica in one of those institutions which Dr Graves would seek to dignify with the slavering epithet of "scientific", would not be able to distinguish a branch of *pinus sylvestris* from a peacock's feather!

one small item in the catalogue of vitality, these empirics, who mistook a shambles for a scientific association, forgot that even this limited object could not be attained without a referential examination of the other departments of animated nature. The evil of this course did not terminate in the injury which it inflicted on the science of botany alone; it fatally extended to the other sciences, and even to the moral character of the profession itself, which was thus taught to resolve all knowledge into what were called 'facts' by this empirical sect; and all those ennobling associations arising out of the proper study of science, into a code of money-making, and humbugging the public about pretended cures. Even at this hour, this body, which presumptuously arrogates the right of dictating the course of study to be pursued by the medical students of Ireland, neither has appointed a professor of botany,* nor requires the slightest information on botanical subjects from those whom they pretend to educate.

The proscription of the science by a profession with which, unfortunately, it was always identified, spread its baneful influence over the public mind, and confirmed its indifference, by example, for botanical pursuits. The obvious but erroneous conclusion that 'a pursuit despised and rejected by persons with whose education it was connected, was not worth the trouble of learning,' became the easy and universal response to every private and public attempt to promote the cultivation of the science. Nothing, indeed, can better illustrate the apathy which everywhere prevailed on this subject, than the results of, and the circumstances attending, the exertions of the Royal Dublin Society to awaken the national mind to a sense of the importance of the science. While we concede to the Society the full merit of the best intentions, we must observe that the means which they adopted of carrying their design into effect, fell far short of the difficulties which they had to surmount; and that to the original error of a bad selection of agents for the accomplishment of their undertaking, they committed the less excusable fault of witnessing and tolerating the incompetency of those agents for years after it had been proved in the most satisfactory manner. It is true they had a refractory soil to operate on; a soil too in which the seeds of positive hostility to the science were extensively sown and fatally prolific. Yet a man enjoying the liberal advantages of the society's patronage, alive to the important trust reposed in him, and armed with a fearless conviction of the utility of his mission, might still have achieved much in the regeneration of the public taste, if he could not entirely succeed in establishing the empire of the science on the wide basis of popular favour. Truth and science might in this, as in many other instances, have proved themselves more infectious and diffusible than error, and showed that they only required to be well planted to ensure their propagation. Unfortunately the choice of the society at that important crisis fell on a man who did not possess a single attribute to qualify him for the restoration of public taste from its vitiated to a healthy condition; and still more culpable and unaccountable was his retention in office after failing in the task they had assigned him. The old-fashioned *prig* whom they appointed Professor of Botany, and continued in that office until death removed him out of the wrong position in which he had been placed, imagined that the wearing of a sky-blue vest, and counting the pistils and stamens of some showy flowers in the theatre of the Society constituted the whole duties of a professor of botany. The burlesque of the science which this old coxcomb annually enacted must be fresh in the recollection of many of our readers. Tricked out in apparel as tawdry as the pie-bald vestiture of the high priest of Flora herself, and thoroughly imbued with that dancing-master style of manner with which Chesterfield and the last century might have been enraptured, Dr Wade always commenced his didactic duties by a preliminary eulogium on the Society. of this truly *Hibernian* specimen of *Oriental* exaggeration it would be impossible to give an adequate description. The Society was duly put forward as the centre of national civilization; the paragon of all learned associations; the sun which blessed the most distant districts of the island with the humanizing influence of science and literature! The waters of the bog of Allan oozed away, in obedience to the improving mandate of the Society, into perfect dryness; the 'purple-blossomed heaths' of the mountains of Cunnemara disappeared in the prospect of the more profitable products of cultivation: *omnis feret omnia tellus*, scarcely described the metamorphose in store for Ireland under

---

* There is, we believe, *one* in a state of incubation in Park street, who, when sufficiently fledged for flight, and the botanical "aerie" in the College is prepared for his reception, will probably wing his way to that elevated station, unless, indeed, the "concours" should in the interim start up and bar his ascent. We beg not to be misunderstood—we mean a *system*, not a *person*, of whom we know nothing.

the ameliorating auspices of his patrons. Dryden in the palmiest days of patronage and dedication, when the refuse flowers of Parnassus were twined at fixed prices into wreaths of immortality for the great, was cold, artful, and penurious of his praise, when compared with the glowing, undisguised and exuberant flattery of this most grateful and generous of all literary and scientific pensioners. During this stage, however, of the doctor's annual *apotheosis* of the institution in Kildare-street, the incense of panegyric was still endurable, because it was general; the members and professors present on these occasions might appropriate just as much of the floating fumes as suited their taste or their vanity; but as the Doctor proceeded to individualize the performance—to apportion his respective quota of the panegyric to each of the Society's professors—it was certainly amusing to see them one by one withdrawing out of the theatre, amidst the laughter of the audience, in anticipated apprehension of the unmerciful brush of this old dauber! One more duty was yet to be fulfilled by the 'little, round, fat, oily man of God', who certainly had 'a roguish twinkle in his eye', with many other characteristics of an occupant of the 'Castle of Indolence'—a hint or address to his 'fair countrywomen' on their peculiar fitness for botanical pursuits, which he usually concluded in drawing their attention to some 'fair inmate' of the garden, and the gallant intimation, to one of the servants, of '*Pulcherrimae detur!*' The prize was accordingly disposed of as the rustic taste of some 'Hodge' of the garden dictated, the decision producing a commotion among the spectators of both sexes which was scarcely exceeded by that of the 'judgement of Paris' himself, and more befitting the celebration of the prurient rites of the *floralia* in ancient Rome than a place devoted to the objects of science and instruction.

Such were the notions which this man entertained of the means to be employed in overcoming the difficulties of his undertaking—such his idea of popularizing a science which has all the prejudices of the learned, and all the apathy of society in general, arrayed against it. We have dwelt with perhaps seeming prolixity and unnecessary severity on the errors committed by this individual and his too indulgent patrons; but when we reflect on the splendid opportunities abused by both these parties, and contrast the melancholy consequences of their conduct with the results of the exertions of other teachers and other learned societies in neighbouring countries, we stand acquitted in our own judgment of having overcharged the wretched picture which we have drawn of the former condition of botany in Ireland.

Yet the evil did not rest here; the decay of the garden at Glasnevin rivalled the buffoonery in the theatre of Kildare-street. Vast numbers of plants, procured with difficulty and at much expense, disappeared from the beds and stoves, through the perverse neglect of the curator who brought up the rear of ruin headed by Dr Wade. In our walks to the garden at this era of barbarism, we not unfrequently saw the thistle flourishing in indigenous luxuriance above the wasted remains of some costly exotic of the green-house; and the label of perhaps one the umbelliferae, outraging the labours and nomenclature of Linnaeus, standing perhaps before one of the konunculaceae.*

Fortunately for the fate of botany in Ireland, and the improvement of the Glasnevin Garden, Dr Wade was 'gathered to his fathers', and some time after his coadjutor in mischief was dismissed. The choice of the Society now fell on a professor who has proved himself equal to the important office which he holds, his success as a popular lecturer being perhaps unrivalled in Dublin. Though it would appear that Dr Litton did not devote more of his attention to botanical pursuits in early life than to the other branches of medical science, he possessed other qualifications in an eminent degree to compensate for this defect. A preliminary education, which may be said to have been truly learned, succeeded by a long and serious application to science and literature in the most extensive acceptation of those terms, prepared him to prosecute with effect the study of his adopted science. To him the sources of knowledge were familiar; the paths to the fountain had been often and observantly trod; and the heights of science, which mutually look into and connect each other's domains, were frequently and successfully scaled. Unlike the Cyclops of a deficient education, who, though possessed of sufficient intellectual power, knows not which way to turn in the darkness of his ignorance when stimulated by choice or necessity to studious exertion, he saw the course distinctly which he ought to pursue,

---

* [Ed. note. This is the spelling used in the original article, undoubtedly this is a typographic error for Ranunculaceae. ECN]

and had only the delightful task of collecting and arranging those materials of instruction with which previous excursions into the fields of literature and science had made him acquainted. His lectures in the theatre of the Society, preparatory to the practical course delivered by him in the garden, present master-pieces of popular composition on the study and advantages of botany, and demonstrate how much can be accomplished by a man of correct taste and general acquirements, when his energies and resources are concentrated on any given subject. In these admirable discourses, the selection of illustration is as judicious and copious as the warmest enthusiast in the cause of the science could desire, and as convincing of the just importance attached to it as would be required by the most sceptical or practical audience.

One of the leading features in these lectures—a feature in them which is highly creditable to the judgment of Dr Litton, is the attempt to awaken in his fellow-citizens a sense of the utility of the science, and to remove, by a correct and philosophic exposition of its objects, and the proper manner of studying it, those vulgar prejudices which existed in the public mind against its cultivation. This design is still further followed up and more fully developed in the course of lectures delivered in the Society's garden, which lectures usually embrace a demonstration of some of the more remarkable natural families of the vegetable kingdom. For proving to a mixed assembly, such as Dr Litton has to address, the advantages which they unconsciously derive from botanical science, and exhibiting to them in the most attractive form the beauty of its philosophy, no better course could be adopted. By examining even but a few of the links in the vast and interesting chain of vegetable existence, the spectator is at once surprised and gratified by the successful evolution of new species, genera, tribes and families, all differing from and yet connected with each other by shades of habit and structure so subtle and yet so distinct, that he knows not which to admire most, the superabundance and perfection of design displayed in the formation of these objects, the wonderful ingenuity of man in the discovery of their organization, connexion, and properties, or the important ends accomplished even by the humblest of them in the grand laboratory of nature. In each of these families he is sure to meet with some individual, the peculiarity of whose qualities, appearance, or structure, will fix his attention on its less remarkable relatives, and induce him to enter on their investigation. He is thus gradually led on to observe the order, harmony, and similitude, which pervade the whole system of the vegetable kingdom, from the humblest moss to the monarch oak, and to arrive at length at the only conclusion for which its study indeed would be worthy of man's attention—the ennobling and instructive conviction of the wisdom, benevolence, and consistency, displayed in its production.

The method of viewing and studying the vegetable kingdom. to which we have alluded, has been prudently made the basis of Dr Litton's lectures; and in carrying his conception into execution he is equally felicitous. Whatever assistance for the illustration of this design can be derived from, or whatever interest can be conferred on, the subject by an extensive acquaintance with the various branches of natural history and general literature, is seized upon and happily applied to the attainment of this object. The general stock of human learning is thus made tributary to the ornament and elucidation of his theme; but with a degree of taste and judgment which shows that he has not ransacked libraries for the purpose of ostentatious extraction, but that his communion with those sources of knowledge has elevated his own ideas and observations to a level with those from whom he borrows. His discourses consequently possess all the freshness and integrity of original composition, carrying the audience along with them without those narratorial interruptions of undisguised plagiarism which so eminently distinguish the clumsy compilations of some of his contemporaries. From the remarks which we have made (and which we could wish were more worthy of the subject), in order to point out the correctness of Dr Litton's views of the nature of his duties and his ample capabilities for their performance, it may be inferred that in the minor details of character he is equally distinguished. In his language and delivery there is indeed a delightful absence of all that charlatanism and mannerism which would of late appear to have found their way from the green-room, not only into the theatres of science, but into every department of social life in the Irish metropolis. His respect for himself and the pursuit which he follows: his convictions of its dignity and importance, unalloyed by that biting scepticism of selfishness and ignorance which may be seen every day in our philosophic institutions, sacrificing the true interests of science by a personal prostitution of manner, to gratify, for private objects, the vicious taste of the polite rabble, have refined his deportment into a dignified simplicity. The objects of instruction are never for a moment merged in obtrusions of vanity, or degraded in clap-traps for vulgar applause. He is widely content to leave to science, which he so purely and faithfully serves, the never-neglected duty of securing to her votaries that portion of admiration and esteem which they respectively deserve. In this he is right; for though such reputation is not easily earned, it is the only kind that wears well. Having secured this prize by honourable and meritorious means, we have only to express a sincere wish that he may long continue to enjoy it.

The Society were not less fortunate in the appointment of a curator. Mr Nevin [sic] has more than realised the expectation of the Society and the public, by his great exertions in the improvement of the garden. The walks, beds, ponds, and stoves, have all undergone the process of repair and alteration; and in little more than twelve months the place presents a degree of order and of elegance almost incredible to those who had witnessed its previous state of ruin. Numerous and valuable additions have been made to the contents of the green-houses, by donations, and by purchases made by Mr Nevin in the English gardens during the last summer. The Society appear to be laudably anxious to afford every facility in their power to the botanical visitors to the garden; and from the success of the exertions which have been hitherto made, a still greater impulse to the cultivation of botany in Ireland may be anticipated.

Our space scarcely justifies an allusion to the localities of this garden, of which it might be expected we should give some account. We should, however, decline the task on other grounds. A botanic garden cannot be *safely* described, being a sort of Calypso's Isle, in which the Circean cup of natural enchantment might over-excite even the apathetic, and betray the most prudent into rhetorical excesses of description. A corner of the earth, in which figures and flowers more beautiful and abundant than ever sprang from Parnassus' sides, or drew vitality from Castalia's dews, present themselves in endless profusion, is surely a dangerous theme in the month of May. We must, therefore, be held excused for not attempting to 'paint the lily',—to give body and form to a scene in which the operations of the matchless pencil of nature must be seen in order to be sufficiently prized and understood. This much however we may venture to aver for the information of those who have not visited the garden, that if all the delightful objects of foreign production which now crowd its space were removed to their native *habitats*, the grounds of Glasnevin would still be a place of which much might be said in their favour. There is wood, water, extent and variety of surface and of soil—not a paltry patch of ground, such as usually devoted to these purposes, but a demesne of upwards of forty acres? To this garden we may further observe there is admission two days in the week without expense; that it lies within less than half an hour's walk of the north side of the city; and that the lectures usually commence there in the month of June, and are delivered at eight o'clock in the morning by Dr LITTON. The citizens of Dublin, and studious strangers in that city, are thus supplied with one of the best sources of amusement and instruction; and the vast number of visitors during the delivery of the lectures, show that this great advantage has at length been duly appreciated. Many observations on the further improvement of the establishment suggest themselves to our mind; but in the present disposition of the Society to promote by all means in their power the true interests of the science, and of the Society's agents to forward their views in diffusing a knowledge of this useful and delightful branch of knowledge, we deem it unnecessary to disturb the general approbation of these efforts, by the publication of suggestions which we have no doubt will be realized in the course of time.

ERINENSIS

Dublin, May 1835.

Appendix 2

Plates from Botanical Magazine
(a) associated with Glasnevin (b) by W. E. Trevithick

Plates associated with Glasnevin

* indicates plates actually using original Glasnevin material

Botanical Magazine

1304*	Fabricia lacvigata	vii.1810	specimen figured from Glasnevin
3521	Fuchsia macrostemma var. recurvata	x.1836	Glasnevin mentioned
3541*	Verbena tweediana	xii.1836	drawing by Ninian Niven of specimen in Glasnevin
3618	Philbertia grandiflora	xi.1837	Glasnevin mentioned
3620	Boussingaultia baselloides	xii.1837	Glasnevin mentioned
3630*	Tweedia versicolor	i.1838	drawings by Ninian Niven of specimens in Glasnevin
3694*	Verbena teucrioides	xii.1838	drawing by Ninian Niven of specimen in Glasnevin
3756*	Aristolochia ciliata	x.1839	specimen figured from Glasnevin
3773*	Passiflora mooreana	i.1840	specimen figured from Glasnevin
3786*	Gonolobus hispidus	iii.1840	specimen figured from Glasnevin
3793*	Miltonia candida var. flavescens	iv.1840	specimen figured from Glasnevin
3828*	Hymenoxys californica	x.1840	specimen figured from Glasnevin
3840	Sida picta	xii.1840	Glasnevin and TCD mentioned
3841*	Grabowskia duplicata	xii.1840	specimen figured from Glasnevin
3907*	Franciscea latifolia	xi.1841	drawing of specimen in Glasnevin sent by David Moore
3929*	Catasetum abruptum	iii.1842	specimen figured from Glasnevin
3932*	Oxalis lasiopetala	iii.1842	specimen figured from Glasnevin
3938*	Oxalis martiana	iv.1842	specimen figured from Glasnevin
3942*	Catasetum globiflorum	v.1842	specimen figured from Glasnevin
3969	Trichocentrum fuscum	ix.1842	Glasnevin mentioned
3971*	Gloxinia tubiflora	x.1842	specimen figured from Glasnevin
4022*	Cestrum viridiflorum	vi.1843	specimen figured from Glasnevin
4039*	Rhipsalis brachiata	ix.1843	specimen figured from Glasnevin
4076*	Phaseolus lobatus	iii.1844	specimen figured from Glasnevin
4120*	Aristolochia ornithocephala	ii.1844	specimen figured from Glasnevin
4200*	Anthocercis ilicifolia	xii.1845	specimen figured from Glasnevin
4423*	Exacum zeylandicum	ii.1849	specimen figured from Glasnevin
4771	Exacum macranthum	iii.1854	Kew plant from Glasnevin
4793*	Buddleia crispa	vii.1854	specimen figured from Glasnevin
4794*	Clematis barbellata	vii.1854	specimen figured from Glasnevin
4796*	Cassiope fastigiata	vii.1854	specimen figured from Glasnevin
4823*	Nymphaea amazonum	xii.1854	specimen figured from Glasnevin
4839*	Brownea grandiceps	iii.1855	specimen figured from Glasnevin
4868*	Thermopsis barbata	viii.1855	specimen figured from Glasnevin
4906*	Banksia victoriae	vii.1856	specimen figured from Glasnevin
5514*	Vellozia candida	vi.1865	specimen figured from Glasnevin
5516*	Acanthus montanus	vii.1865	specimen figured from Glasnevin
5533*	Lankesteria barteri	ix.1865	specimen figured from Glasnevin
5578*	Ceropegia sororia	v.1866	specimen figured from Glasnevin
5617*	Combretum micropetalum	xii.1866	specimen figured from Glasnevin
5696*	Dicentranthera macrophylla	iii.1868	specimen figured from Glasnevin
5719*	Paeonia emodi	vii.1868	specimen figured from Glasnevin
5788*	Steriphoma paradoxum	viii.1869	specimen figured from Glasnevin
5990*	Masdevallia lindenii	ix.1872	specimen figured from Glasnevin
6040	Greyia sutherlandii	vi.1873	Glasnevin mentioned
6053	Hibbertia baudouinii	ix.1873	Glasnevin mentioned
6113*	Crinum moorei	viii.1874	specimen figured from Glasnevin
6356	Fevillea moorei	iv.1878	Kew plant from Glasnevin
6469*	Brownea ariza	i.1880	specimen figured from Glasnevin

7026*	Calanthe striata	xi.1888	specimen figured from Glasnevin
7123*	Lueddemannia pescatorei	vii.1890	specimen figured from Glasnevin
7143*	Acineta densa	xi.1890	flowers figured from Glasnevin, foliage from Kew
7189	Pleurothallis immersa	viii.1891	Kew plant from Glasnevin
7233	Resterpia striata	v.1892	Kew plant from Glasnevin
7262*	Moorea irrorata	xi.1892	specimen figured from Glasnevin
7428	Saccolabium mooreanum	viii.1895	named for Moore
7431*	Pleurothallis scapha	viii.1895	specimen figured from Glasnevin
7464	Dipodium paludosum	iii.1896	Kew plant from Glasnevin
7476	Masdevallia corniculata var. inflata	v.1896	Glasnevin mentioned
7488	Cyrtanthus huttonii	viii.1896	Glasnevin mentioned
7499	Lathyrus undulatus	x.1896	Kew plant from Glasnevin
7677	Lonicera hildebrandiana	x.1899	Glasnevin mentioned
7766	Masdevallia dorsum	iii.1901	Glasnevin mentioned
7781	Nymphaea flavo-virens	vi.1901	Glasnevin mentioned
7859	Masdevallia schroederiana	x.1902	Kew plant from Glasnevin
7928*	Agapetes mooreana	xii.1903	specimen figured from Glasnevin
7958*	Bulbophyllum weddelii	vi.1904	specimen figured from Glasnevin
7968*	Vanda pumila	viii.1904	specimen figured from Glasnevin
7994	Phyllostachys nigra	i.1905	Glasnevin mentioned
8000*	Bulbophyllum crenulatum	ii.1905	specimen figured from Glasnevin
8017*	Nepenthes rajah	vi.1905	specimen figured from Glasnevin
8020	Lycaste locusta	vi.1905	Kew plant from Glasnevin
8031	Odontoglossum ramulosum	viii.1905	Kew plant from Glasnevin
8033	Cirrhopetalum breviscapum	ix.1905	Kew plant from Glasnevin
8041*	Momordes buccinator var. aurantiacum	x.1905	specimen figured from Glasnevin
8056*	Wittmackia lingulata	i.1906	specimen figured from Glasnevin
8062*	Arachnanthe annamensis	iii.1906	specimen figured from Glasnevin
8103	Lycaste dyeriana	xi.1906	Kew plant from Glasnevin
8109	Vanda watsonii	xii.1906	Glasnevin mentioned
8131*	Cymbidium erythrostylum	iv.1907	specimen figured from Glasnevin
8160*	Bulbophyllum dichronum	x.1907	specimen figured from Glasnevin
8161*	Paeonia cambessedesii	x.1907	specimen figured from Glasnevin
8203	Coelogyne perakensis	vi.1908	Kew plant from Glasnevin
8216	Bulbophyllum galbinum	ix.1908	Glasnevin mentioned
8221	Polystachya lawrenceana	x.1908	Kew plant from Glasnevin
8229	Eria hyacinthoides	xii.1908	Kew plant from Glasnevin
8262	Coelogyne venusta	vi.1909	Glasnevin mentioned
8266*	Mahonia arguta	vii.1909	specimen figured from Glasnevin
8273*	Megaclinium purpureorachis	ix.1909	specimen figured from Glasnevin
8297*	Coelogyne mooreana	ii.1910	specimen figured from Glasnevin
8305	Pittosporum colensoi	iii.1910	Glasnevin mentioned
8318*	Fouqueria splendens	vi.1910	specimen figured from Glasnevin
8543*	Ribes laurifolium	iii.1914	specimens figured from Glasnevin, Kew and Aldenham
8631	Promenaea microptera	x.1915	Glasnevin mentioned
8636*	Anemone obtusiloba f. patula	xi.1915	specimen figured from Glasnevin
8662*	Eria tomentosa	v.1916	specimen figured from Glasnevin
8711*	Plagiospermum sinense f. brachypoda	v.1917	specimen figured from Glasnevin
8721*	Rhododendron cuffeanum	viii.1917	specimen figured from Glasnevin
8734	Megacarpa polyandra	xi.1917	Glasnevin mentioned
8765*	Ramonda serbica	vii-ix.1918	specimen figured from Glasnevin
8785	Bulbophyllum hamelinii	x-xii.1918	Glasnevin mentioned
8791*	Primula chasmophila	i-iii.1919	specimen figured from Glasnevin
8792	Bulbophyllum robustum	i-iii.1919	Kew plant from Glasnevin
8809*	Odontoglossum cristatum	vii-ix.1919	specimen figured from Glasnevin
8852*	Nuphar polysepalum	vii.ix.1920	specimen figured from Glasnevin
8925*	Veratrum wilsonii	1938	specimen figured from Glasnevin
8949*	Maxillaria fletcheriana	ii.1923	specimen figured from Glasnevin
8954*	Cirrhopetalum tripudiana	ii.1923	specimen figured from Glasnevin
8979	Maxillaria chrysantha	xi.1923	Kew plant from Glasnevin
8999	Cotoneaster salicifolia	iii.1924	Headfort plant from Glasnevin

9002	Boykinia tellimoides	vi.1924	Glasnevin mentioned
9004*	Magnolia wilsonii	vi.1924	specimen figured from Glasnevin
9015	Dendrochilum uncatum	ix.1924	Glasnevin and TCD mentioned
9063	Viburnum grandiflorum	vi.1925	Headfort plant from Glasnevin
9068*	Chinodoxa siehei	vi.1925	specimen figured from Glasnevin
9107	Primula inayatii	ix.1926	Glasnevin mentioned
9131	Abelia triflora	i.1928	Glasnevin mentioned
9146*	Schizandra rubriflora	ii.1928	specimens figured from Glasnevin and Aldenham
9160*	Hypericum leschenaultii	x.1928	specimen figured from Glasnevin
9175	Cyrtanthus rhododactylis	x.1929	Glasnevin mentioned
9179	Encyclia mooreana	x.1929	named for F. Moore
9183	Euonymus grandiflorus f. salicifolia	x.1929	Glasnevin mentioned
9200*	Rheum palmatum var. dissectum f. rubriflora	v.1930	specimen figured from Glasnevin
9247	Arctotis speciosa var. hayana	vii.1931	Glasnevin mentioned
9281	Berberis francisci-ferdinandii	v.1932	Headfort plant from Glasnevin
9325	Physosiphon lindleyi	vii.1935	Glasnevin mentioned
9345*	Hypericum kouytchense	i.1934	specimen figured from Glasnevin
9362*	Deutzia rubens	vii.1934	specimen figured from Glasnevin
9376*	Maxillaria fuscata	x.1934	specimen figured from Glasnevin
9389	Cotoneaster hebephylla var. monopyrena	ii.1935	Headfort plant from Glasnevin
9390*	Phlox maculata x P. paniculata	ii.1935	specimen figured from Glasnevin
9400*	Lycaste longiscapa	v.1935	specimen figured from Glasnevin
9418	Pelexia maculata	xi.1935	Glasnevin mentioned
9459	Bulbophyllum orthoglossum	xi.1936	Glasnevin mentioned
9478	Cotoneaster cooperi var. microcarpa	ii.1937	Headfort plant from Glasnevin
9561	Rhododendron ravum	vii.1939	Glasnevin mentioned
9563	Ptychogyne flexuosa	vii.1939	Glasnevin mentioned
9564	Buddleia fallowiana	vii.1939	Headfort plant from Glasnevin
9581*	Codonopsis convolvulacea var. forrestii	xii.1939	specimen figured from Glasnevin
9621*	Agapanthus inapertus	vi.1942	specimen figured from Glasnevin

Botanical Magazine, new series

83	Eria globifera	x.1949	Kew plant from Glasnevin
199	Selago galpinii	viii.1953	Glasnevin mentioned
374	Greyia sutherlandii	ii.1962	(see 6040 above)
506	Solandra maxima	ix.1957	Glasnevin mentioned

Kew Magazine

18	Geranium orientalitibeticum	viii.1984	Glasnevin mentioned

Notes

1 Ninian Niven sent at least three drawings of plants to W. Hooker for use in *Botanical Magazine*. Two were used and reproduced although Niven's name does not appear on the published plate.

   3541 Niven sent a drawing—plate bears no artist's name

   3630 Niven sent specimen 'here requested' and also a 'beautiful drawing', implying that the drawing was not used. W. Fitch is credited on the plate.

   3694 Niven's drawing was used after reduction, but W. Fitch is credited as artist on the plate.

2 3907 The artist's name cannot be discovered, the painting was displayed at the Royal Horticultural Society of Ireland's show in 1841 (*Gardeners' Chronicle*, 1 (1841)).

3 4200 *Anthocercis ilicifolia*—then plant was killed by the potato blight fungus (*Phytophthora infestans*) in 1846.

4 6113 *Crinum moorei*—the descendants of the original (type) plant still grow at Glasnevin.

5 7262 *Moorea irrorata*—now *Neomoorea wallisii*. As pointed out elsewhere (E. C. Nelson, 'Orchid paintings at Glasnevin', *Orchid Review*, 89 (1981), 373–377, 384) part of this plate may have been copied from an original watercolour by Lydia Shackleton.

Plates from Botanical Magazine drawn or lithographed by William Edward Trevithick

del. = delineavit (= drew this)
lith. = lithographed

9030	Kleinia stapeliiformis	R. M. Adam & W. E. T. del., L. Snelling lith.
9041	Aeginetia indica	W. E. T. del., L. Snelling lith.
9046	Chelonistele pusila	W. E. T. del., L. Snelling lith.
9048	Tulipa humilis	W. E. T. del. & lith.
9052	Arnia melanocarpa	L. Snelling & W. E. T. del., L. Snelling lith.
9054	Campomanesia thea	W. E. T. del., L. Snelling lith.
9055	Begonia manicata	W. E. T. del. & lith.
9056	Pulsatilla regeliana	W. E. T. del. & lith.
9062	Prunus yedoensis	W. E. T. del. & lith.
9064	Primula edgeworthii	W. E. T. del. & lith.
9065	Maxillaria lepidota	W. E. T. del. & lith.
9067	Smilax excelsa	G. Atkinson & W. E. T. del., W. E. T. lith.
9069	Arctotis roodae	W. E. T. del. & lith.
9070	Rosa mirifica	W. E. T. del. & lith.
9071	Dendrobium victoriae-reginae	L. Snelling & W. E. T. del., L. Snelling lith.
9074	Staphylea holocarpa	W. E. T. del. & lith.
9075	Roscoea humeana	W. E. T. del. & lith.
9076	Berberis replicata	W. E. T. del. & lith.
9078	Tulipa australis var. arvorum	W. E. T. del. & lith.
9079	Salix bockii	W. E. T. del. & lith.
9081	Aster staticifolius	W. E. T. del. & lith.
9082	Clerodendron colebrookianum	W. E. T. & G. Atkinson del., W. E. T. lith.
9084	Roscoea cautleoides	W. E. T. del. & lith.
9086	Catasetum tenebrosum	W. E. T. del. & lith.
9087	X Malvastrum hypomadarum	W. E. T. del. & lith.
9088	Aconitum angelicum	W. E. T. del. & lith.
9089	Berberis vernae	L. Snelling & W. E. T. del., L. Snelling lith.
9092	Geranium farreri	W. E. T. del. & lith.
9100	Primula siamensis	W. E. T. del. & lith.
9101	Mamillaria conospea	W. E. T. del. & lith.
9102	Berberis lycioides	W. E. T. del. & lith.
9105	Rhodospatha forgetii	W. E. T. del. & lith.
9108	Fritillaria libanotica	W. E. T. & G. Atkinson del., W. E. T. lith.
9109	Cirropetalum miniatum	W. E. T. del. & lith.
9110	Brunnera macrophylla	W. E. T. del. & lith.
9111	Sargentodoxa cuneata	W. E. T. del. & lith.
9112	Sargentodoxa cuneata	W. E. T. del. & lith.
9113	Diplomera hirsuta	W. E. T. del. & lith.
9114	Anemone glaucifolia	G. Atkinson del., W. E. T. lith.
9123	Aster forrestii	W. E. T. del. & lith.
9126	Habranthus robustus	L. Snelling & W. E. T. del., L. Snelling lith.
9127	Venidium fastuosum	W. E. T. del., L. Snelling lith.
9141	Halium umbellatum	W. E. T. del., L. Snelling lith.
9143	Trollius yunnanensis	W. E. T. del., L. Snelling lith.
9148	Fritillaria acmopetala	W. E. T. del. & lith.
9158	Cestrum psittacinum	W. E. T. del., L. Snelling lith.
9166	Primula vulgaris f. heterochroma	W. E. T. del., L. Snelling lith.
9168	Corokia macrocarpa	W. E. T. del., L. Snelling lith.
9176	Siphonosmanthus suavis	W. E. T. del., L. Snelling lith.
9178	Salvia caerulea	G. Atkinson & W. E. T. del., L. Snelling lith.
9179	Encyclia mooreana	W. E. T. del., L. Snelling lith.
9180	Cosmos diversifolius	W. E. T. del., L. Snelling lith.
9181	Bridgesia amabilis	W. E. T. del., L. Snelling lith.
9187	Acradenia frankliniae	W. E. T. del., L. Snelling lith.
9188	Ranunculus psilostachys	W. E. T. del., L. Snelling lith.
9191	Sarcopodium lyonii	W. E. T. & L. Snelling del., L. Snelling lith.
9192	Prunus pilouscula	L. Snelling & W. E. T. del., L. Snelling lith.
9193	Tsuga chinensis	W. E. T. & L. Snelling del., L. Snelling lith.

APPENDIX 2

9198	Phalaenopsis violacea	W. E. T. del., L. Snelling lith.
9201	Abies faberi and A. forrestii	W. E. T. del., L. Snelling lith.
9205	Lilium formosanum	W. E. T. del., L. Snelling lith.
9206	Maxillaria elatior	W. E. T. del., L. Snelling lith.
9209	Ligustrum quihoui	W. E. T. del., L. Snelling lith.
9215	Veronica teucrium f. thracica	W. E. T. del., L. Snelling lith.
9217	Rhododendron didymum	L. Snelling & W. E. T. del., L. Snelling lith.
9223	Aster farreri	W. E. T. del., L. Snelling lith.
9224	Tremacron forrestii	W. E. T. del., L. Snelling lith.
9231	Lycaste suaveolens	W. E. T. del., L. Snelling lith.
9250	Salvia juriscii	W. E. T. del., L. Snelling lith.
9264	Ceanothus fendleri	W. E. T. del., L. Snelling lith.
9278	Calceolaria acutifolia	W. E. T. del., L. Snelling lith.
9286	Coprosma propinqua	L. Snelling & W. E. T. del., L. Snelling lith.

Appendix 3

Personnel associated with Glasnevin

Members of the Royal Dublin Society's Committee of Botany and Horticulture

This Committee, whose function was to run the Botanic Gardens, was made up of members of the (Royal) Dublin Society. It was set up a few years after the Gardens was established, probably soon after 1800 and was certainly in existence in 1810. Normally it was composed of eleven members, but on occasions there were less than this number.

Official lists of committee members have not survived, or may never have existed. The list to follow has been compiled from the extant records of the (Royal) Dublin Society. The longest serving member was Simon Foot J.P., elected to the Committee in 1825, served as chairman of the Committee from 25 June 1866 to 6 February 1871 and attended his last Committee meeting on 18 November 1872. Foot also served on a committee set up by the Royal Dublin Society in June 1823 to consider the future management of the Botanic Gardens.

Name	Known dates served (*elected +resigned)	Name	Known dates served (*elected +resigned)
Adair, Samuel P.	1869*-73	Kilmaine, Lord	1815
Betham, Sir William	1825*	Lanigan, Dr L.	1825*
Boyd, John	1815	La Touche, P. D.	1815
Brady, Francis W.	1855	Lentaigne, Dr John	1848-54, 1869
Brennan, Edmund P.	1852-55		
Bruce, Haliday	1830, 1831+	Michin, Dr Humphry	1873-75
Butler, Hon. Cavendish	1851-54	Mulvany, Henry W.	1830-31, 1846-55
Cane, Arthur B.	1846-47, 1850-51	Neligan, Dr John M.	1846-49
Carey, Richard	1875-77		
Chamney, George	1825*-31	O'Grady, Dr M. M.	1846-55
Clarke, Dr	1825*, 1830-31	O'Lier, J.M.	1825*
Croften, Arthur B.	1847-48	Orpen, Dr T. H.	1815
Croker, Dr C. P.	1846-54		
Crosthwaite, Leland	1846-52	Pasley, Joseph	1815
		Pasley, Joshua	1815
Darley, Dr Benjamin	1869	Paterson, Dr H.	1850-54
De Ricci, Dr Herman R.	1869	Peele, R.M.	1825* (d. 1849)
D'Olier, I. M.	1825*, 1830-31, 1846-49		
D'Olier, John R.	1869	Roe, George	1846-55
Draper, Stephen	1815		
Dunn, John	1811*	Shaw, Lieut. Colonel	1846-57
		Shaw, Sir Robert	1830-31, 1851-52, 1855
Eustace, Dr John	1869	Slacke, John	1811*
Farrar, William	1815	Smith, Walter G.	1877
Foot, Simon, J.P.	1825*-72		
Fox, R.	1815	Thornhill, Dr William	1869
Frazer, Dr William	1855	Toler, Dr John	1869
Giffard, John	1815	Verschoyle, Richard	1826*
Guinness, R. R.	1825	Vereker, Amos	1869
Hamill, Hugh	1811*, 1815	Walsh, Dr Albert J.	1855
Harty, Dr	1830-31	West, Dr	1830-31
Houghton, Edward	1825*, 1830-31	Wybrants, Robert	1846-48, 1851-55

APPENDIX 3

Chronological table of senior staff 1795-1985

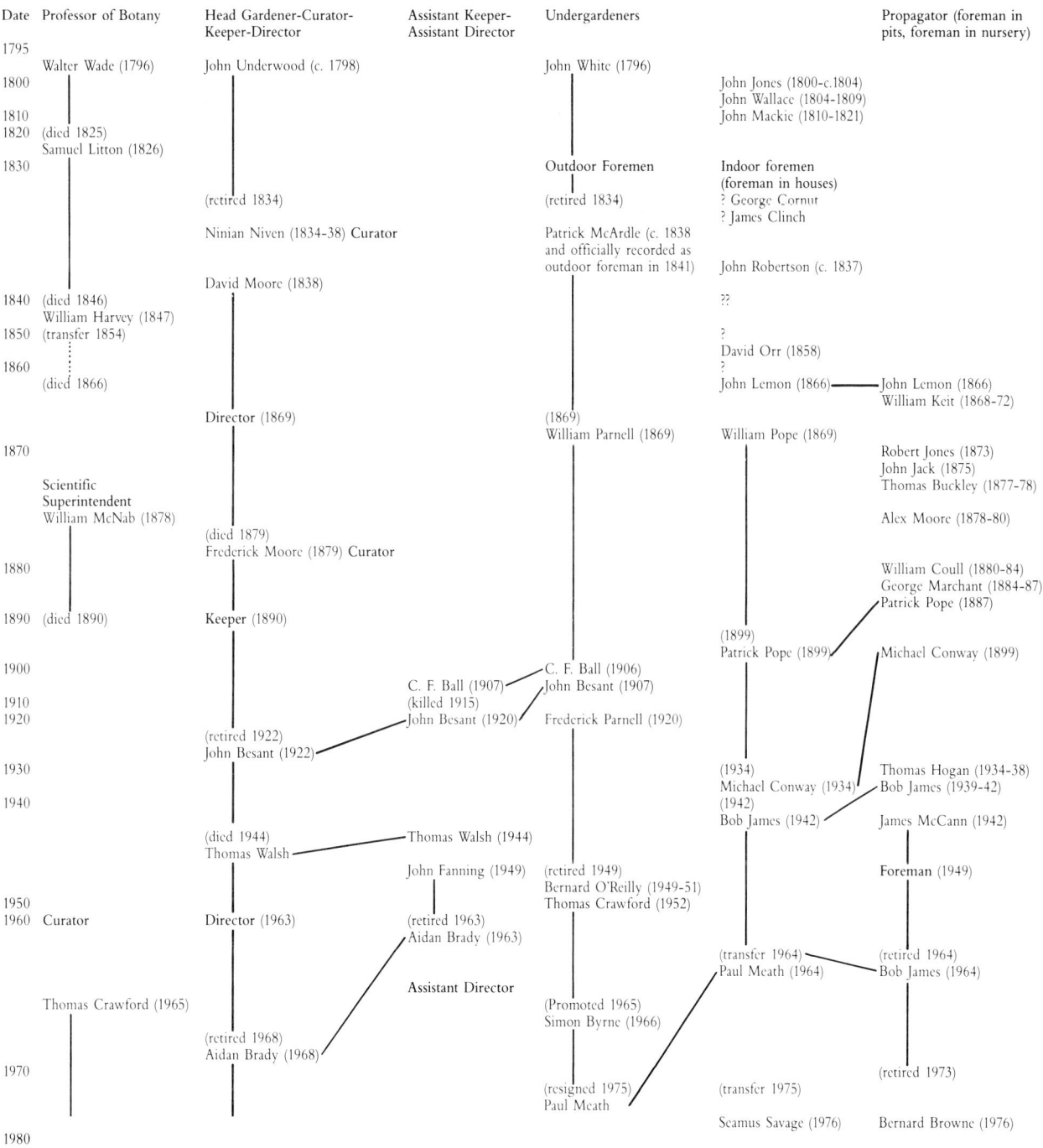

BIBLIOGRAPHY

In writing this history we have consulted published and unpublished sources which are acknowledged in the following notes. However, to avoid endless repetitions and to reduce the length of these notes, we have not referenced all the official reports and proceedings of the Royal Dublin Society and the government departments which controlled the Botanic Gardens between 1795 and the present day. All unacknowledged materials in the main text will be derived from at least one of the following sources

1. The manuscript minute books of the Royal Dublin Society, Ballsbridge, comprising
   (a) Minutes of the (Royal) Dublin Society, 1790-1877.
   (b) Minutes of proceedings of the council of the Royal Dublin Society, 1837-1874.
   (c) Minutes of the Committee of Botany, (Royal) Dublin Society, 1816-1845, 1857-1877. (The minute book for 1846-1856 is missing from the archive).

2. Proceedings of the (Royal) Dublin Society, 1790-1877. These were published annually, and contain transcripts of the minutes of the council, as well as the annual reports of the curator (later director) of the Glasnevin Botanic Gardens, and of the Professor of Botany.

3. Annual reports of the Keeper/Director of the Royal (later National) Botanic Gardens, Glasnevin. Published annually, these form part of the annual report of the government departments which controlled the Gardens between 1878 and the present day.

NOTES AND REFERENCES

Prologue

1. The title 'National Botanic Gardens, Glasnevin' is derived directly from 'Royal Botanic Gardens, Glasnevin' which was officially decreed in July 1885. The plural 'Gardens' was sanctioned by Viscount Sandon (Vice-president of the Board of Education) because it was common usage in Dublin, and because the original promoters of the establishment formed several distinct gardens within the grounds (see e.g. pp. 54-56). Glasnevin Mss.; V. Ball to W. R. McNab, 10.vii.1885; V. Ball to W. R. McNab, 14.vii.1885.
2. The Dublin Society, now the Royal Dublin Society, assumed the prefix 'Royal' on 29 June 1820 when George IV became a patron. H. F. Berry, *A history of the Royal Dublin Society* (London, 1915), p. 241.

Chapter 1

1. A. G. Morton, *History of botanical science* (London, 1981). W. T. Stearn, 'From Theophrastus and Dioscorides to Sibthorp and Smith: the background and origin of Flora Graeca', *Biological Journal of the Linnean Society, London*, 8 (1976), pp. 285-298.
2. W. Blunt, *The art of botanical illustration* (London, 1950). W. Blunt and S. Raphael, *The illustrated herbal* (London, [1980]).
3. A. W. Hill, 'The history and function of botanic gardens', *Annals of the Missouri Botanic Garden*, 2 (1915), pp. 185-223 (see especially pp. 188-189). R. Willis, 'Description of the ancient plan of the monastery of St Gall in the ninth century', *Archaeological Journal*, 5 (1848), pp. 85-117.
4. F. Shaw, 'Medicine in Ireland in mediaeval times', in *What's past is prologue. A retrospect of Irish Medicine* (Dublin, 1952). E. O'Brien and A. Crookshank, *A portrait of Irish medicine. An illustrated history of medicine in Ireland* (Dublin, 1984).
5. A. Chiarugi, 'Le date di fondazione dei primi orti botanici del mondo: Pisa (estate 1543); Padova (7 iuglio 1545); Firenze (1 dicembre 1545)', *Nuovo Giornale Botanico Italiano*, 60 (1953), pp. 785-839. W. T. Stearn, 'Botanical gardens and botanical literature in the eighteenth century', *Catalogue of botanical books in the collection of Rachel McMasters Miller Hunt* (Pittsburgh, 1961) vol. 2, pp. xli-cxl. F. A. Stafleu, 'Botanical gardens before 1818', *Boissiera*, 14 (1969), pp. 31-46. S. M. Walters, *The Shaping of Cambridge Botany*, (Cambridge, 1981). L. T. Tomasi, 'Projects for botanical and other gardens: a 16th century manual', *Journal of Garden History*, 3 (1983), pp. 1-34. A. G. Morton (1981 no. 1 above).
6. J. Gerard, *The herball or generall historie of plantes* (London, 1597), p. 1185. N. Colgan, 'The shamrock in literature: a critical chronology', *Journal of the Royal Society of Antiquaries of Ireland*, 6, series V, (1896), pp. 211-226, 349-361.
7. Commendatory letter by George Baker, in J. Gerard (1597, no. 6 above), preface. See also B. Henrey, *British botanical and horticultural literature before 1800* (Oxford, 1975), p. 39, footnote 4.
8. S. Raphael, 'The Oxford Botanic Garden', in *Of Oxfordshire Gardens* (Oxford, 1982), pp. 3-23. R. T. Gunther, *Oxford Gardens* (Oxford, 1912)
9. Thomas Baskerville, 'Account of Oxford, c. 1670-1700', quoted by B. Henrey (1975, no. 7 above), p. 996.
10. H. R. Fletcher and W. H. Brown, *The Royal Botanic Garden, Edinburgh, 1670-1970* (Edinburgh, 1970).
11. R. H. Semple, *Memoirs of the Botanic Garden at Chelsea* (London, 1878). W. T. Stearn (1961, no. 5 above), pp. lxxiii-lxxx.

Chapter 2

1. See for example P. W. Joyce, *A social history of ancient Ireland* (London, 1903), vol. 2, pp. 148-158, 354-363. J. G. D. Lamb, 'The apple in Ireland; its history and varieties', *Economic Proceedings of the Royal Dublin Society*, 4 (1951), pp. 1-63. E. C. Nelson, *An Irish Flower Garden*, (Kilkenny, 1984), pp. 3-7.
2. D. M. Synnott, 'Folk-lore, legend and Irish plants', in E. C. Nelson and A. Brady (eds), *Irish gardening and horticulture* (Dublin, 1979), pp. 36-43.
3. Patrick Bowe kindly drew my (E.C.N.) attention to this—see E. C. Nelson (1984, no. 1 above).
4. A. Zettersten, 'The Virtues of Herbs in the Loscombe manuscript; a contribution to Anglo-Irish language and literature', *Acta Universitatis Lundensis* (1965), section 1. For general discussion of mediaeval gardens, see J. H. Harvey, *Mediaeval Gardens* (London, 1981). J. H. Harvey, 'The first English garden book. Mayster Jon Gardener's treatise and its background', *Garden History 13* (1985), pp. 83-101.
5. L. Walsh, *Richard Heaton of Ballyskenagh 1601-1660* (Roscrea, 1978). For comments on several inaccuracies in this book, see E. C. Nelson, in *The Naturalist, 105* (1980), pp. 162-163.
6. K. T. Hoppen. 'Sir William Petty: polymath 1623-1687', *History Today*, 15 (1965), pp. 126-134. E. O'Brien and A. Crookshank, *A portrait of Irish medicine. An illustrated history of medicine in Ireland* (Dublin, 1984).
7. E. C. Nelson, 'The influence of Leiden on botany in Dublin in the early eighteenth century', *Huntia*, 4 (1982), pp. 133-146. T. C. Barnard, 'The Hartlib circle and the origin of the Dublin Philosophical Society', *Irish Historical Studies*, 19 (1974), pp. 56-71.
8. J. D. H. Widdess, *A history of the Royal College of Physicians of Ireland 1654-1963* (Edinburgh and London, 1963). E. C. Nelson (1982, no. 7 above).
9. K. T. Hoppen, *The common scientist in the seventeenth century: a study of the Dublin Philosophical Society, 1683-1708* (London, 1970).
10. E. C. Nelson, 'Sir Arthur Rawdon (1662-1695) of Moira: his life and letters, family and friends, and his Jamaican plants', *Proceedings and Reports of the Belfast Natural History and Philosophical Society, 10*, Second series, (1983), pp. 30-52.
11. TCD Mss; Register vol. 5/1, f. 264 (see E. C. Nelson (1982, no. 7 above) and references therein).
12. E. C. Nelson (1982, no. 7 above).
13. TCD Mss; Deed 16a.
14. TCD Mss; Deed 51b.
15. TCD Mss; Bursar's Vouchers P4/2/7.
16. TCD Mss; Bursar's Vouchers P4/4/1.
17. TCD Mss; Register vol. 5/1, f. 276.
18. J. D. H. Widdess (1963, no. 8 above).
19. Royal College of Physicians of Ireland Mss; Minute book (entry dated 6.ii.1692/3).
20. TCD Mss; Bursar's Vouchers P4/6/8; P4/12/12; P4/16/11; P2/57/1 f. 234; P2/57/1 f. 386; P2/57/1 f. 282.
21. TCD Mss; Register vol. 5/1, f. 282.
22. E. C. Nelson (1982, no. 7 above—see especially footnote 28).
23. TCD Mss; Archbishop King Papers, 1995-2008 (1384).
24. Details of the marriage settlement are recorded in the Registry of Deeds, King's Inns, Dublin (deed 42718, dated 7.vii.1711).
25. E. St. J. Brooks, 'Henry Nicholson, first lecturer in botany and the earliest physic garden', *Hermathena*, 83 (1954), pp. 3-15. Details of the manuscript sources quoted here are given by Brooks, and also by E. C. Nelson (1982, no. 7 above).
26. H. F. Berry, *A history of the Royal Dublin Society* (London, 1915), p. 10. W. Maple, *A method of tanning without bark* (Dublin, 1729).
27. The manuscript catalogue (Botany Library, British Museum (Natural History), London) has a manuscript title page, reading 'Catalogus plantarum in Horto Dublinensis', and the fly-leaf bears an annotation reading, in part, 'This catalogue of the plants growing in the Public Garden at Dublin was sent by Dr William Stephens...' (see E. C. Nelson (1982, no. 7 above), footnote 58).
28. E. C. Nelson, ' "In the Contemplation of Vegetables"—Caleb Threlkeld (1676-1728), his life, background and contribution to Irish botany', *Journal of the Society for the Bibliography of Natural History*, 9 (1979), pp. 257-273. E. C. Nelson, *The first Irish flora: Synopsis Stirpium Hibernicarum by Caleb Threlkeld (1726)* (To be published 1987).
29. H. F. Berry (1915, no. 26 above), p. 13.
30. M. Gahan, 'The development of crafts', in J. Meenan and D. Clarke (eds), *The Royal Dublin Society 1731-1981* (Dublin, 1981) pp. 239-240.
31. TCD Mss; Register vol. 5/1 f. 601. The Register records Chemys' election as 'Professor of Botany', but there was no consistency in the use of the terms professor and lecturer at that time—William Clement was appointed 'Lecturer', although his status within the University was the same as that held by his predecessor. The Register of the University does not officially record the foundation of a Chair in Botany, but it is generally accepted that when the School of Physic was formed in 1711 a professorship in botany was included.

    For discussion of William Clement (his surname is often spelled Clements but this appears to be incorrect), see R. B. McDowell and D. A. Webb, *Trinity College Dublin 1592-1952. An academic history* (Cambridge, 1982). T. P. C. Kirkpatrick, *History of the medical teaching in Trinity College Dublin and the School of Physic in Ireland* (Dublin, 1912).
32. J. Gilbourne, *The Medical Review, a poem being a panygeric on the Faculty of Dublin...* (Dublin, 1775).
33. E. Hill, *An address to the students in physic* (Dublin, 1803) p. 7. E. Hill, *An address to the President and Fellows of the...College of Physicians in Ireland* (Dublin, 1805), pp. 44-45.
34. E. P. Wright, *Notes from the Botany School, Dublin*, 1 (1896), pp. 1-4. See also *Doctor Hill's Library...which will be sold by auction...1st of July 1816 by Thomas Jones...* (Dublin, 1816) (TCD call number I.M.5).
35. TCD Mss; 1770-1774b (J. Hely Hutchinson's unpublished history of the University of Dublin). See T. P. C. Kirkpatrick (1912, no. 31 above) p. 148.
36. G. O. Sayles, 'Contemporary sketches of the members of the Irish parliament in 1782', *Proceedings of the Royal Irish Academy*, 56 C3 (1954), p. 240.
37. E. Hill (1803, no. 33 above) p. 17—Hill seems to confuse the sequence of events—the testimonial for Wade is dated 1791.

38 *The Parliamentary Register [of House of Commons, Ireland]*, 13 (1793), pp. 367, 391, 435-437, 452.
39 We can trace no record of this botanical garden, nor of the site leased by Hill; there is no deed registered in the Registry of Deeds, King's Inns, Dublin. The only accounts of it are in Hill's own published version of these events (see no. 33 above).
40 E. Hill (1803, no. 33 above) p. 56.

Chapter 3

1 J. Rutty, *A spiritual diary and soliloquies* (London, 1776) – quoted by J. D. H. Widdess, *A history of the Royal College of Physicians of Ireland* (Edinburgh and London, 1963), pp. 73-75.
2 A. B. Lambert, 'Anecdotes of the late Dr Patrick Browne, author of the Natural History of Jamaica', *Transactions of the Linnean Society, London*, 4 (1798), pp. 31-34.
3 The original manuscript is in the Linnean Society, London.
4 *Burke's Family Records* (London, 1976) p. 1172.
5 W. D. Moore, 'An outline of the history of pharmacy in Ireland', *Dublin Quarterly Journal of Medical Science*, 6 (1848), pp. 90-94. H. F. Berry, *A history of the Royal Dublin Society* (London, 1915) p. 144.
6 G. S. Bougler, 'Walter Wade', *Dictionary of National Biography* (Oxford, 1975) pp. 2166-2167 (Compact edition). C. A. Cameron (*History of the Royal College of Surgeons in Ireland and of the Irish Schools of Medicine* (Dublin, 1886) pp. 511-512) suggested that Wade graduated from a Scottish university, but it is also possible that, like his uncle, Walter Wade studied in a French university (see no. 7 below).
7 R. Laslier (Bibliothèque Municipale, Reims, France) to E. C. Nelson, 8.ii.1980. Burke (1976, no. 4 above).
8 W. Wade, *Carta a humanigo sobre o estado actual da inoculacao...* (Lisboa, 1768).
9 Information from the records of the Religious Society of Friends, Historical Library, Eustace Street, Dublin.
10 C. A. Cameron (1886, no. 6 above) p. 512.
11 Royal College of Physicians in Ireland, Dublin, Mss; Minutes Book, entries dated 30.x.1796 and 23.iv.1787.
12 E. C. Nelson, 'Walter Wade's "Flora Dubliniensis", an enigmatic Irish botanical publication', *Long Room*, 20-21 (1981), pp. 16-20.
13 A. Malcomson, *John Foster: the politics of the Anglo-Irish Ascendancy* (Oxford, 1978).
14 A. Young, *A tour in Ireland* (Dublin, 1780) vol. 1, p. 113.
15 S. Savage, *Catalogue of the manuscripts in the library of the Linnean Society of London; part IV – Calendar of the Ellis manuscripts* (London, 1948). (This catalogue gives précis of letters from the Fosters to Ellis).
16 Ellis Mss; F. Foster to J. Ellis 12.x.1768.
17 Ellis Mss; J. Foster to J. Ellis 27.vii.1769.
18 Ellis Mss; J. Foster to J. Ellis 10.ix.1770.
19 Ellis Mss; J. Foster to J. Ellis 15.x.1770.
20 Information on books purchased by John Foster is contained in letters and manuscripts in the Ellis Mss (see no. 15 above) and in the Foster/Massereene Papers (Public Record Office of Northern Ireland, Belfast). A sale catalogue of John Foster's library is extant; the library was sold on 6 April 1843, by Charles Sharpe, Anglesea Street, Dublin (Foster Collection, Library (Special Collections), Queen's University, Belfast).
21 H. F. Berry (1915, no. 5 above), pp. 91-92.
22 Quoted by A. Malcomson (1978, see no. 13 above), p. 362.
23 A. Malcomson (1978, see no. 13 above), p. 415.
24 E. Hill, *An address to the students of physic* (Dublin, 1803), pp. 16-17.
25 *The Parliamentary Register [for the House of Commons, Ireland]*, 10 (1790), p. 159.
26 H. F. Berry (1915, see no. 5 above), pp. 136-140.
27 A. Young (1780, see no. 5 above) vol. 1, p. 20.
28 *Proceedings of the Dublin Society*, 26 (1790), p. 141 (entry for 22.vii.1790).
29 Foster/Massereene Papers; D562/7828.
30 Foster/Massereene Papers; D562/7828.
31 W. Wade, *A statement of the progress...made for the purpose of instituting a public Botanic Garden near...Dublin* (Dublin, 1793), pp. 6-7 (Foster/Massereene Papers; D562/7828).
32 E. Hill (1803, no. 24 above), p. 17.
33 Foster/Massereene Papers; D562/7824.
34 E. Hill (1803, no. 24 above), p. 17.
35 *The Parliamentary Register [for House of Commons, Ireland]*, 13 (1793), p. 452.
36 W. Wade (1793, no. 31 above), p. 9.
37 Foster/Massereene Papers; D562/7810. No other record of this proposal has been traced.
38 Smith Mss; A. Caldwell to J. E. Smith 23.ix.1793 (see W. R. Dawson, *Catalogue of the manuscripts in the library of the Linnean Society of London; part 1: The Smith papers* (London, 1934)).
39 Smith Mss; A. Caldwell to J. E. Smith 23.ix.1793.
40 Smith Mss; W. Wade to J. E. Smith 20.ix.1794.
41 Smith Mss; A. Caldwell to J. E. Smith 23.ix.1793.
42 Foster/Massereene Papers; D562/7837.
43 RDS Mss; Minute books; entry dated 5.iii.1795.
44 RDS Mss; Minute books; entry dated 5.iii.1795.
45 Smith Mss; W. Wade to J. E. Smith 17.iii.1795.

Chapter 4

1 The origin of the name Glasnevin presents Irish scholars with many difficulties. The derivation and translation given here was kindly supplied by the Irish Placenames Commission, Phoenix Park, Dublin (C. Bale to E. C. Nelson 25.ii.1982).

Other suggestions about the origin of Glasnevin are
  (a) *Glas-Naeidhen* meaning Naeidhe's streamlet after '...some old pagan chief named Naeidhe [who] must have resided...' by the banks of the River Tolka—see P. W. Joyce, *The origin and history of Irish names of places* (Dublin, 1871), p. 441.
  (b) *Glaisin aoibhinn* meaning a pleasant little field—this was cited as early as 1818 by J. Warburton, J. Whitelaw and R. Walsh, *History of the city of Dublin* (London, 1818), p. 1285. It was repeated by P. E. M. Clinch, 'Botany and the Botanic Gardens', in J. Meenan and D. Clarke (eds) *The Royal Dublin Society 1731-1981* (Dublin, 1981)—Clinch's contribution to this collection of essays is unreliable and inconsistent.
2 T. K. Cromwell, *Excursions through Ireland* (London, 1820), vol. 2, pp. 17-26.
3 Lord Killanin and M. V. Duignan, *The Shell Guide to Ireland* (London, 1967), p. 251 (Second edition). J. D'Alton, *The history of County Dublin* (Dublin, 1838), pp. 173-182.
4 W. Stokes, *Lives of saints from the Book of Lismore* (Dublin, 1890), p. 174.
5 J. D'Alton (1838, no. 3 above) p. 177.
6 We are grateful to the Rector of Glasnevin, the Rev. Desmond Harmon, and the Select Vestry for access to this map.
7 Tickell Mss; deed dated 16.iv.1736.
8 G. D. Burchataell and T. V. Sadleir, *Alumni Dubliniensis...* (Dublin, 1935).
9 B. M. Mansfield, 'Dr Richard Helsham, "The most eminent physician of this city and kingdom" (Dean Swift)', *Old Kilkenny Review, 3(1)* (Series 2) (1984), pp. 24-31. It is possible that John Putland's father leased the land originally and that later the lease passed to his son.
10 H. F. Berry, *A history of the Royal Dublin Society* (London, 1915), p. 84.
11 G. A. Aitkin, 'Thomas Tickell', *Dictionary of National Biography* (Oxford, 1975), p.2088 (Compact edition).
12 P. Smithers, *The life of Joseph Addison* (Oxford, 1968), pp. 311-312 (Second edition).
13 P. Smithers (1968, no. 12 above).
14 H. Wood, 'Addison's connexion with Ireland', *Journal of the Proceedings of the Royal Society of Antiquaries of Ireland, 34* (1904), pp. 133-158.
15 The legend of Tickell and Addison walking together is repeated in innumerable sources, although as early as 1903 it was rejected—see H. Wood (1904, no. 14 above), Lord Killanin and M. V. Duignan (1967, no. 3 above).
16 H. F. Berry (1915, no. 8 above), pp. 188-189.
17 This is stated by several authors, for example, J. Warburton, J. Whitelaw and R. Walsh (1818, no. 1 above) and J. D'Alton (1838, no. 3 above). See also R. E. Tickell, *Thomas Tickell and eighteenth century poets 1685-1740* (London, 1931), p. 127.
18 J. O'Keeffe, *Recollections of the life of John O'Keeffe* (London, 1826), vol. 1, p. 212.
19 Tickell Mss; P. Delany to T. Tickell 14.viii.1736.
20 Quoted by E. Malins and the Knight of Glin, *Lost demesnes, Irish landscape gardening 1660-1845* (London, 1976), p. 44.
21 Dr Hurd, quoted by R. Hayden, *Mrs Delany, her life and her flowers* (London, 1980), p. 170.
22 Quoted by J. D'Alton (1838, no. 3 above), pp. 174-175.
23 M. Hadfield, *A history of British gardening* (London, 1985), pp. 177-178. (Revised edition, published by Penguin Books Ltd). C. Thacker, *The history of gardens* (London, 1979), pp. 181-182.
24 J. Swift, 'A description of Mother Ludwell's Cove', quoted by E. Malins and the Knight of Glin (1976, no. 20 above).
25 J. Addison, *The Spectator* (24.vi.1712) quoted by e.g. C. Thacker (1979 no. 23 above), p. 182.
26 R. Hayden, (1980, no. 21 above), p. 86.
27 E. C. Nelson, 'Some records (c. 1690-1830) of greenhouses in Irish gardens', *Moorea, 2* (1983), pp. 21-28.
28 R. Hayden (1980, no. 21 above), p. 104.
29 E. Darwin, 'The loves of the plants', *The Botanic Gardens* (Dublin, 1796), part II (Fourth edition)—see E. C. Nelson, '[review of] R. Hayden, *Mrs Delany, her life and her flowers*', *Archives of Natural History, 10* (1981), pp. 361-362.
30 L. Pilkington, quoted by E. Malins and the Knight of Glin (1976, no. 20 above), p. 37.
31 J. C. Walker, 'Essay on the rise and progress of gardening in Ireland', *Transactions of the Royal Irish Academy, 4* (1791), pp. 3-19.
32 Smith Mss; A. Caldwell to J. E. Smith 23.ix.1793.
33 T. K. Cromwell (1820, no. 2 above), p. 18.
34 Tickell Mss; deed dated 24.x.1794.
35 H. F. Berry (1915, no. 10 above), p. 86.

Chapter 5

1 Foster/Massereene Papers D562/7837; J. Kiernan to J. Foster 1.ii.1794 [i.e. 1795]—the year date is certainly an error as the Society minutes indicate acquisition of the land at Glasnevin in 1795.
2 Foster/Massereene Papers; D207/29/32. RDS Mss; Minutes book (entries from 1795 to 1800 inclusive).
3 J. Warburton, J. Whitelaw and R. Walsh, *History of the city of Dublin* (London, 1818), pp. 1279-1305.
4 For details of the finances of the Gardens, see E. M. McCracken, 'The finances of the Glasnevin Botanic Gardens under the Royal Dublin Society, 1795-1877', *Glasra, 5* (1981), pp. 45-50.
5 The only copy book known to us is in the Foster/Massereene Papers D562/7829c.
6 RDS Mss; Minutes book (entry dated 19.xii.1793). A bibliography of the National Botanic Gardens is given by E. C. Nelson, 'A select, annotated bibliography of the National Botanic Gardens, Glasnevin, Dublin', *Glasra, 5* (1981), pp. 1-20.
7 RDS Mss; Minutes book (entry dated 12.iii.1795).
8 E. C. Nelson, 'Walter Wade's "Flora Dubliniensis", an enigmatic Irish botanical publication', *Long Room, 20-21* (1980), pp. 16-20.
9 Public Record Office, London, HO100/62 (ff 131-132); J. Foster to Lord Camden 21.vii.1796.
10 Camden's letter endorsing the request from J. Foster (no. 9 above) is dated 23.vii.1796 (PRO, London, HO100/62 (ff 129-130)).
11 We have been unable to trace any contemporary documents confirming that John Underwood was Scottish, nor have any records been found providing details of his early career in horticulture.
12 Linnean Society, London, Mss; nomination paper of J. Underwood.

13 For a detailed account of Lee and Kennedy, see E. J. Willson, *James Lee and the Vineyard Nursery* (Fulham, 1963), and E. J. Willson, *West London nursery gardens* (Fulham, 1982).

14 Parish registers, St Mobhi's Church, Glasnevin. See also A. E. Vicars, *Index to the prerogative wills of Ireland 1536-1810* (Dublin, 1897).

15 Parish registers, St Mobhi's Church, Glasnevin—the children recorded were (i) Mary, b. 21.ii.1805; (ii) John Newcommen, b. 4.ii.1807; (iii) Sarah, b. 21.i.1809; (iv) Eliza, b. 25.v.1811; (v) Charlotte, b. 26.xii.1812; (vi) William Newcommen, b. 11.v.1815; (vii) Jane, b. 6.vii.1818. The Whites also worshipped in St Mobhi's Parish Church at Glasnevin.

16 John White's competence in Irish was slight; he used Irish names in his monograph of grasses (see E. C. Nelson (1981, no. 6 above)) but does not demonstrate skill in the citation of those names. John White himself stated that he was a native of County Louth in a letter addressed to the Grand Jury of the county dated July 1814 (Foster/Massereene Papers D2519/4/1483).

17 Foster/Massereene Papers D207/29/30.

18 Plans and elevations of these buildings were published in the early catalogues and in the *Transactions of the Dublin Society* (see E. C. Nelson (1981, no. 6 above), p. 8, item 10). An estimate dated 20.v.1799 for these glasshouses is in Foster/Massereene Papers D562/7830.

19 [W. Wade], *Catalogue of plants in the Dublin Society's Botanic Garden...* (also in *Transactions of the Dublin Society for 1799*, 1 (1800)).

20 [W. Wade], 'A short description of the...botanical and agricultural garden, at Glasnevin.',*Transactions of the Dublin Society for 1799*, 1 (1800). T. K. Cromwell, *Excursions through Ireland*, (London 1820), vol. 2, pp. 19-25.

21 J. Warburton, J. Whitelaw and R. Walsh (1818, no. 3 above), p. 1284.

22 The Dublin Society published a prospectus of the Gardens as a preface to the catalogue (see no. 20 above) and this was also published in *The Dublin Magazine and Irish Monthly Register*, 5 (July 1800), pp. 1-8. (See E. C. Nelson (1981, no. 6 above)).

23 Foster/Massereene Papers D207/29/94; J. Underwood to J. Foster 9.xii.1809.

24 There are underground pipes linking the pond and the River Tolka; the channels are closed by sluices but these are very rarely opened. The pond thus fills and empties by natural seepage. One consequence of this is that the pond may flood in very wet weather, as the water levels are not precisely regulated. A major flood occurred in December 1954, when the Tolka overflowed and washed away hundreds of plants. Two large cedars of Lebanon were toppled.

25 The mixture of Linnean and Jussieuian systems persisted until the 1830s. Jussieu's system of classification was the forerunner of the modern method of ranking plants in families; Jussieu used the term 'Natural Orders'. For discussion of these systems see, e.g. F. A. Stafleu, *Linnaeus and the Linnaeans* (Utrecht, 1970).

26 RDS Mss; Minutes book. Foster/Massereene Papers D207/46/8.

27 H. F. Berry, *A history of the Royal Dublin Society* (London, 1915), p. 191.

28 Foster/Massereene Papers D207/29/70. H. F. Berry (1915, no. 27 above), p. 191. A mill existed at Glasnevin as early as 1535, and one is marked on the map dated 1720 (see Fig. 17, p. 38). In 1812 the Dublin Society renewed a lease on the mill site and garden on the western side of the road—this is the land presently called the Mill Field. The mill was in a ruinous state but was repaired as it was thought that the Society could use it to demonstrate methods of dressing hemp and making cement (see H. F. Berry (1915, no. 36 above) p. 191-192). But the mill proved costly. By 1817 the Society had spent £1,184 on it and in that year the mill was leased to John Hill.

The mill, a T-shaped building, is just visible in Fig. 36 (p. 73); it is clearly shown on deeds relating to 'the estate of the Rt Revd. the Dean of Christ Church let to William John Gore...' dated 1807 (these deeds are now in the National Library of Ireland (Mss 2789 and 2790)). We are extremely grateful to J. A. O'Doherty of Glasnevin for his help and interest, especially in unravelling the history of the mill.

29 Foster/Massereene Papers D207/29/63; Bishop of Kildare to J. Foster 15.xi.1807: D207/29/66; J. Foster to Bishop of Kildare 6.xii.1807: D207/29/67; Bishop of Kildare to J. Foster 8.xii.1807.

30 Foster/Massereene Papers D207/46/34; C. Vallency to J. Foster 30.iii.1807.

31 Foster/Massereene Papers D562/7842; J. Underwood to J. Foster 26.v.1803.

32 Foster/Massereene Papers D562/7842.

33 Foster/Massereene Papers D207/29/102; J. Underwood to C. Vallency 7.ix.1807.

34 Foster/Massereene Papers D207/29/80: D207/29/83; J. Underwood to J. Foster 10.v.1809. *Telopea speciosissima* was figured in *Botanical Magazine* in 1808 (tab. 1128) from a plant belonging to E. J. A. Woodford of Hertfordshire.

35 *Proceedings of the Dublin Society...1814-1815*, 51 (1815), p. 142.

36 H. F. Berry (1915, no. 27 above), pp. 236-237. E. M. McCracken, 'The origins of the library at Glasnevin Botanic Garden', *Irish Booklore*, 2(1) (1972), pp. 82-88.

37 TCD Mss, Sirr Papers, 868/2 f. 270; J. Templeton to T. Russell 19.iii.1797. We are grateful to Colán MacArthur for information about this letter and to the Keeper of Manuscripts, TCD, for providing a copy.

38 British Museum (Natural History), London, Botany Library. Dr David Mabberley (School of Botany, University of Oxford) kindly brought this manuscript to our attention. The descriptions of grasses are also contained in one of Robert Brown's manuscript notebooks. For a detailed biography of this most important botanist see D. J. Mabberley, *Jupiter Botanicus, Robert Brown of the British Museum* (London, 1985). Brown probably visited the Glasnevin Botanic Gardens again during a brief tour in Ireland in August and September 1810; he certainly visited the College Botanic Garden at Ballsbridge on this later occasion. He also visited Ireland in 1843.

39 E. Wakefield, *An account of Ireland, statistical and political* (London, 1812).

40 Diary of Sir Vere Hunt, entry dated May 1813. (This information was kindly supplied by Dr C. O'Mahony, Regional Archivist, Limerick).

41 A. Plumtre, *Narrative of a residence in Ireland during the summer of 1814...* (London, 1817), p. 28. We are grateful to F. E. Dixon for bringing this to our notice.

42 R. Walsh, 'Botanic Gardens', in J. Warburton, J. Whitelaw and R. Walsh, (1818, no. 3 above), p. 1287.

43 The attribution of this fine portrait to the Irish artist Allen Ramsay is generally accepted by art historians in Ireland as being improbable. Elsewhere, ECN questioned that the portrait could be of Dr Wade, but he has revised his opinion

and now accepts that the identity of the sitter, as indicated in a manuscript note attached to the painting, should be accepted (see *Archives of Natural History* 13 (1986) — Newsletter for fuller discussion). The portrait is reproduced in colour in *Irish Arts Review* 3(1)(1986), p. 3.

Chapter 6

1 A. Malcomson, *John Foster: the politics of the Anglo-Irish Ascendancy* (Oxford 1978).
2 Foster/Massereene Papers D207/29/130; B. McCarthy to J. Foster 4.x.1813.
3 Foster/Massereene Papers D207/29/113.
4 Foster/Massereene Papers D207/29/94; J. Underwood to J. Foster 9.xii.1809.
5 This punning phrase may have been current in Dublin in the early nineteenth century, for Wade used a phrase approaching it in a letter dated 29.ix.1821 (Foster/Massereene Papers D207/59/32) '...under the auspices, assistance and *Fostering* care of Lord Oriel, [the Botanic Gardens] has heretofore flourished...'. The only printed item containing it that we can trace is in a letter written by Joseph Hamilton to the editor of the *Irish Farmer's and Gardener's Magazine*, 1 (1834), pp. 568-569.
6 Foster/Massereene Papers D207/29/82; W. Wade to J. Foster 6.v.1809.
7 Foster/Massereene Papers D207/59/2; W. Wade to J. Foster 14.iii.1811.
8 See E. C. Nelson, 'A select annotated bibliography of the National Botanic Gardens, Glasnevin, Dublin', *Glasra*, 5 (1981), pp. 1-20.
9 This volume is in the National Botanic Gardens, Glasnevin.
10 These are listed by E. C. Nelson (1981, see no. 8 above)
11 Both of these publications were printed in the *Transactions of the Dublin Society*, and also issued separately — see E. C. Nelson (1981, no. 8 above).
12 Bedford Mss, Irish Correspondence, Book A, p. ii; Duke of Bedford to W. Wade 6.iv.1806. We are grateful to Paul Smith for providing information on the Bedford-Wade correspondence and to Mrs M. Draper, Archivist, Bedford Office, London, for making available copies of the letters.
13 Foster/Massereene Papers D207/29/12; W. Wade to J. Foster 13.vii.1807.
14 Foster/Massereene Papers D207/46/8; C. Vallency to J. Foster 15.vi.1807.
15 Foster/Massereene Papers D207/59/3; W. Wade to J. Foster 13.vi.1811.
16 Foster/Massereene Papers D207/59/3; W. Wade to J. Foster 13.vi.1811.
17 Foster/Massereene (Chilham) Papers T2519/4/1386; W. Wade to J. Foster 12.ix.1812.
18 Foster/Massereene (Chilham) Papers T2519/4/1494; W. Wade to J. Foster 19.ix.1814.
19 Foster/Massereene Papers D207/59/24; W. Wade to J. Foster 4.x.1819.
20 Foster/Massereene Papers D207/59/4; W. Wade to J. Foster 29.vii.1811.
21 Foster/Massereene Papers D207/59/5; W. Wade to J. Foster 21.x.1811.
22 Foster/Massereene Papers D207/59/1; W. Wade to J. Foster 12.ii.1811.
23 Foster/Massereene Papers D207/59/1; W. Wade to J. Foster 12.ii.1811.
24 Foster/Massereene Papers D207/59/22; W. Wade to J. Foster 3.v.1819.
25 Foster/Massereene Papers D207/59/27; W. Wade to J. Foster 23.i.1821.
26 Foster/Massereene Papers D207/59/28; W. Wade to J. Foster 10.ii.1821.
27 Royal Society, London. Mss; Certificates IV, 181.
28 W. Wade, 'Catalogus plantarum rariorum in comitatu Gallovidiae, praecipue Cunnamara inventarum...' *Transactions of the Dublin Society*, 2 (11) (1801), pp. 103-127.
29 W. Wade, *Plantae rariores in Hiberniae inventae...* (Dublin 1804); also published in *Transactions of the Dublin Society*, 4 (1804). Wade gave specimens of this moss to Robert Brown during the latter's visit to Glasnevin in September 1800. See also D. Turner, *Muscologiae Hibernicae Spicilegium* (Yarmouth 1804), p. 104.
30 W. Wade, *Quercus or oaks, from the French of Michaux...with notes and an appendix* (Dublin 1809).
31 W. Wade, *Salices, or an essay towards a general history of sallows, willows and osiers...* (Dublin 1811).
32 W. Wade, *De Buddlea globosa et de Holco odorato...with two coloured plates* (Dublin 1804).
33 W. Wade, *Hortus gramineus Glasnevinensis; or, the Glasnevin Grass Garden* (Dublin 1818). (For bibliographic information on this, see E. C. Nelson (1981, no. 8 above, item 22)).
34 W. Wade, *De Buddlea globosa...* (Dublin 1804) — it also appeared in *Transactions of the Dublin Society*, 4 (1804), pp. 215-221.
35 C. Schulze, 'A memoir on the great advantage...of cultivating bees in Ireland', *Transactions of the Dublin Society*, 1(1) (1800).
36 Foster/Massereene Papers D207/29/94; J. Underwood to J. Foster 9.xii.1809.
37 Foster/Massereene Papers D207/59/1; W. Wade to J. Foster 12.ii.1811.
38 Foster/Massereene Papers D562/16097; J. White to J. Foster 9.i.1822.
39 Foster/Massereene Papers D562/16147; J.White to J. Foster 3.v.1822.
40 Foster/Massereene Papers D562/16165; J. White to J. Foster 29.vi.1822. See also D562/16172; J. White to J. Foster 18.vii.1822.
41 This Latin name is not validly published as it appears without any description in W. Wade, *Prospectus of lectures on botany* (Dublin 1823). (See E. C. Nelson, *An Irish Flower Garden* (Kilkenny 1984) pp. 15-17.)
42 J. White, *An essay on the indigenous grasses of Ireland* (Dublin 1808).
43 This catalogue (see no. 33 above) contained pressed specimens of the plants noted, so a substantial amount of material had to be prepared for the publication. It is probable that an insufficient number of specimens was available, and that this accounts for the extraordinary rarity of this publication. (See E. C. Nelson (1981, no. 8 above)).
44 [K. Baily], *The Irish Flora* (Dublin 1833). Katherine Sophia Baily was the compiler of this book — she was 21 years old at the time. Although published anonymously, several writers named her as the author in the 1830s (for example *Irish Farmer's and Gardener's Magazine*, 1 (1834), p. 262).
45 Foster/Massereene Papers D207/59/25; W. Wade to J. Foster 29.xi.1819.
46 Foster/Massereene Papers D207/59/25; W. Wade to J. Foster 29.xi.1819.
47 Foster/Massereene Papers D207/59/25; W. Wade to J. Foster 29.xi.1819.

48 *Proceedings of the Royal Dublin Society 1820-1821*, 57 (1821), p. 109.
49 Foster/Massereene Papers D207/59/26; J. Foster to W. Wade 30.iii.1820.
50 John Underwood's memorial was printed in full in the *Proceedings of the Royal Dublin Society*, 56 (1820), pp. 172-177.
51 Foster/Massereene Papers D207/29/156; J. Foster to J. White 23.vii.1820.
52 Foster/Massereene Papers D207/59/27; W. Wade to J. Foster, 23.i.1821.
53 Foster/Massereene Papers D207/59/158; J. White to J. Foster 23.i.1821.
54 Foster/Massereene Papers D207/59/29; W. Wade to J. Foster 10.iii.1821.
55 Foster/Massereene Papers D207/59/32; W. Wade to J. Foster 29.x.1821.
56 Foster/Massereene Papers D207/29/191; J. White to J. Foster 27.iv.1825.
57 M. Fallon, *The sketches of Erinensis: selections of Irish medical satire 1824-1836* (London, 1979). Fallon did not republish the article on Wade and Litton which originally appeared in *The Lancet* (16.v.1835), pp. 253-258. This is reprinted here as Appendix 1.
58 *Proceedings of the Royal Dublin Society...1825-1826*, 62 (1826) pp. 9-11. In the Foster/Massereene Papers (D562/7867) is a draft of a letter from Lord Oriel to B. McCarthy, Secretary of the Royal Dublin Society, dated 20.xi.1825. Clearly this was intended as a reply to a letter received by Oriel from the Society. Oriel wrote:
> I am much honoured and gratified by the favourable view which the Botany Committee has taken of the letter I ventured to trouble the Society with and I will hope, that both it and the College [i.e. Trinity College, Dublin] will make trial at least, whether a union of the two gardens under the same person may not prove the most economical to both, and most effectual to answer the scientific purpose of the College, and the practical objects of the Society...

This shows a remarkable change in Foster's attitude to the University with which he had refused to co-operate in the 1790s.
59 C. A. Cameron, *History of the Royal College of Surgeons in Ireland...* (Dublin 1888) pp. 618-619.
60 Contemporary commentators, including *Erinensis*, take care to note the different lecturing styles of Wade and Litton,
61 S. Litton, *Syllabus of a course of lectures on botany...* (Dublin 1830). S. Litton, *Syllabus of a course of lectures on practical botany...* (Dublin 1833). E. C. Nelson (1981, no. 8 above), items no. 26, 27.
62 A. Malcomson (1978, no. 1 above).
63 *Proceedings of the Royal Dublin Society...1830-1831*, 67 (1831) Appendix II, pp. xix-xxi.
64 Z, 'Botanic Garden, Glasnevin', *Dublin Penny Journal* (1833-1834), pp. 107-108.
65 J. Hamilton, 'On the improvements in Glasnevin Garden', *Irish Farmer's and Gardener's Magazine*, 1 ((1) 1834), pp. 568-569 (see no. 5 above).

Chapter 7

1 Moore Mss; J. T. Mackay to the Rev. Dr Sadleir 1.ii.1835.
2 It is generally, but incorrectly, stated that Ninian Niven was the son of James Niven, who worked as a botanical collector in South Africa—the source of this error appears to be in the first edition of J. Britten and G. S. Boulger *A biographical dictionary of British and Irish Botanists* (London 1893). However, Ninian Niven's father was also called Ninian Niven and he was a well-known gardener in Scotland. We are very grateful to Paddy and Jennifer Woods, Royal Botanic Garden, Edinburgh, for their help in unravelling the Niven family tree through work in Scottish archives. The brief biography of Ninian Niven published in *The Garden*, 7 (1875), pp. xi-xii, is accurate.
3 D. Beaton, 'My autobiography', *The Cottage Gardener*, 13 (1854), p. 156.
4 N. Niven, 'Account of a method of cultivating the grape vine', *Memoirs of the Caledonian Horticultural Society*, 4 (1829), pp. 234-243.
5 N. Niven 'Fuschia [sic] macrostemma var. recurvata', *Irish Farmer's and Gardener's Magazine*, 1 (1834), p. 139. *Botanical Magazine* (tab. 3521).
6 This 'remarkable hybrid raised in Glasnevin Garden many years ago...' is mentioned by W. R. McNab, *A guide to the Royal Botanic Gardens, Glasnevin* (Dublin 1885), p. 22. We can trace no other account of this plant. It is unlikely to be a hybrid and is perhaps a cultivar, selected by Niven. There is a plant labelled *Strelitzia* × *nivenii* in the National Botanic Gardens (July 1985) but it has not flowered in recent years and appears to be only a small form of *S. reginae*.
7 N. Niven, 'On the villa plantations in the neighbourhood of Dublin', *Irish Farmer's and Gardener's Magazine*, 1 (1833), pp. 6-8.
8 E. Murphy, 'Observations on the Stirling Agricultural Exhibition', *Irish Farmer's and Gardener's Magazine*, 1 (1834), pp. 345-346.
9 N. Niven, 'Observations of the Glasnevin Garden', *Irish Farmer's and Gardener's Magazine*, 1 (1834), pp. 406-408.
10 J. Hamilton, 'On the improvements in Glasnevin Garden', *Irish Farmer's and Gardener's Magazine*, 1 (1834), pp. 568-569.
11 J. C., 'Observations on Mr Niven's proposed alterations in the Glasnevin Botanic Garden', *Irish Farmer's and Gardener's Magazine*, 1 (1834), pp. 517-520.
12 *Proceedings of the Royal Dublin Society*, 73 (1837).
13 N. Niven, *The visitor's companion to the Botanic Garden, Glasnevin* (Dublin 1838), pp. 155-162.
14 *The Literary Gazette and Journal of the Belles Lettres* (1834), p. 532. J. Morrell and A. Thackray, *Gentlemen of science* (London 1981), pp. 175-186.
15 Kew Mss, English Letters 8/99; N. Niven to W. J. Hooker 18.iii.1836.
16 *Irish Farmer's and Gardener's Magazine*, 2 (1835), p. 432. *Loudon's Gardeners' Magazine*, 12 (1836), pp. 118-120.
17 *Proceedings of the Royal Dublin Society*, 73 (1837).
18 Glasnevin Mss; Accessions book 1834-1889 (see Ch. 10, note 12 for details of these records).
19 *Proceedings of the Royal Dublin Society*, 73 (1837), pp. 6-8.
20 None of these specimens has been traced today in the herbarium in the National Botanic Gardens, Glasnevin.
21 No copy of this circular has been traced; its publication was noted by J. Hamilton, *Irish Farmer's and Gardener's Magazine*, 1 (1834), p. 569.
22 N. Niven (1838, no. 13 above), p. 71.
23 *Irish Farmer's and Gardener's Magazine*, 5 (1838), pp. 357-363.

24 *Proceedings of the Royal Dublin Society*, 72 (1836), Appendix III, pp. xvi-xvii. *Irish Farmer's and Gardener's Magazine*, 3 (1837), p. 185. *Magazine of Zoology and Botany*, 2 (1837), pp. 378-379. Niven's experiments on the translocation of fluids in plants were not novel; these internal currents had been described as early as 1727 by Stephen Hales (see A. G. Morton, *A history of botanical science* (London 1981), pp. 250-251).

25 N. Niven, 'On the formation of a geological arrangement of the rocky strata of Ireland, with the plants indigenous to the same', *Irish Farmer's and Gardener's Magazine*, 2 (1835), pp. 531-532.

26 N. Niven, *Essay on the recent failure of the potato crop* (Dublin 1835). This was also published in the *Irish Farmer's and Gardener's Magazine*, 2 (1835).

27 N. Niven, *The potato epidemic* (Dublin 1846). E. C. Nelson, 'David Moore, Miles J. Berkeley and scientific studies of potato blight in Ireland, 1845-1847', *Archives of Natural History*, 11 (1983), pp. 249-261.

28 W. Robinson, 'The Garden-Farm and New Winter Garden, Dublin', *Gardeners' Chronicle* (1864), pp. 1226-1227.

29 E. Malins and P. T. P. Bowe, *Irish gardens and demesnes from 1830*, (London 1980), pp. 36-42. N. Niven, 'A general outline...of the new public garden', *Prospectus of the proposed public gardens at Monkstown Castle* (Dublin 1839), pp. 17-25.

30 E. Malins and P. T. P. Bowe (1980, no. 29 above).

31 *Irish Farmer's Gazette* (October 1853), p. 308.

32 *Irish Farmer's and Gardener's Magazine*, 5 (1838), p. 225.

33 J. A. Robins, 'History of the Royal Horticultural Society of Ireland', in E. C. Nelson and A. Brady (eds) *Irish Gardening and Horticulture* (Dublin 1979), pp. 71-92.

34 *Irish Farmers' Gazette*, 8 (1849), p. 441.

35 E. C. Nelson (1983, no. 27 above).

36 N. Niven, 'Observations on the Glasnevin Garden', *Irish Farmer's and Gardener's Magazine*, 1 (1834), p. 408.

37 N. Niven, *Redemption thoughts* (Dublin 1869).

38 *Gardeners' Chronicle*, 11 (1879), p. 277.

39 For a discussion of this subject see J. Prest, *The Garden of Eden; the botanic garden and the recreation of Eden* (New Haven and London 1981).

40 Moore Mss; List of plants dated 10.xi.1838.

41 *Proceedings of the Royal Dublin Society*, 75 (1839), pp. 51-53.

Chapter 8

1 The Register of Dundee Parish recording David's baptism used the form Moir. A document among the Moore family papers, dated 1829, used the form Muir (see no. 5 below). It is noteworthy that all members of the family changed to using the surname Moore, and yet some commentators have stated that David made the alteration to disguise his Scottish origins. This latter suggestion is seemingly contradicted by several of his obituaries which noted that he retained his Scottish accent (see *Daily News*, 10.vi.1879. N. Moore, 'David Moore', *Dictionary of National Biography* 13, p. 792. S. A. Stewart and T. H. Corry, *Flora of the north-east of Ireland* (Belfast 1888), pp. xviii-xx).

2 E. C. Nelson, 'David Moore's date of birth, a correction', *Glasra*, 7 (1983), p. 24.

3 A. Eliot, 'Forfarshire Naturalists, 1. William Gardiner, jun., author of "The Flora of Forfarshire", &c', *People's Journal* (5.iv.1879). M. R. D. Seaward, E. C. Nelson and B. J. Coppins, 'David Moore and Isaac Carroll: some lichenological and bryological correspondence', *Glasra*, 6 (1982), pp. 73-96.

4 A. J. Warden, *Angus or Forfarshire* (Dundee 1880), pt. 14, ch. 35, pp. 200-201.

5 Moore Mss; testimonial for D. Muir, 27.iii.1829.

6 J. C. Loudon, *Encyclopaedia of gardening* (London 1834), pp. 354-355.

7 D. Moore, 'An account of a new Catasetum from Brazil, with observations on the importation and treatment of "Orchidaceoeus" plants', *Irish Farmer's and Gardener's Magazine*, 1 (1834), pp. 190-192.

8 Moore Mss; Helen Moore to D. Moore 14.viii.1832.

9 Ordnance Survey, Phoenix Park, Dublin, Mss; (records of payments to staff).

10 Kew Mss, English Letters 6/58; D. Moore to W. J. Hooker 2.iv.1834. English Letters 11/89; D. Moore to W. J. Hooker 26.i.1838. Ordnance Survey, Phoenix Park, Dublin, Mss; (records of payments to staff).

11 Kew Mss, South African Letters (1830-1844) 58/39; W. H. Harvey to W. J. Hooker 16.iv.1835.

12 Moore Mss; J. T. Mackay to D. Moore 13.xi.1834.

13 It is not known what fern Harvey meant—David Moore did not discover *Asplenium billotii* (= *A. lanceolatum*) although he may have mistaken a form of *A. adiantum-nigrum* for the rarer species. (D. M. Synnott to E. C. Nelson pers. comm.).

14 These plants were found '...growing near each other' in June 1836 (Kew Mss, English Letters 8/93; D. Moore to W. J. Hooker 1.vii.1836). The sedge survives in cultivation; the cultivated plants originated from the Lough Neagh population. The grass still grows at Lough Neagh.

15 The memoir was originally published in a small edition for circulation to members of the British Association for the Advancement of Science at the meeting in Dublin, August 1835 (see J. Morrell and A. Thackray, *Gentlemen of science* (London 1981), pp. 333-334). A second edition, expanded and revised, was published in 1837; only this second edition contains George du Noyer's illustrations and the detailed botanical appendix by David Moore.

16 Kew Mss; English Letters 13/89; D. Moore to W. J. Hooker, undated (c. 1837). English Letters 13/87; D. Moore to W. J. Hooker 9.x.1837.

17 The original watercolours are in the National Botanic Gardens, Glasnevin. M. J. P. Scannell and C. I. Houston, 'George V. du Noyer (1817-1869), a catalogue of plant paintings at the National Botanic Gardens, Glasnevin, with aspects of his scientific life', *Journal of Life Sciences, Royal Dublin Society*, 2 (1980), pp. 1-13. M. J. P. Scannell, 'The apple paintings of George du Noyer', *The Garden*, 108 (1983), pp. 458-459.

18 Kew Mss, English Letters 13/89; D. Moore to W. J. Hooker, undated (c. 1857). The rose illustrations were not published.

19 E. C. Nelson, 'William McCalla, a second panegyric for an Irish phycologist', *Irish Naturalist's Journal*, 20 (1981), pp. 271-283.

20 Charles Moore joined the Irish Ordnance Survey on 2 May 1837, according to records preserved in the Ordnance Survey, Phoenix Park, Dublin. He worked for at least part of his time in County Donegal. In June 1838, at a *fête champêtre* in Belfast Botanical Garden, Charles Moore was awarded one of the Templeton prizes for a display of native plants. The event was reported in *The Northern Whig*, 30 June 1838:

> The first Templeton Prize was awarded to Mr Charles Moore, Assistant in the Geological department of the Irish Ordnance Survey, for the following plants...[there follows a list of native species, including ferns, orchids, sundews and 'Pyrola Secunda, Media and Minor']...Two additional medals were awarded, for the rarest single indigenous plants — one to Miss Richards of Bangor Castle for Ophrys Apifera; the other to J. A. Whitla, Esq., of Gobrana for Epipactis Grandiflora...'

(A photostat copy of *The Northern Whig* was kindly provided by the Librarian, Linen Hall Library, Belfast). In E. M. McCracken, *The palm house and Botanic Garden, Belfast* (Belfast 1971), p. 20, the Templeton Prize was incorrectly attributed to David Moore.

21 Moore Mss; Marriage certificate dated 7.iv.1836.
22 This was the practice at that time — none of the members of the Moore family known to us have any knowledge of David Moore's first or second families.
23 Kew Mss, English Letters 11/89; D. Moore to W. J. Hooker 26.i.1838. English Letters 11/90; D. Moore to W. J. Hooker 16.iii.1838.
24 Kew Mss, English Letters 11/92; D. Moore to W. J. Hooker 6.v.1838.
25 Kew Mss, English Letters 11/93; D. Moore to W. J. Hooker 19.vi.1838.
26 Kew Mss, English Letters 11/93; D. Moore to W. J. Hooker 16.vii.1838.
27 Moore Mss; both the original letters and a printed compilation of these testimonials survive.
28 Moore Mss; N. Niven 'Copy of Daniel Clifford's engagement May 1838'. Moore Mss; D. Moore to N. Niven 19.xi.1838.
29 H. F. Berry, *A history of the Royal Dublin Society* (London 1915), p. 263.
30 Moore Mss; David Moore, 'Memorandum for the young men employed in the Royal Dublin Society's botanic Garden', 1.xii.1838.
31 Moore Mss; 'Terms on which apprentices are taken in the Royal Dublin Society's Botanic Garden', 15.iii.1845.
32 E. M. McCracken, 'The arboretum, Dublin Botanic Garden', *Quarterly Journal of Forestry*, 63 (1969), pp. 242-253.
33 E. M. McCracken, 'Origins of the library at Glasnevin', *Irish Booklore*, 2(2) (1972), pp. 82-88.
34 Glasnevin Mss; Account book [for] library, 24.vii.1852 to 30.xii.1941.
35 Kew Mss, English Letters 16/268; D. Moore to W. J. Hooker 2.iv.1841.
36 H. F. Berry (1915, no. 29 above), pp. 258-266.
37 E. Murphy, 'Royal Dublin Society, the Botanic Garden', *Irish Farmer's and Gardener's Magazine*, 8 (1841), pp. 141-143.
38 J. Lindley, [editorial dated 17 April 1841], *Gardeners' Chronicle*, 1 (1841), p. 243.
39 Kew Mss, English Letters 16/273; D. Moore to W. J. Hooker 4.x.1841.
40 W. J. Bean, *The Royal Botanic Gardens, Kew* (London 1908), pp. 28-33. R. G. C. Desmond, 'The Hookers and the development of the Royal Botanic Gardens, Kew', *Biological Journal of the Linnean Society, London*, 7 (1975), pp. 173-182. W. Blunt, *In for a penny* (London 1978).
41 E. J. Diestelkamp, 'The design and building of the Palm House, Royal Botanic Gardens, Kew', *Journal of Garden History*, 2 (1982), pp. 233-272. E. J. Diestelkamp, 'The conservatories and hot-houses of Richard Turner', *Historic greenhouse and Royal Kew* (Kew 1982), pp. 6-11.
42 Kew Mss, English Letters 18/54; D. Moore to W. J. Hooker 26.x.1842.
43 No foundation stone was laid and no promenade took place.
44 For detailed discussions of the history of the design and construction of the Curvilinear Range, we are indebted to Dr Edward J. Diestelkamp who kindly made available to us a copy of the relevant chapter from his unpublished doctoral thesis, *The iron and glass architecture of Richard Turner* (University College, London, 1982). The following paragraphs are based on that work, and on E. J. Diestelkamp and E. C. Nelson, 'Richard Turner's Legacy, the Glasnevin Curvilinear Glasshouse', *Taisce Journal 3(1)*, (1979), pp. 4-5. E. J. Diestelkamp, 'Richard Turner (c. 1798-1881) and his glasshouses', *Glasra*, 5 (1981), pp. 51-53.
45 Meyler was a member of the Royal Dublin Society (H. F. Berry (1915, no. 29 above) p. 238). See A. Meyler, *Observations on ventilation...* (London c. 1818).
46 E. J. Diestelkamp (1981, no. 44 above).
47 E. M. McCracken (1971, no. 20 above).
48 Moore Mss; W. McNab to D. Moore 16.v.1846.
49 Kew Mss, English Letters 26/330; J. Lyons to W. J. Hooker 4.v.1848. Lyons' other comments in this letter are interesting for they concern the heating of the Glasnevin curvilinear range. He remarked, with reference to Turner's newly completed Palm House at Kew that '...I [Lyons] hope the heating will be complete, he [Turner] has failed in the new houses in Glasnevin Garden. They cannot in the stove obtain more heat than about 45[°F] in winter...'.
50 E. J. Diestelkamp (1982, no. 44 above).
51 Kew Mss, English Letters 22/273; R. Turner to W. J. Hooker 23.v.1844.
52 A central pavilion with a circular floor-plan was included by Turner in a conservatory designed for Killakee House, Co. Dublin — the building is not extant but an engraving was published by C. M'Intosh, *The book of the garden* (Edinburgh and London, 1855), vol. 1, plate 20;
53 E. C. Nelson, *John Lyons and his orchid manual* (Kilkenny 1983) — this volume includes a facsimile of Lyons' manual with a detailed biographical introduction. See also E. C. Nelson, 'Two centuries of orchids at Glasnevin', *Orchid Review*, 89 (1981), pp. 136-141, 191-194. E. C. Nelson, 'John Lyons and the first orchid manual', *Orchid Review*, 91 (1983), pp. 74-77.
54 J. Arditti, 'A history of orchid hybridization, seed germination and tissue culture', *Botanical Journal of the Linnean Society, London*, 89 (1984), pp. 359-381.
55 For a review of this subject, see E. C. Nelson, 'David Moore, Miles J. Berkeley and scientific studies of potato blight in Ireland, 1845-1847', *Archives of Natural History*, 11 (1983), pp. 249-261.

56 Shortly after the death of his first wife, Hannah, in the middle of December 1840, David Moore wrote to a friend Isabella Morgan at Cookstown, Co. Tyrone, that he had lost '...the only source of my earthly happiness...' (Moore Mss; D. Moore to I. Morgan 21.xii.1840). In 1843, he married Isabella and, as far as we can ascertain, they had two children before her death (see no. 22 above).
57 Erinensis, 'Dr Samuel Litton and the Botanic Garden', *The Lancet, 16* (16.v.1835) (see Ch. 6, no. 57) — see Appendix 1.
58 W. J. Hooker, 'Littonia modesta', *Botanical Magazine*, tab. 4723. E. C. Nelson, *An Irish Flower Garden* (Kilkenny 1984), pp. 184-185.

Chapter 9

1 Kew Mss, English Letters 25/216; W. H. Harvey to W. J. Hooker 10.vi.1847.
2 Kew Mss, English Letters 25/220; W. H. Harvey to W. J. Hooker 11.vi.1847.
3 Kew Mss, English Letters 25/221; W. H. Harvey to W. J. Hooker 22.vi.1847.
4 No limit was placed on the professorship. The Royal Dublin Society gradually lost control of its professors, and in 1854 they were transferred to the Museum of Irish Industry (see p. 139)
5 The other candidates were W. E. Steele, O'Bryen Bellingham, P. Clinton, J. Morissy, A. Mitchell and W. Andrews.
6 Kew Mss, English Letters 25/231; W. H. Harvey to W. J. Hooker 10.xii.1847.
7 Kew Mss, English Letters 25/232; W. H. Harvey to W. J. Hooker 13.xii.1847.
8 Kew Mss, English Letters 25/233; W. H. Harvey to W. J. Hooker 21.xii.1847
9 Kew Mss, English Letters 26/185; W. H. Harvey to W. J. Hooker 4.i.1848.
10 Kew Mss, English Letters 26/187; W. H. Harvey to W. J. Hooker 27.i.1848.
11 [L. Fisher], *Memoir of W. H. Harvey...* (London, 1869), pp. 1-9.
12 James White of Ballitore has been described as an excellent and prominent Irish botanist in several unreliable secondary sources. He must not be confused with John White, undergardener at Glasnevin. This error is made by the otherwise reliable R. L. Praeger, 'William Henry Harvey 1811-1866', in F. W. Oliver (ed) *Makers of British botany* (Cambridge 1913). At best, James White was a good teacher and keen amateur (see L. Fisher (1869, no. 11 above, p. 2).
13 W. H. Harvey to a cousin, 1827 — quoted by L. Fisher (1869, no. 11 above), pp. 4-5.
14 E. C. Nelson, 'Thomas Coulter (1793-1843) in North America, some bibliographic problems and some solutions', *Contributions to the history of North American Natural History* (London 1983), pp. 59-71. A full biography of Thomas Coulter is in preparation by E. C. Nelson and the late A. Probert.
15 Kew Mss, English Letters 26/191; W. H. Harvey to W. J. Hooker 11.iv.1848.
16 Kew Mss, English Letters 29/191; W. H. Harvey to W. J. Hooker, undated (about 24.v.1850).
17 Kew Mss, English Letters 29/341; W. H. Harvey to W. J. Hooker, undated (see no. 16 above): English Letters 29/330; W. H. Harvey to W. J. Hooker 31.v.1840.
18 *Irish Farmer's and Gardener's Magazine,* 5 (1838), p. 226.
19 Glasnevin Mss; Accessions book for 1834-1889.
20 *Freeman's Journal*, 24 June 1853.
21 E. M. McCracken, *The palm house and Botanic Garden, Belfast* (Belfast 1971), p. 46.
22 *Proceedings of the Royal Dublin Society,* 87 (1850). See also M. Craig, 'The Society's Buildings', in J. Meenan and D. Clarke (eds) *Royal Dublin Society 1731-1981* (Dublin 1981), p. 66.
23 W. R. McNab, *A guide to the Royal Botanic Gardens, Glasnevin* (Dublin 1885). This building is now used as a workshop and store, and stands in the nursery area. It is perhaps one of the earliest prefabricated buildings in existence anywhere in the world. The cast-iron ribs bear the maker's name 'J. H. Porter, London'. We are grateful to Dr Edward Diestelkamp for information on this building. (See *Examples of iron buildings and roofing, manufactured by John H. Porter, Southwark* (London 1849) RIBA London, ref. no. Q8).
24 W. Robinson, 'Notes on gardens XIX [-XXII], Royal Botanic Gardens, Glasnevin', *Gardeners' Chronicle* (1864), pp. 988-989, 1011, 1035-1036, 1083-1084, 1131. It should be noted that there is no record of William Robinson having trained as a gardener under David Moore at Glasnevin, despite the assertion by M. Allen (*William Robinson 1838-1935* (London, 1982)) — see E. C. Nelson, 'An Irish bachelor as Father of the English Flower Garden', *Moorea*, 2 (1983), pp. 54-56.
25 Seed and plant donations are recorded in the Accessions Book for 1834 to 1889. Extracts from this will be published by E. C. Nelson, 'The sources of plants for the Botanic Gardens at Glasnevin 1795-1879'.
26 E. M. McCracken (1971, no. 21 above).
27 *The Royal Dublin Society and the citizens of Dublin* (Dublin 1860). 'Issued by a preliminary committee formed for promoting the freer opening of the Glasnevin Botanic Garden' (Copy in the Royal Irish Academy, Dublin).
28 W. Gregory, *Sir William Gregory, K.C.M.G.,...an autobiography edited by Lady Gregory* (London 1894), pp. 210-211 (Second edition).
29 E. J. Diestelkamp, *The iron and glass architecture of Richard Turner.* Unpublished doctoral thesis, University College, London, 1982.
30 Turner built other fine glasshouses in Ireland but none was as extensive and elaborate as that at Glasnevin — see E. J. Diestelkamp, 'Richard Turner (c. 1789-1881) and his glasshouses', *Glasra*, 5 (1981), pp. 51-53.
31 E. C. Nelson, *An Irish Flower Garden* (Kilkenny 1984), pp. 83-84. (There is an error in this account of Wilhelm Keit; the Brussels nursery where he worked belonged to Linder, not Sander.) Further material on Keit is contained in D. P. McCracken, 'William Keit and the Durban Botanic Garden', *Bothalia, 16(1)* (1986). B. D. Schrire, 'Centenary of the Natal Herbarium, Durban, 1882-1982', *Bothalia,* 14 (1983), p. 225. M. Gunn and L. E. Codd, *Botanical Exploration of southern Africa* (Capetown 1981).
32 Kew Mss, English Letters 95/336; D. Moore to J. D. Hooker 27.vii.1872.
33 R. G. Strey to E. C. Nelson, 30.x.1978. Strey's information came from Keit's letters presently in a private collection in South Africa. (See E. M. McCracken and E. C. Nelson, 'Julius Wilhelm Keit in Dublin' (in preparation)
34 Kew Mss, English Letters 95/338; D. Moore to J. D. Hooker 6.ix.1872.

35 Glasnevin Mss; Accessions Book for 1834-1889.
36 F. M. Leighton, 'The genus *Agapanthus* L'Heritier', *Journal of South African Botany*, Supplementary vol. 4 (1965). E. C. Nelson (1984, no. 31 above).
37 An obituary of Alexander Moore was published in *The Garden*, 25 (1884), p. 378. Glasnevin Mss; Letters vol. 1 (November 1877-November 1881), item no. E11. Letters vol. 1, item no. E43: memorandum dated 17.vi.1879.
38 Kew Mss, English Letters 95/341; D. Moore to J. D. Hooker 15.viii.1873.
39 This volume is in National Botanic Gardens, Glasnevin.
40 This volume is in National Botanic Gardens, Glasnevin.
41 Glasnevin Mss; Letters vol. 1(November 1877-November 1881), item no. E46.
42 Glasnevin Mss; Letters vol. 1 (November 1877-November 1881), W. Parnell to W. Steele 22.vi.1879.
43 Glasnevin Mss; Letters vol. 1 (November 1877-November 1881), item no. E46 (Testimonials of William Parnell) — this comment is in a testimonial from David Moore dated 21.ii.1852.
44 David Orr's date of death is given incorrectly in standard biographical dictionaries (e.g. R. G. C. Desmond, *British and Irish botanists and horticulturists* (London 1977)). We are grateful to Mary Davies for providing this correct date. Orr refused to give his age or date of birth to the authorities at Glasnevin (Glasnevin Mss; Letters vol. 1 (November 1877-November 1881), item no. E55) and there are other difficulties in providing an accurate account of the early years of his life. However it is certain that he was demoted from indoor foreman to assistant foreman in 1866.
45 D. A. Webb, 'William Henry Harvey 1811-1866 and the tradition of systematic botany', *Hermathena, 103* (1966), pp. 32-45.
46 Moore Mss; N. MacLeod to D. Moore 29.xi.1869.
47 R. D. Meikle, *Willows and poplars of Great Britain and Ireland* (London 1984), pp. 126-129. E. C. Nelson (1984, no. 31 above), pp. 22-23.
48 A list of David Moore's papers on mosses is contained in W. Lett, 'Census report on the mosses of Ireland', *Proceedings of the Royal Irish Academy, 32, B7* (1915), pp. 84-85. See also D. Moore, 'Report on Irish Hepaticae', *Proceedings of the Royal Irish Academy, 2* (series 2) (1877), pp. 591-672. M. R. D. Seaward, E. C. Nelson and B. J. Coppins, 'David Moore and Isaac Carroll: some lichenological and bryological correspondence', *Glasra, 6* (1982), pp. 73-86.
49 A second edition of *Cybele Hibernica* was published in October 1898, edited by N. Colgan and R. W. Scully, but David Moore's name was not printed on the title-page — this omission prompted an apology to Frederick Moore from Miss Frances More (Moore Mss; F. M. More to F. W. Moore 19.xi.1898).
50 For discussion see E. C. Nelson, 'A select annotated bibliography of the National Botanic Gardens, Glasnevin', *Glasra, 5* (1981), pp. 1-20. By the late 1870s these guide-books had been out-of-print for several years, and in 1885, Professor W. R. McNab produced what was described on the title-page as 'a revised and enlarged' edition of David Moore's work. Twenty years later, a *Short Guide* was prepared under the direction of Col. G. Plunkett. In its first year of issue, 1906, 3,900 copies were sold at one halfpenny each. Further 'editions' of this were printed — a list of these has not been published as the exact bibliographic status of each one needs careful investigation. This was replaced in April 1971 by a pamphlet (a folded sheet) with coloured photographs and a map. This pamphlet was revised in 1980 and the current 'A short guide to the Gardens' was published in June 1981, costing 25p (the equivalent of 5 shillings).
51 David Moore published a work on grasses which was produced in at least four editions. The first edition, titled *Concise notices of the indigenous grasses of Ireland best suited for agriculture with dried specimens of each kind* was published in 1843 and was printed by R. Purdie of Mullingar. A second edition, with the amended title *Concise notices of British grasses best suited for agriculture...* was published in 1850, by James McGlashan of Dublin. A third edition was at least planned in 1851, and another edition, also numbered the third edition, was published in Dublin by McGlashan and Gill and is dated 1856. Moore's publications on other horticultural and agricultural subjects are innumerable and scattered through journals and periodicals — no complete bibliography of Moore has been compiled but a partial bibliography, extracted from Irish journals, is included in R. J. Hayes, *Sources for the history of Irish civilization, Articles in Irish periodicals* (Boston 1970), pp. 711-712. His publications on indigenous Irish plants are listed by R. L. Praeger, *Irish Topographical Botany* (Dublin, 1903), pp. cxxxi-cxxxiii.
52 D. Moore, 'On the successful establishment of *Loranthus europaeus*...in the Botanic Garden, Glasnevin', *Journal of the Royal Dublin Society, 6* (1873), pp. 383-387.
53 Kew Mss; English Letters 95/325; D. Moore to J. D. Hooker 29.vii.1870.
54 D. Moore, 'On a hybrid Sarracenia...', *Atti del Congresso Internazionale Botanico, Firenze* (1874), pp. 174-179 — this was preprinted as a booklet and published separately in Dublin in 1874 (see E. C. Nelson, '*Sarracenia* hybrids raised at Glasnevin Botanic Gardens, Ireland; nomenclature and typification', *Taxon 35* (1986)). The two certificates presented with the medals are in the archive at the National Botanic Gardens, Glasnevin. The two medals, which are identical, are extant; one is in the possession of Dr Moore's descendants, and the second is in the National Museum of Ireland, Dublin. A description of the latter medal was published by M. J. P. Scannell and J. Teahan, 'The gold medal of the Royale [sic] Societa Toscana di Orticultura awarded to David Moore in 1874 for studies on *Sarracenia*', *Irish Numismatics, 96* (1983), pp. 191-193. This account contains several invalid statements. There is no way of telling which of the medals was awarded for the display of *Sarracenia* and which for the display of *Ouvirandra*. The medals were cast in quantity, and the name of each of the many recipients was later engraved on the reverse face. The medal was not awarded, as suggested by Scannell and Teahan, for Moore's research on *Sarracenia*, but for the display of plants at the International Congress, in the same manner as gold medals are given at horticultural shows in Britain and Ireland today.
55 D. Moore (1874, no. 54 above). The extremely delicate leaves of this plant were wrapped in banana leaves for the journey to Florence.
56 E. C. Nelson and M. R. D. Seaward, 'Charles Darwin's correspondence with David Moore of Glasnevin on insectivorous plants and potatoes', *Biological Journal of the Linnean Society, London, 15* (1981), pp. 157-164.
57 Kew Mss; English Letters 95/343; D. Moore to J. D. Hooker 22.vii.1874.
58 *Science and Revelation: a series of lectures in reply to the theories of Tyndall, Huxley, Darwin, Spencer, etc.* (Belfast 1875).
59 *Belfast Newsletter*, Tuesday 8.xii.1874. (Original in Moore Mss)

60 D. Moore, 'Design in the structure and fertilization of plants: a proof of the existence of God', in *Science and revelation* (see no. 58 above). (This was reprinted and issued as a pamphlet).
61 *Botanical Magazine*, tab. 6113. F. W. Moore, 'Moore's Crinum', *The Garden, 19* (1881), p. 260 and plate 274.
62 F. W. Burbidge, *The Narcissus, its history and culture* (London, 1875), plate 22. Moore Mss; F. W. Burbidge to D. Moore 21.vii.1874.
63 Kew Mss; English Letters 95/327; D. Moore to J. D. Hooker 6.xii.1870.
64 D. Moore, 'On a supposed new species of *Ceratozamia*', *Scientific Proceedings of the Royal Dublin Society, 2* (1878), pp. 112-114.
65 Kew Mss; English Letters 95/327; D. Moore to J. D. Hooker 6.xii.1870.
66 A. Gray, [Obituary for W. H. Harvey] *American Journal of Science and Arts, 42* (1866), pp. 273-277 (quoted by R. L. Praeger (1912, no. 12 above)).
67 We are grateful to J. White, Department of Botany, University College, Dublin, for his helpful comments about the early history of the Royal College of Science. See also R. A. Jarrell, 'The Department of Science and Art and control of Irish Science 1853-1905', *Irish Historical Studies, 23* (1983), pp. 330-347.
68 Glasnevin Mss; Letters vol. 1 (November 1877-November 1881), N. MacLeod to W. Steele i.xi.1877.
69 Glasnevin Mss; Letters vol. 1 (November 1877-November 1881), D. Moore to W. Steele 7.xi.1877.
70 Glasnevin Mss; Letters vol. 1 (November 1877-November 1881), Col. Donnelly to W. Steele 26.xi.1877.
71 Glasnevin Mss; Letters vol. 1 (November 1877-November 1881), Col. Donnelly to W. Steele 26.xi.1877.
72 Glasnevin Mss; Letters vol. 1 (November 1877-November 1881), item no. E6.
73 Moore Mss; draft of a letter undated and unaddressed but certainly David Moore's response to the memoranda from his superiors.
74 Glasnevin Mss; Letters vol. 1 (November 1877-November 1881), item no. E39, W. R. McNab to W. Steele 27.iii.1879.
75 Glasnevin Mss; Letters vol. 1 (November 1877-November 1881), item no. E21. D. Moore to W. Steele 7.vii.1878.
76 Kew Mss, English Letters 95/366; D. Moore to W. T. Dyer 16.xi.1878.
77 Kew Mss, English Letters 95/367; D. Moore to W. T. Dyer 21.xi.1878.
78 Kew Mss, English Letters 95/371; D. Moore to W. T. Dyer 12.iii.1879.
79 Kew Mss, English Letters 95/382; F. W. Moore to J. D.Hooker 10.vi.1879.
80 *Dundee Advertiser*, 9.vi.1879.
81 *Daily News*, 10.vi.1879.
82 *Gardeners' Chronicle, 11* (1879), p. 756.

Chapter 10

1 *Botanical Magazine*, tab. 1304.
2 Foster/Massereene Papers D207/59/30; W. Wade to J. Foster, undated (c. 1821).
3 E. J. Willson, *James Lee and the Vineyard Nursery* (Fulham 1961). E. C. Nelson 'Australian plants cultivated in England before 1788', *Telopea, 2* (1983), pp. 347-353.
4 J. Ewan and N. Ewan, 'John Lyon, nurseryman and plant hunter, and his journal, 1799-1814', *Transactions of the American Philosophical Society, 53* (new series) (1963), pp. 5-69.
5 Foster/Massereene Papers D207/29/83; J. Underwood to J. Foster 10.v.1809.
6 A. Coats, *The quest for plants* (London 1969), pp. 296-299.
7 Foster/Massereene Papers D207/59/3; W. Wade to J. Foster 13.vi.1811.
8 Foster/Massereene Papers D207/59/17; J. Foster to W. Wade, undated (c. vi.1811).
9 E. C. Nelson, 'Tropical drift fruits and seeds on the coasts in the British Isles and Western Europe II: History (1560-c. 1860) and folklore', *Scottish Naturalist*, 1983, pp. 11-63.
10 Records of donations are contained in the manuscript accessions books (see no. 12 below) and were published, more or less regularly, in the *Proceedings of the (Royal) Dublin Society*.
11 See Ch. 9, note 14.
12 The earliest register of donations includes entries dated from 12 March 1830 to July 1833. This was followed by a more substantial register dated from 3 April 1834 to 31 December 1889. Registers for the years 1890 to the present day are also extant. Extracts from the early registers are being prepared for publication by E. C. Nelson (see Ch. 9, note 29 above).
13 See Ch. 7, p. 91.
14 A. Castellanos, 'Las exploraciones botanicas en la epoca de la independencia 1810-1853', *Hombergia, 4* (1945), pp. 3-14. R. Gorer, *The growth of gardens* (London 1978), pp. 108-109 (Gorer's book contains information on Tweedie's work, based on records supplied by Dr Brian Morley from Glasnevin sources).
15 Peritus, 'On emigration, with reference to gardeners and on the prospects of botanical collectors', *Loudon's Gardeners' Magazine, 16* (1840), pp. 115-116.
16 Kew Mss. English Letters 20/137; D. Moore to W. J. Hooker 9.xii.1843.
17 R. Gorer (1978, no. 14 above), p. 108.
18 Kew Mss, English Letters 9/284; N. Niven to W. J. Hooker 24.vii.1837.
19 *Botanical Magazine*, tab. 3773 (David Moore's christian name is given incorrectly as Dugald in the text).
20 Kew Mss, English Letters 18/54; D. Moore to W. J. Hooker 24.x.1842.
21 E. C. Nelson, 'In honour of David and Frederick Moore', *Moorea, 1* (1982), pp. 1-4. E. C. Nelson, *An Irish Flower Garden* (Kilkenny 1984), pp. 93-94.
22 B. D. Morley, 'Edward Madden and Glasnevin', *Stream and Field, 2* (June-July 1971), pp. 18-19. B. D. Morley, 'Edward Madden (1805-1856)', *Journal of the Royal Horticultural Society, 97* (1972), pp. 203-206. R. Gorer (1978, no. 14 above), pp. 19-20, 119-122. E.C. Nelson (1984, no. 21 above).
24 E. Madden, 'Diary of an excursion to the Shatool and Boorun Passes over the Himalaya in September 1845', *Journal of the Asiatic Society of Bengal, 15* (1846), pp. 79-135.
25 D. Moore to N. Ward 1.ii.1842, quoted by N. Ward, *On the growth of plants in closely glazed cases* (London 1842), pp. 91-92.

26 *Botanical Magazine*, tab. 3929, 3942. Kew Mss, English Letters 15/84; D. Moore to W. J. Hooker 4.xi.1840.
27 *Irish Farmer's and Gardener's Magazine*, 3 (1836), pp. 140-141.
28 Ordnance Survey, Phoenix Park, Dublin, Mss; (staff records including accounts of wages payments).
29 The most thorough account of the 'Grand Bungle' is contained in L. A. Gilbert, *Botanical investigation of New South Wales 1811-1840*. Unpublished doctoral thesis, University of Sydney, 1971, pp. 484-493, and L. A. Gilbert, *The Royal Botanic Gardens, Sydney, a history 1816-1985* (Sydney 1986).
30 Kew Mss, English Letters 33/310; D. Moore to W. J. Hooker
31 Moore Mss; W. G. Milne to D. Moore 28.xi.1865.
32 Moore Mss; W. G. Milne to D. Moore 28.vi.1865.
33 Moore Mss; W. G. Milne to D. Moore 26.vii.1865.
34 D. Moore, 'The late Mr Grant Milne', *Gardeners' Chronicle* (1866), p. 731. The original letter is not among the Moore papers but it is clear that the dates given by Moore in the *Gardeners' Chronicle* are incorrect. The letter from Lauche must have been dated 4.iv.1866, not 4.iv.1864.
35 D. Moore (1866, no. 34 above).
36 Glasnevin Mss; Accessions book for 1834-1889.
37 E. C. Nelson (1984, no. 21 above), pp. 76-79.
38 A. A. Dorrien-Smith, 'An attempt to introduce *Olearia semidentata* into the British Isles', *Bulletin of Miscellaneous Information (Kew)* 1910, pp. 120-126.
39 Based on H. Travers' correspondence with F. W. Moore (Glasnevin Mss), dated from 13.vii.1906 to 13.x.1911. *Olearia semidentata* is listed among living plants received at Glasnevin from New Zealand in October 1908.
40 E. C. Nelson, 'Charlotte Isabel Wheeler Cuffe 1867-1967', *Rhododendrons 1981-1982 with Magnolias and Camellias* (1981), pp. 33-41.
41 R. Farrer, 'Second exploration in Asia, no. 25, Maymyo', *Gardeners' Chronicle*, 69 (third series) (1921), p. 54.

Chapter 11

1 Moore Mss; Col. Donnelly to F. W. Moore 23.vi.1879. W. Steele to F. W. Moore 27.vi.1879. On 22 June 1879, the day before Moore was told of his appointment, William Parnell wrote to Steele applying for the vacant keepership; Parnell's letter is in Glasnevin Mss; Letters vol. 1 (November 1877-November 1881).
2 Register of Scot's Church, Mary's Abbey, Parnell Square, Dublin.
3 F. W. Moore, 'Some reminiscences', *New Flora and Sylva*, 12 (1940), pp. 180-187.
4 Kew Mss, English Letters 95/337; D. Moore to J. D. Hooker 27.vii.1872.
5 F. W. Moore (1940, no. 3 above), p. 181.
6 Moore Mss; H. Cleghorn to F. W. Moore 11.xii.1874. See also *Gardeners' Chronicle*, 11 (1879), p. 756, where it is stated that Frederick Moore took '...an examination for the Indian Forest Service with much credit'. In letters to J. H. Balfour, Regius Keeper of the Royal Botanic Garden, Edinburgh, David Moore provided some additional information on his son's progress. In a letter dated 14 December 1874 he wrote: '[Frederick] only got 4th place and there being two vacancies he is of course out for the present...He did well in the natural sciences, but failed in the Algebra and free hand drawing.' On 7 August 1876, David Moore noted that '...my son who stood the examination for the Indian Forestry and only came out fourth last, is this year at the University of Leiden studying under Dr Suringar.' (Balfour Mss, Royal Botanic Garden, Edinburgh).
7 Frederick Moore was following in the tradition of Irish botany — E. C. Nelson, 'The influence of Leiden on botany in Dublin in the early eighteenth century', *Huntia*, 4 (1982), pp. 133-146.
8 Moore Mss; D. Moore to F. W. Moore 6.vi.1876.
9 Michael Dowd's appointment to the College Botanic Garden was noted in *The Garden*, 8 (1875), p. 341, and his resignation is recorded in Kew Mss, English Letters 95/351; D. Moore to J. D. Hooker 18.x.1876.
10 Kew Mss, English Letters 95/551; D. Moore to J. D. Hooker 18.xi.1876.
11 Moore Mss; A. Hart to D. Moore 24.iv.1876 and 28.x.1876.
12 Moore Mss; T. Stack to F. W. Moore 29.xi.1876, and F. W. Moore to A. Hart 24.xi.1876.
13 Kew Mss, English Letters 95/322; D. Moore to J. D. Hooker 28.iii.1868. Moore wrote '...Poor Bain who is unable to write himself...gives up the garden today...The College people have behaved very handsomely to him...Dr Dickson takes charge of the garden, with two foreman, one for the outside departments, the other for houses.' Bain was only 53 years of age when he retired, but it is clear that he did vacate the curatorship — in *The Garden*, 1 (1872) p. 153, he was described as '...late of the College Gardens...'.
14 David Moore had also had a row with Bain (in 1857) and James Townsend Mackay acted as mediator (Moore Mss; J. T. Mackay to D. Moore 19.vi.1857). Moore Mss; A. Hart to F. W. Moore 29.x.1877.
15 This phrase appears in a typescript, probably written by J. W. Besant, title 'Sir Frederick Moore M.A., M.R.I.A., F.L.S., V.M.H.' (Glasnevin Mss).
16 Moore Mss; D. Moore to F. W. Moore 23.vii.1878.
17 B.M., 'Glasnevin', *Gardeners' Chronicle*, 13 (1878), pp. 555-556.
18 Moore Mss; E. P. Wright to F. W. Moore 11.vi.1879 and T. Stack to F. W. Moore 18.vi.1879.
19 This was taken up and quoted in English horticultural journals, e.g. *The Garden*, 15 (1879), p. 486.
20 Kew Mss, English Letters 95/382-3; F. W. Moore to J. D. Hooker 10.vi.1879.
21 Moore Mss; Mrs A. M. L. Lindsay to Mrs M. Moore 16.vi.1879.
22 Moore Mss; T. Stack to F. W. Moore 18.vi.1879.
23 Moore Mss; Col. Donnelly to F. W. Moore 23.vi.1879.
24 'The Glasnevin Botanic Garden', *Gardeners' Chronicle*, 12 (1879), p. 18.
25 *The Garden*, 16 (1879), p. 6.

26 Glasnevin Mss; Letters vol. 1 (November 1877-November 1881); N. MacLeod to F. W. Moore 26.vii.1879. *The Garden*, 16 (1879), pp. 25-26 (quoted from *Daily News*).
27 'The College Botanic Garden, Dublin', *The Garden*, 16 (1879), p. 25.
28 Moore Mss; N. MacLeod to F. W. Moore 26.vi.1879.
29 *The Garden* 16 (1879), pp. 25-26.
30 F. W. Moore, 'The late Mr Edward Woodall', *Gardeners' Chronicle, 101* (1937), pp. 164-165.
31 B. D. Morley and E. C. Nelson, 'William Edward Gumbleton (1840-1911) connoisseur and bibliophile', *Garden History*, 7(3), (1979), pp. 53-65.
32 Glasnevin Mss; Letters vol. 17 (1898), F. W. Moore to G. T. Plunkett.
33 Glasnevin Mss; Letters vol. 1 (November 1877-November 1881), item no. E71, W. R. McNab to W. Steele 23.i.1880.
34 Glasnevin Mss; Letters vol. 1 (November 1877-November 1881), item no. 81, Science and Art Department to W. Steele 25.iii.1880.
35 Glasnevin Mss; Letters vol. 5 (1886) — a series of letters dated February and March 1886.
36 W. R. McNab, *A guide to the Royal Botanic Gardens, Glasnevin* (Dublin 1885). The engraving originally appeared in *Gardeners' Chronicle*, 22 (1884). One of the plates in the *Chronicle* shows the old Palm House which was demolished in 1883 (Fig. 63, this volume), and its replacement — the present Great Palm House — is also depicted. There is a plan of the old Palm House in the archives of the Office of Public Works, St Stephen's Green, Dublin.
37 Glasnevin Mss; Letters vol. 3 (1885), W. Ball to W. R. McNab 14.vii.1885.
38 For a bibliography of McNab, see *Annals of Botany*, 3 (1890), pp. 477-479.
39 Glasnevin Mss; Letters vol. 9 (1890) — series of letters.
40 This microscope was manufactured by Powell and Lealand, 170 Euston Road, London. Some of the lenses are marked with the name R. & J. Beck. On the inside of the box lid in pencil is the inscription 'W. R. McNab March 4th 1880'.
41 The McNab herbarium has been the subject of study in recent years and catalogues of some of the specimens have been published — B. D. Morley and Sr M. Powell, 'The Robert Brown material in the National Herbarium, Glasnevin, Dublin', *Glasra (Contributions from the National Botanic Gardens, Glasnevin)*, 1 (1979), pp. 12-39.
42 Kew Mss, English Letters 95/413; F. W. Moore to W. T. Dyer 3.iv.1890.
43 Glasnevin Mss; Letters vol. 9 (1890), item no. 42.
44 Glasnevin Mss; Letters vol. 10 (1891), item no. 53.
45 See note 15 above.
46 T. Crawford and E. C. Nelson, 'Irish horticulturists 1, W. H. Crawford', *Garden History*, 7(2) (1979), pp. 23-26.
47 F. W. Moore, 'Lachenalia', *Journal of the Royal Horticultural Society*, 13 (1891), p. 216.
48 R. D. Trotter, 'Sir Frederick Moore, D.Sc., F.L.S., V.M.H. 1857-1949', *Journal of the Royal Horticultural Society*, 74 (1949), pp. 473-475.
49 See note 15 above.
50 F. W. Moore, 'Orchids', *Orchid Review*, 11 (1903), pp. 163-164.
51 F. H. Woolward, *The genus Masdevallia* (London 1896). E. C. Nelson, 'Two centuries of orchids at Glasnevin', *Orchid Review*, 89 (1981a), pp. 136-140, 191-194. E. C. Nelson, 'Orchid paintings at Glasnevin', *Orchid Review*, 89 (1981b), pp. 373-377, 384. Two of Miss Woolward's paintings are in the National Botanic Gardens, Glasnevin.
52 'Orchids at Glasnevin', *Orchid Review*, 11 (1903), pp. 237-240.
53 *Handlist of orchids cultivated in the National Botanic Gardens, Glasnevin, Dublin* (Dublin 1960). This catalogue was privately printed but was not widely distributed; there are several copies in the Gardens' library.
54 Glasnevin Mss; Letters vol. 18 (1899), F. W. Moore to Director, Science and Art Museum 1.xi.1899. R. James, 'Windows on the past', *Vitis*, 1(2) (1983), 8. (The date given for Conway's retirement in this article is not correct — it should read 31 January 1942).
55 E. C. Nelson (1981, no. 51 above).
56 R. E. Arnold, 'Neomoorea irrorata', *Orchid Review*, 57 (1949), pp. 184-185. R. A. Rolfe, 'Neomoorea irrorata', *Orchid Review*, 23 (1915), pp. 80-81. *Botanical Magazine*, tab. 7262.
57 In a document approving the commissioning of paintings from Alice Jacob, H. B. White stated that 'I beg to say that this service [i.e. painting flowers] has hitherto been always charged to G. P. Purchase of seeds etc. at the Royal Botanic Gardens and I presume it will still remain a charge upon that subhead of the vote'. (Glasnevin Mss; Letters vol. 28 (recte 27) (1908), H. B. White to Professor Campbell 18.xi.1908). There are, however, no other explicit references to this expenditure in official or private correspondence between 1884 and 1908.
58 Shackleton Mss; by kind permission of the late George Shackleton. An account of Lydia Shackleton's work as an artist is in preparation (July 1985) by E. C. Nelson based on this material, and especially on her early sketch-books and diaries.
59 B. D. Morley, 'Lydia Shackleton's paintings in the National Botanic Gardens, Glasnevin', *Glasra*, 2 (1979), pp. 25-36. (This paper includes a catalogue of her paintings of *Sarracenia*, *Lachenalia*, *Paeonia* and *Helleborus*, as well as a few miscellaneous genera, but the items numbered 98 and 99 are not by Shackleton — they are pencil sketches by G. P. Baker). E. C. Nelson, 'Lydia Shackleton (1828-1914)', *The Garden*, 107 (1982), pp. 233-235. M. J. P. Scannell and H. Lahert, 'Lydia Shackleton 1828-1914 botanist and artist', *Journal of County Kildare Archaeological Society*, 16 (1984), pp. 331-339. The latter paper contains a partial catalogue of Lydia Shackleton's paintings of native Irish plants, again with a few miscellaneous exotic subjects. These were executed for the National Museum and not for the Botanic Gardens. Some of the native plants were copied from sketches in Miss Shackleton's sketch-books. Scannell and Lahert noted that Lydia Shackleton visited the United States of America '...but the year of the visit and the reason for it are not known'. Miss Shackleton went to America on two occasions, firstly between September 1873 and July 1876, and later between May 1888 and October 1889. On both occasions she visited friends and relations, including the Yarnall family of Pennsylvania and the Davis and Suliot families in Ohio. (J. Shackleton, pers. comm. 1985).
60 F. C. Stern, *A study of the genus Paeonia* (London 1946), pp. 94-95. The plate of *Paeonia emodi* in *Botanical Magazine* (tab. 5719) was prepared by W. Fitch using specimens supplied by Dr David Moore in 1868.

61 The paintings of orchids bearing the monogram 'J. H.' (see E. C. Nelson (1981b, no. 51 above)) have presented difficulties, but Josephine Humphries' work and her monogram closely resemble the Glasnevin material. Miss Humphries was another woman of Quaker origin, and the Humphries and Shackleton families are related by marriage. We are grateful to Mrs Valerie White, Helen's Bay, for her help.
62 F. C. Hayes, *A handy book of horticulture* (London 1900). Miss Jacob's name is on the title page.
63 Glasnevin Mss, Letters vol. 28 (recte 27) (1908); Department of Agriculture and Technical Instruction to H. B. White 17.xi.1908. We are grateful to Philip Jacob, Shankill, for his help in tracing information on Alice Jacob, and also to Professor John Turpin, National College of Art and Design, Dublin, for his most valuable comments. See J. Turpin, 'The South Kensington System and the Dublin Metropolitan School of Art 1877-1900', *Dublin Historical Record, 36* (1983), pp. 42-64. J. Turpin, 'The Metropolitan School of Art 1900-1923', *Dublin Historical Record, 37* (1984), pp. 59-78; *Dublin Historical Record, 38* (1985), pp. 42-52, 86-102.
64 Lists of *Nerine* cultivars with details of parentage are in Glasnevin Mss (these were compiled about 1925).
65 Moore Mss; E. Wilmott to F. W. Moore, undated.
66 F. W. Moore (1891, no. 47 above), pp. 216-231.
67 The couple were married in 1901. Several other plants were named for Phylis Moore, including *Cypripedium* 'Mrs F. W. Moore', *Chaenomeles* 'Phylis Moore', and *Cytisus* 'Lady Moore'.
68 *Botanical Magazine*, tab. 8017. Records of the Glasnevin hybrids are contained in various manuscripts.
69 The plate published in vol. 30 of *The Garden* (16.x.1886) facing p. 366 is stated to have been '...drawn in the Glasnevin Botanic Garden, Dublin'. It depicts *Sarracenia* × *moorei* and *S.* × *popei* (one pitcher and one flower of each). The present location of the original watercolour is not known — it is not among the Shackleton paintings in Glasnevin, but undoubtedly was painted by her.
70 Kew Mss, English Letters 131/DBN/298; F. W. Moore to A. Hill 24.iv.1918.
71 F. W. Burbidge, 'Glasnevin', *Gardeners' Chronicle, 22* (1884), pp. 487-488, 525-526.
72 R. Farrer, *My rock-garden* (London 1908), p. 8.
73 B. M., 'Glasnevin', *Gardeners' Chronicle, 10* (1878), p. 556.
74 Office of Public Works, St Stephen's Green, Dublin. The plans are contained in folder with other plans of buildings at the National Botanic Gardens, Glasnevin.
75 'Proposed Palm House, Royal Botanic Gardens, Glasnevin' (no. 74 above).
76 B. D. Morley and E. C. Nelson (1979, no. 31 above).
77 Kew Mss, English Letters 131/DBN/174; F. W. Moore to A. Hill 10.ix.1911.
78 Glasnevin Mss; Letters vol. 9 (1890), F. W. Moore to Director, Science and Art Museum, 20.x.1890.
79 P. Moore, 'Obituary, Miss H. M. White, LL.D. (Dublin)', *Gardening Illustrated, 58* (25.vii.1936), p. v. E. C. Nelson, *An Irish Flower Garden* (Kilkenny, 1984), pp. 124-125.
80 Glasnevin Mss; Letters vol. 17 (1898), F. W. Moore to G. E. Plunkett 23.vi.1898. In a letter dated 18.ii.1898, Miss Graves indicated that she had been told of her acceptance at Glasnevin for the training course.
81 The women who came to Glasnevin were always known as 'lady gardeners'. Their names are recorded in various manuscripts, including a list giving the names of all the women who worked between 1898 and 1936. Lady pupils were accepted for training at the National Botanic Gardens in the 1940s and 1950s. Girls are now entitled to apply, on equal terms with young men, for the three-year training course in amenity horticulture.
82 We are grateful to Elsie Miller for her assistance with the history of lady gardeners.
83 M. A. Crosbie, 'The training of a lady gardener', *Irish Gardening, 7* (1912), pp. 44-45.
84 Rowallane Mss; P. Moore to H. Armytage Moore 8.i.1908.
85 E. V. Miller, 'Reminiscences of gardening sixty years ago', *Moorea, 4* (1985), pp. 21-22. For examples of Miss Miller's photographs, see e.g. *Irish Gardening, 16* (1921), pp. 62-63, 87, 92, 126, and E. H. Walpole (no. 111 below).
86 S. Pim, *The wood and the trees, a biography of Augustine Henry* (Kilkenny 1984) (second edition).
87 Glasnevin Mss; Letters vol. 27 (recte 26) (1907), W. Watson to F. W. Moore 20.ix.1906.
88 F. W. Moore, 'Obituary for C. F. Ball', *Irish Gardening, 10* (1915), pp. 161-162. W. Watson to F. W. Moore (no. 87 above).
89 Glasnevin Mss; Letters vol. 27 (recte 26) (1907), A. K. Bulley to F. W. Moore 21.ix.1906.
90 Glasnevin Mss; Letters vol. 27 (recte 26) (1907), F. W. Moore to Director, Science and Art Institutions 29.xi.1906 and F. W. Moore to H. B. White 4.xii.1906.
91 Glasnevin Mss; Letters vol. 27 (recte 26) (1907), memorandum H. B. White to T. P. Gill 30.xi.1906.
92 Glasnevin Mss; Letters vol. 27 (recte 26) (1907), F. W. Moore to H. B. White 4.xii.1906.
93 F. W. Moore (1915, no. 88 above).
94 E. C. Nelson (1984, no. 78 above), pp. 128-129. W. F. Walsh, R. I. Ross and E. C. Nelson, *An Irish Florilegium* (London 1983), plate 32.
95 C. F. Ball, 'Botanizing in Bulgaria', *Journal of the Royal Horticultural Society, 39* (1913), pp. 1-7. C. F. Ball, 'Botanizing in Bulgaria', *Gardeners' Chronicle, 51* (1912), pp. 252-253, 274-275.
96 C. F. Ball, 'A search for alpines in the granite region of the Maritime Alps', *Irish Gardening, 9* (1914), pp. 28-29, 43-44.
97 Amos Perry
98 Glasnevin Mss; W. E. Gumbleton to C. F. Ball 6.xi.1910. See B. D. Morley and E. C. Nelson (1979, no. 31 above).
99 Glasnevin Mss; W. E. Gumbleton to C. F. Ball 10.xi.1910.
100 Glasnevin Mss; W. E. Gumbleton to C. F. Ball 14.xi.1910. C. F. Ball, 'Olearias in Ireland', *Gardeners' Chronicle, 49* (1911), pp. 52-53.
101 *Irish Gardening, 10* (1915), p. 154 (extract from a letter dated 24.viii.1915).
102 *Irish Gardening, 10* (1915) (extract from a letter dated 20.viii.1915).
103 F. W. Moore (1915, no. 88 above).
104 'Roll of Honour, Stephen George Rose', *Irish Gardening, 12* (1917), pp. 89-90.
105 Kew Mss, English Letters 131/265; F. W. Moore to D. Prain 1.v.1916.

106 Kew Mss, English Letters 131/265; F. W. Moore to D. Prain 1.v.1916.
107 Kew Mss, English Letters 131/266; F. W. Moore to A. Hill 12.v.1916.
108 W. Watson, 'Retirement of Sir Frederick Moore', *Gardeners' Chronicle, 71* (1922), p. 252.
109 M. Allen, *William Robinson 1838-1935: Father of the English flower garden* (London 1982), p. 213 (Fig. 32). Photographs showing the Moores at Gravetye also appeared in *Gardening Illustrated, 60* (23.vii.1938), p. 459.
110 F. Cabot, 'As it was in the beginning', *Bulletin of the American Rock Garden Society, 42* (1984), pp. 22-48.
111 J. Robins, 'History of the Royal Horticultural Society of Ireland', in E. C. Nelson and A. Brady (eds) *Irish gardening and horticulture* (Dublin 1979), pp. 70-92.
112 The following Latin oration was delivered on the occasion:

> Virum stirpe huic Collegio familiarissima ortum, amicum nobis omnibus, quotquot adsumus, bene notum, bene dilectum, in oculis nostris viventem ad vos duco, EQUITEM, FREDERICUM WILLELMUM MOORE. Huic, credo, nascenti dea quae vere semper fruitur Ovidianum illud:—
>
> *Arbitrium tu mihi floris habe*
>
> in aurem dixit; rerum enim naturae pulcherrimum nactua provinciam in arte herbaria regnum sibi haud dubium asseruit. Et mehercule, cum ei coronum apricam laudis meae nectere iubeor, eloquentiam florum fabulosam orationi aptiorem esse liquet; quod si consuetudo nostra me non nisi Latine loqui sinit, in lucem claram illud ante omnia proferre volo quod ipsius nomine florum genus quoddam pretiosissimum nuncupatur. Porro tot plantarum species sibi vindicavit, quippe quae in sua tutela primum florescere inceperint, ut facere vix possim quin, verbis paulum immutatis, cum scriptore sacro exclamem: 'omnes plantas terrae nominibus suis appellavit.' Praesento vobis custodem Regiis Hortis Dublinensibus per multos annos, olim nostris praepositum, Florae antistitem numismate Victoriano inter primos donatum, in certamine campestri Hiberniae propugnatorem fortem, quem rogo plausibus effusissimis prosequamini.

This translation is included in the Moore Mss.

> I bring before you a man sprung from a stock very familiar to this College, a friend well-known to all of us who are here today, living with our affection and esteem, Sir Frederick William Moore. In his ear, I fancy, when he was born, the goddess who always enjoys the delights of spring whispered Ovid's command:
>
> *Hold the judgement of flowers for me*
>
> For he sought the most beautiful province in nature and gained for himself a recognised kingdom in the art of gardening. And by Jove, when I am ordered to weave him a gay garland of praise, it is clear that the fabled language of flowers would be best fitted for my speech. Since, however, our custom allows me to speak only in Latin, I would now, before all else, bring to light the fact that a certain very valuable genus of flowers is called by his name. In fact he has made so many species of plants his own, insofar as they first bloomed and flourished under his care, that I am driven to exclaim with the sacred writer, but with a very slight change: 'He has called all the plants of the earth by their names'.
>
> I present to you the Keeper for many years of the Royal Gardens in Dublin, formerly the Head of our own Gardens, one of the first chief priests of Flora to be awarded the Victoria Medal, and also an outstanding champion of Ireland in field games, whom I now ask you to honour with unrestrained applause.

113 F. W. Moore, 'Recollections of Mount Usher', in E. H. Walpole, *Mount Usher*, pp. 16-24 (see Ch. 12, no. 22).
114 R. L. Praeger, 'Sir Frederick William Moore', *Proceedings of the Linnean Society of London, 162* (1950), pp. 108-109.
115 J. M. Watson, 'The late Sir Frederick Moore', *Gardening Illustrated, 71* (1949), p. 284.

Chapter 12

1 F. W. Moore, 'John W. Besant A.H.R.H.S.', *Gardeners' Chronicle, 116* (1944), p. 124. For additional information about her father, we are most grateful to Mrs Margaret Mansfield, Little River, New Zealand.
2 F. W. Moore (1944, no. 1 above).
3 Glasnevin Mss; Letters vol. 27 (recte 26) (1907), A. K. Bulley to F. W. Moore 21.ix.1906 and 23.ix.1906.
4 Moore Mss; W. Watson to F. W. Moore 25.ix.1906. There is an edited typescript of this letter among the official papers at Glasnevin (Letters vol. 27 (recte 26) (1907))—Moore softened some of Watson's plain language. We have quoted the original unedited manuscript.
5 Glasnevin Mss; Workman's day-book, April 1907-April 1917. Margaret Mansfield to E. C. Nelson 28.xi.1981.
6 Margaret Mansfield to E. C. Nelson 28.xi.1981.
7 Glasnevin Mss; J. W. Besant to Secretary of Department of Agriculture 7.x.1927.
8 The table below shows the age at entrance of the 215 young men who entered Glasnevin for instruction between 1902 and 1932.

Age	Number	% of total
under 17	21	10
17	22	10
18-20	98	45·5
21-25	57	26·5
over 25	17	8

The apprentices were of a relatively mature age, one half being between 18 and 20 years old, and a quarter between 21 and 25. As the school-leaving age for the greater part of the period was 14 years, they had spent some time gaining practical experience in private gardens before entering the Gardens. The map (Fig. 128) shows the home area of the young men who came to Glasnevin in the first third of the present century. By far the greatest number, 71 (approximately one-third), came from the Dublin area, approximately a half came from with 60 km of Dublin, and three-quarters within 120 km. Outside of the 120 km radius only counties Cork and Waterford were the home of any appreciable number of apprentices. None came from counties Londonderry, Longford, Monaghan and Leitrim.

9 J. Hutchinson, 'W. E. Trevithick', *Journal of the Kew Guild, 7(63)* (1958), pp. 592-593. M. J. P. Scannell, 'William Edward Trevithick (1900-1958) gardener and botanic artist', *Vitis, 1(2)* (1983), pp. 28-29. There are many inaccuracies in biographical accounts of Trevithick. Trevithick was born on 30 April 1899, became a student gardener at Glasnevin on 4 April 1918

(he applied on 9 January 1917), and left for Kew on 31 March 1920 on completion of his apprenticeship. An incomplete list of his *Botanical Magazine* plates was given by Scannell—a full list is given in Appendix II. His photographs were occasionally published in *Irish Gardening* (e.g. vol. 15 (1920), pp. 21, 84, 85).

10 'Obituary, the late William Pope', *Irish Gardening, 11* (1916), p. 12.
11 Glasnevin Mss; M. Hornibrook to J. W. Besant 10.x.1922.
12 Glasnevin Mss; 'Shrubs...sent to Glasnevin from Knapton' 10.x.1922.
13 S. Pim, *The wood and the trees, a biography of Augustine Henry* (Kilkenny 1984) (second edition).
14 For accounts of some of these manuscripts, see J. Morley, 'Notes on some manuscript data for "The trees of Great Britain and Ireland" ', *Yearbook of the International Dendrology Society 1974* (1975), pp. 55-59. B. D. Morley, 'Augustine Henry; his botanical activities in China 1882-1890', *Glasra, 3* (1979), pp. 21-81. E. C. Nelson, 'Augustine Henry and the exploration of the Chinese flora', *Arnoldia, 43* (1983), pp. 21-38.
15 J. W. Besant, 'The late Professor Henry, V.M.H.', *Gardeners' Chronicle, 87* (1930), pp. 264-275.
16 J. W. Besant, 'Plantae Henryanae', *Gardeners' Chronicle, 96* (1935), pp. 334-335. For a more recent assessment of Dr Henry's contribution to horticulture see E. C. Nelson, 'The garden history of Augustine Henry's plants', in S. Pim (1984, no. 13 above), pp. 217-236.
17 Glasnevin Mss; Accessions books.
18 Earl of Rosse, 'Dr H. H. Hu's plant-hunting expedition 1937-1940', *Yearbook of the International Dendrology Society 1970* (1971), pp. 7-11; *Yearbook 1972* (1973), pp. 7-12; *Yearbook 1973* (1974), pp. 11-18. Earl of Rosse, 'Plant introductions to Ireland from the Himalayas and the Far East', in E. C. Nelson and A. Brady (eds) *Irish gardening and horticulture* (Dublin 1979), pp. 137-150.
19 Glasnevin Mss; J. W. Besant, Annual report for the year ended 31 March 1938.
20 Glasnevin Mss; miscellaneous papers.
21 J. W. Besant, 'Standardized composts', *Journal of the Royal Horticultural Society, 67* (1942), pp. 163-165.
22 E. H. Walpole, *Mount Usher 1868-1928* (this book has no imprint; presumably it was privately published about 1928).
23 J. W. Besant, 'Botanic Gardens: origin, history and development', *Journal of the Department of Agriculture, Ireland, 33(2)* (1935), pp. 173-182. Also J. W. Besant, 'The Botanic Gardens, Glasnevin, Dublin', *Journal of the Royal Horticultural Society 65* (1940), pp. 349-353.
24 A. W. Hill, 'Kew's tribute to Glasnevin', *Gardeners' Chronicle* (1931).
25 R. L. Praeger, 'An account of the genus *Sedum* as found in cultivation', *Journal of the Royal Horticultural Society, 46* (1921), pp. 1-314. R. L. Praeger, *An account of the Sempervivum group* (London 1932).
26 F. C. Stern, *A study of the genus Paeonia* (London 1946) (see Ch. 11, no. 60).
27 W. T. Stearn, '*Epimedium* and *Vancouveria* (Berberidaceae), a monograph', *Journal of the Linnean Society, Botany, 51* (1938), pp. 409-535.
28 F. W. Moore (1944, no. 1 above).
29 B. Mansfield, 'The late Mr J. W. Besant A.H.R.H.S.' (source not traced, supplied by Mrs M. Mansfield).
30 Margaret Mansfield to E. C. Nelson 28.xi.1981.
31 For information on the career and family background of Dr Walsh we are most grateful to Dr Maureen Walsh, Howth.
32 T. Walsh, *The Augustine Henry Forestry herbarium at the National Botanic Gardens, Glasnevin, Dublin, a catalogue of the specimens* (Dublin 1957). T. Walsh, *The Augustine Henry Forestry Herbarium at the Botanic Gardens, Glasnevin, Dublin*. Unpublished doctoral thesis, University of Dublin, 1950.
33 A plaque commemorating John Fanning is in the canteen at the National Botanic Gardens, Glasnevin.
34 Information on Parnell's time at the Royal Botanic Garden, Edinburgh, was kindly provided by Jennifer Woods.
35 Glasnevin Mss; R. James to A. Brady 1.ix.1982. J. Mageean, 'The business behind the Botanic Gardens', *Feature Magazine* (October 1947), pp. 12-13—this contains a photograph of a group of 'grand old men', Mick Doherty, Tommy Goggins, Freddie Parnell, Andy Doyle and Tommy Brady, who, in 1947, had between them served a total of more than 220 years in Glasnevin.
36 Glasnevin Mss.
37 Glasnevin Mss; T. Walsh to Turner 25.vi.1962 (ref. BD 974).
38 B. D. Morley, '*Todea barbara* at Glasnevin', *Journal of the Royal Horticultural Society, 95* (1970), pp. 415. E. C. Nelson, '*Todea barbara* in National Botanic Gardens, Dublin', *The Garden, 103* (1978), p. 418.
39 E. C. Nelson, 'The influence of Leiden on botany in Dublin in the early eighteenth century', *Huntia, 4* (1982), pp. 133-146.

Index

Abelia triflora, 160
Abies forrestii, 218
Abutilon vitifolium, 155
Acacia, 100
Acer (maple), 101
　pentaphyllum, 223
ACOT, 237, 239
Acton, John, 91
Addison, Joseph, 40-41, 43, 44
　legend of, at Glasnevin, 40, n. 4/15
Adiantum capillus-veneris (maidenhair fern), 101
Aesculus (horse-chestnut), 53, 190
Agapanthus, 136
　campanulatus, 136
　mooreanus, 136
　umbellatus var. *mooreanus*, 136
Agave, 194
Aiton, William, 49
Ajuga pyramidalis, 140
Albert Agricultural College, Glasnevin, 201
Aldridge, John, 117-8
Alexander the Great, 3
Alexandra College, Dublin, 201
Allman, George, 117-8, 127, 148
Allman, William, 78
Alnus (alder), 190
Alocasia, 166
Aloe, 10, 53, 173, 194
Alopecurus pratensis, Plate 3
Alps, 210-11
Amaryllis, 58
Amazon, river, 122
Amazon water-lily: see *Victoria amazonica*
America (North), 26, 120, 153-4
America (South), 91, 101, 112, 153-8, 161, 184
Andrews, William, 114, 117
Andromeda, 56
Anemone obtusiloba, 169-171, 188
　vernalis, 209
Anguloa virginalis, Plate 13
Annesley, Richard Grove, 222
Anthocercis ilicifolia, 115
Anthoxanthum odoratum (syn. *Holcus odoratus*), 64, 69, 71
Anthurium, 166
Antrim (Ireland), 72, 98-99, 140
Apothecaries Hall (Dublin), 147
Araucaria, 72, 76
　araucana, 157
　excelsa (Norfolk Island pine), 73-76, 81, 121
Araujia sericofera, 157
Arbutus unedo, 7, 56
Ardilaun, Lord, 190
Argentina, 155-158
Aristotle, 3
Armstrong, Margaret, 9
Arran, Earl of, 91, 156-7
Asplenium billottii, n. 8/13
　lanceolatum, 98, n. 8/13
Aster, 219
Astilbe, 224
Australia, 49, 73, 97, 127, 146, 152-3, 161-2, 179, 219, 225, 227, 237

Bacon, Robert, 230
Baily, Katherine Sophia, n. 6/14
Bain, John, 175
Baker, John Wynne, 31, 36

Baker, Margaret (Mrs David Moore), 173, 175
Baldwin, Sarah (Mrs Henry Nicholson), 10
Balfour, Andrew, 5
Balfour, Isaac Bayley, 151
Balfour, James Hutton, 149, 163, n. 11/6
Ball, Mrs Alice, 209
Ball, Charles Frederick, 181, 205-11, 214-5, 224, 236, Plate 14
Ball, Valentine, 179, 181
Ballitore (Kildare), 118, 120, 184, n. 9/12
Ballybough (Dublin), 15
Banks, Sir Joseph, 21, 49, 152, 154
Banksia, 49, 73, 100, 153
Barr and Sons Ltd (nurserymen), 205
Barralet, John J., 28
Barron, William (nurseryman), 205
Beare, Philip O'Sullivan, 6
Bedford, Duke of, 65, 194
Bees Ltd (nurserymen), 206
Begonia, 174
Bellingham, O'Bryen, 114
Bennett-Poë, John, 176-7, 188
Bentham, George, 138
Berberis, 209, 223
　vernae, 218
Berkeley, Revd Miles, 114
Bermuda, 79
Berry, H. F., 31
Besant, John William, 206-210, 214-226, 229, 232, 236
Besant, Margaret: see Mrs Brendan Mansfield
Betham, Sir William, 75, 138
Betula (birch), 190
Bidwill, John Carne, 162
Birr (Offaly), 7, 224
Bobart, Jacob (junior), 10
Bobart, Jacob (senior), 5, 8
Boerhaave, Herman, 10-11
Bologna (Italy), 4
Bolton, James, 44
Bonpland, Aimé, 122
Books and manuscript works:
　Annals of the Four Masters, 37
　Book of Lismore, 37
　Botanalogia universalis Hibernica (Keogh), 21
　Botanical Dictionary (Milne), 27
　Botanical Elements (Stephens), 11
　Catalogus plantarum (Stephens), 13
　Catalogus systematicus (Wade), 48
　Causae Plantarum (Theophrastus), 3
　Codex Vindobonensis, 3
　Cybele Hibernica (Moore and More), 140, 237
　Dissertatio...de corpore (Nicholson), 10
　Encyclopaedia (Loudon), 173
　English Rock Garden (Farrer), 193
　Essay on the indigenous grasses... (White), Plate 3
　Fasciculus plantarum Hiberniae (Browne), 21-22
　Flora Capensis (Harvey), 120-121
　Flora Dubliniensis: see *Specimen Florae Dubliniensis*
　Flora Hibernica (Mackay), 99
　Flora Londinensis (Curtis), 24-27
　Gardener's Dictionary (Miller), 27
　Genera of South African Plants (Harvey), 120
　Genera Plantarum (Linnaeus), 27

　Guide to the Royal Botanic Gardens (McNab), 179, 180, n. 9/59
　Hand-book for the Botanic Gardens...Glasnevin (Moore), 112, 140, n. 9/50
　Historia plantarum (Theophrastus), 3
　Hortus Britannicus (Sweet), 173
　Icones Plantarum Sinicarum, 223
　Loves of Plants, The (Darwin, E.), 44
　Menthae Britannicae (Sole), 103
　Methodus Plantarum...Dubliniensis (Nicholson), 10-11
　Monograph on...Lilium (Elwes), 204
　Mount Usher (Walpole), 213, 224
　My Rock Garden (Farrer), 193
　Natural History (Pliny), 104
　Natural History of Dublin (Rutty), 21
　Parliamentary Register, 19
　Prospectus of the proposed public garden...Monkstown (Niven), 93
　Quercus (Michaux, ed. Wade), 69
　Redemption Thoughts (Niven), 82, 95
　Sketch(es) of lectures on...grasses (Wade), 65
　Specimen Florae Dubliniensis (Wade), 18, 23-24, 48
　Statement of the progress... (Wade), 31
　Synopsis Stirpium Hibernicarum (Threlkeld), 11, 21
　Theatrum botanicum (How), 7
　The herball or generall histories of plants (Gerard), 5
　The names of herbes (Turner), 5
　The potato epidemic (Niven), 94
　Thesaurus Capensis (Harvey), 121
　The Visitor's Companion (Niven), 86-88, 92-93, 140
　Traité des arbres (Duhamel du Monceau), 27
　Trees of Great Britain and Ireland (Elwes and Henry), 26, 204, 220
　Yew trees of Britain and Ireland (Lowe), 40
Boshill, Paul, 135
Botanical gardens:
　Adelaide (Australia), 236
　Arnold Aboretum (USA), 201
　Ballsbridge, Dublin: see under Dublin, University
　Belfast (Ireland), 109, 127, 153-4, 158, n. 8/20
　Birmingham (England), 91
　Bologna (Italy), 35
　Calcutta (India), 154-5
　Chelsea Physic Garden, London (England), 5, 49
　Chiswick, London (England), 112, 122
　Christchurch (New Zealand), 217
　Cork (Ireland), 146
　Dublin (Ireland):
　　College of Physicians' proposal (1693), 9
　　National Botanic Gardens: see under Dublin, National Botanic Gardens
　　Petty's proposal (1653), 7
　　Philosophical Society (1680s), 8
　　Royal Dublin Society: see under Societies
　　Trinity College: see under Dublin, University
　Durban (Natal), 137
　Edinburgh (Scotland), 5, 91, 99, 109, 124, 151, 155, 161-2, 181, 220, 227, 236, n. 11/6

Florence (Italy), 4
Galway (Ireland), 147
Ghent (Belgium), 229
Glasgow (Scotland), 83, 91, 99, 153-4, 162, 214
Glasnevin: see Dublin, National Botanic Gardens
Gothenburg (Sweden), 229
Halle (Germany), 237
Harold's Cross, Dublin: see under Dublin, University
Heidelberg (Germany), 5
Hull (England), 83
Kew, London (England), 29, 34, 39, 49, 61-62, 105, 109, 122, 124, 136, 138, 145, 151, 158, 161-2, 179, 181, 184, 188, 190, 206, 209, 211, 214, 217, 220, 226, 236
Kingston (Jamaica), 53
Leiden (Netherlands), 4-5, 8, 10-11, 174, 181
Leningrad: see St Petersburg
Liverpool (England), 75, 91, 138, 153-4
Manchester (England), 91
Maymyo (Burma), 171
Melbourne (Australia), 235
Montpellier (France), 5
Montreal (Canada), 229
New York (USA), 218
Oxford (England), 5, 8, 10, 123
Padua (Italy), 4
Pisa (Italy), 4
Regent's Park, London (England), 123
St Gall (Switzerland), 2
St Louis (USA), [153], 229
St Petersburg (Leningrad, USSR), 147
Sydney (Australia), 97, 99, 154, 162, 179
Trinity College, Dublin: see under Dublin, University
Zurich (Switzerland), 194
Botanical Institute, Beijing (China), 223
Boyd, James and Co (Paisley), 193, 195
Bradbury, John, 153-4
Brady, Aidan, 232, 234
Brady, Thomas, 206, 223, 230
Bride, Patrick, 33
Bridgford, Hannah (Mrs David Moore), 99, n. 8/56
Bridgford, Sarah, 99
Bridgford, Thomas (nurseryman), 99
British Association for the Advancement of Science, 88-90, 128, 143, 190, n. 8/15
British Museum (Natural History), (London), 179, 184
Brongniart, Adolphe, 117
Brown, Robert, 62, n. 5/38
Browne, J., 83
Browne, Patrick, 21-23
Brownea, 182
Brunsvegia, 58
Buckley, Thomas, 136
Buddleja crispa, 160
 davidii, 224
 globosa, 56, 71
Bulgaria, 209-210
Bulley, Arthur Kilpin, 206-7, 214
Burbidge, Frederick William, 145, 176-177, 179, 183, 188, 190, Plate 12
Burchall, William, 154
Burgh, Margaretta (Mrs Foster, Baroness Ferrard), 28
Burgh, Thomas, 19
Burma, 168-171
Burnet, Richard (nurseryman), 48
Burren, The (Clare), 7
Burton, Decimus, 109
Butler, Isaac, 21
Butler, Lady Rachel, 175
Butler, Thomas, 83
Buxbaumia aphylla, 69

Cacti, 73, 101, 111, 154, 159, 194, 219
Caesalpina bonduc, 154
Calabar (West Africa), 163-4
Caladium, 166

Calamagrostis stricta, 98
Calceolaria, 209
 × *ballii*, 209
 deflexa, 209
 forgeti, 209
Caldwell, Andrew, 33-34, 36
Camden, Lord, 48, 65
Camellia, 73, 126, 174
Cameroons (West Africa), 164, Plate 8
Campanula, 209
Campbell, Alexander (nurseryman), 138
Campbelltown (New South Wales), 136
Camperdown, Earl of, 97
Candolle, Augustin-Pyramus de, 117
Cape of Good Hope (South Africa), 49, 53, 73, 120, 153, 181
Cardiocrinum giganteum, 159, 173
Carex buxbaumii, 98, Plate 5
Carpenter, James, 73-4
Carpinus (hornbeam), 190
Carroll, Henry, 138
Carroll, Isaac, 187
Carroll, Thomas, 71
Carteret, Lord, 41
Carthamnus tinctoria, 15
Cassiope tetragona, 160
Catasetum, 98
 abruptum, 161
 globiflorum, 161
Cattleya forbesii, 113
Cavit, J., 155
Cedrus atlantica, 191
 libani, 222
Cephalotus follicularis, 142-3
Ceratozamia fusca-viridis, 147
 longifolia, 147
Chaenomeles 'Phylis Moore', n. 11/67
Chambers, Mary (Mrs Walter Wade), 23
Chatham Islands (New Zealand), 166-168, 240
Chemys, Charles, 15, n. 2/31
Cheyne, Charles (Viscount Newhaven), 5
China, 53, 169, 220, 222-223
Cineraria, 202
Citrus aurantiifolia, 161
Clancy, William, 107, 111
Clarendon, Frederick V., 131
Clematis, 56, 224
 obtusiuscula, 224
 orientalis, 224
 tangutica, 224
Clement, William, 15-16. n. 2/31
Clinch, James, 85
Cloncurry, Lady, 175
Coelogyne mooreana, 184-5
Cole, John (nurseryman), 9
Colleges: see under
 Albert Agricultural College
 Dublin, University of
 Royal College of Physicians
 Royal College of Science
 Royal College of Surgeons
 Veterinary College
Columnea gloriosa, 236
Congreve, Ambrose, 229
Connemara (Galway), 69, 91, 99, 158
Convolvulus, 26
Conway, Michael, 184, 206, 229
Cook, Captain James, 15, 152
Cooper, Roland, 171, 220
Coprosma, 166
Cornut, George, 85
Cortaderia selloana, 157-8, 238
Cottingham, Edward, 155
Cotyledon, 53
Coulter, Thomas, 120, 154, 194
Courtney and Stephens Ltd, 109
Cowley, Herbert, 209
Crassula, 53
Crataegus altaica, 209
Crawford, Thomas, 221, 229, 235
Crawford, William, 182
Crinum moorei, 145, 190, 238
Crocker, Emmeline, 203
Croker, J. D., 154

Cromwell, Oliver, 7
Cromwell, T. K., 37, 45
Crosbie, May, 202-3
Cuffe, Charlotte: see Wheeler Cuffe, Lady Charlotte
Cunningham, Allan, 88, 99
Cunningham of Liverpool, 91
Cunningham, James (nurseryman), 97, 160
Cupressus lusitanica, 26
Currey, Florence (Fanny), 201
Curtis, William, 24, 27, 49, 51, 59, 152
Cuscuta, 142
Cuthbert, Mrs Margaret, 89, 116-7
Cyathea dealbata, 166
 medullaris, 146
Cycads, 111, 126, 146-7, 179
Cymbidium mooreanum, 184
Cyphomandra betacea, 157
Cypripedium burbidgeanum (= *Paphiopedilum*) 'Mrs F. W. Moore', n. 11/67
Cytisus 'Lady Moore', n. 11/67

Daboecia cantabrica, 91
Dactylorhiza elata, 185, 186
 maculata, 185
Dahlia, 96
Daly, Hon. Denis Bowes, 71
Daly, Peter, 204, 206
Danby, Earl of: see Danvers, Henry
Danvers, Henry Lord, 5
Darley, Frederick, 107, 109-10
Darlingtonia californica, 142
Darwin, Charles, 143-5, 219
Darwin, Erasmus, 44
Darwin, Francis, 144
Dawson, Thomas, 27
Dawson and Mitchell, 107
Decaisne, Joseph, 117
Delany, Mrs Mary, 42-45
Delany, Patrick, 34, 36-7, 42-4, 239
Denham, Captain, 162
Devonshire, Duke of, 91
Dianthus microlepis, 209
Dicentra chrysantha, 190
Dickson, Alexander, 148
Dickson, James (nurseryman), 49, 51
Dicksonia, 49
Dicksons of Hawlmark (nurserymen), 230
Dictamnus fraxinifolius, 215
Dionaea muscipula (Venus's fly-trap), 26, 127, 143
Dioscorides, Pedianos, 3-4, 6
Dixon, Henry, 211
Doherty, Michael, 206, 223, 230
Donegal (Ireland), 8
Donegan, Patrick (nurseryman), 48
Donnelly, Colonel, 150-1, 176
Dorrien-Smith, Arthur, 168
Douglas fir: see *Pseudotsuga menziesii*
Dowd, Michael, 175
Dowling, Michael, 203, 206
Down (Ireland), 8, 72
Downes, Bishop, 41
Doyle, Andrew, 206, 219, 223, 230
Doyle, Joseph, 181
Doyle, Robert, 73-4
Dracaena draco (dragon tree), 121
Dracophyllum, 166
Dresden (Germany), 135
Drimys: see *Pseudowintera*
Drosera (sundew), 7
Drosophyllum lusitanicum, 143-4
Drummond, James, 146
Drummond, Dr James Lawson, 99
Drummond, Thomas, 128, 154
Drummonds of Stirling (nurserymen), 91
Dryas octopetala, 91
Dublin (Ireland):
 Christchurch Cathedral and Chapter, 37-39, 58
 Exhibition Palace and Winter Garden Co., 135
 General Dispensary, 69
 Institution (Sackville Street), 78

INDEX

National Botanic Gardens, Glasnevin: see also Society, Royal Dublin
  Area of Gardens, 19, 43, 46
  Admissionn to Gardens, 53, 71, 122, 129
  Apprentices (and students), 71, 75, 89, 101, 104, 107, 174, 181, 197, 202, 217, 232, 237, 239, n. 12/8
  Accessions records, 127, 154, 156, n. 10/12
  Aquarium (marine), 57
  Arboretum, 53, 56, 76, 79, 88, 96, 103, 190 (see also Far Grounds)
  Bicycles, 198
  Big Wind, 103, 115
  Buildings:
    Acacia House, 91
    Alpine House 219, 230
    Aquatic House, 120, 123, 194, 230
    Bothy, 197
    Cactus and Succulent House, 194, 196, 219
    Camellia House, 194
    Curvilinear Range, 104, 111, 112, 115, 120, 121, 124, 126, 131-134, 145-46, 194
    Doric Cottages, 61
    Epiphyte House, 8, 58, 91, 106, 171
    Fern House, 121, 194, 196-197, 230-2, 234
    Gate Lodges, 51, 60-63, 85, 138, 151, 195, 197-98
    Iron House: see Museum
    Kemmis House, 107, 111
    Lecture room, 103, 116, 121, 126
    Long Range, 73, 80, 86, 88, 91, 105, 106, 122, 124, 194
    Museum (Iron House), 123-125, 149, 150, n. 9/23
    Model Cottage, 81, 86
    Mill, 58, 198, n. 5/28
    Octagon House, 73-74, 91-92, 106, 112, 120-121, 126, 146, 194, 230
    Original Range, 52-54, 93, 105
    Orchid House, 126, 182, 188, 194
    Palm Houses, 58, 73, 123-125, 143, 146, 179, 193-4, n. 11/36
    Pelargonium House, 55, 91, 126
    Propagating pits, 111, 150
    Professor's (Director's) House, 46, 54-55, 58, 61, 65, 112, 115, 117, 126, 128, 179, 198, 215
    Reading Room, 104
    Shelters, 198
    Tea Room, 198, 201
    Turner's Iron Glasshouse, 195
    Victoria House: see Aquatic House
    Walls, 103, 134
  Catalogues of collections, 53, 59, 72, 76, 100, 102, 153, 204, 225, 228, 233
  Cattle Garden, 56-57, 79, 86
  Chain Tent, 87-88
  Clerk, 135, 217
  Drinking fountains, 198
  Dwarf conifers, 192, 220, 222
  Dyers' Garden, 56-57, 79
  Esculent Garden, 56
  Far Grounds, 190, 229, 235
  Fêtes, 122, 128
  Finances, 18-20, 29, 31, 33, 46, 48-49, 51, 58-9, 71, 73, 79, 85, 96, 99-101, 104-7, 109-10, 123-4, 139, 147, 175-6, 181-2, 190
  Fireman and stoker, 135, 226
  Flower Garden, 88
  Foremen, 85, 135, 138, 150, 184, 217, 230, 237, Appendix III
  Fruticetum, 53, 76, 190 (see also Arboretum)
  Gate-keeper, 135
  Gentlemen Gardeners, 203
  Grass Garden, 88
  Guide books, 56-7, 79, 86-8, 92-3, 110, 112, 140, 179-80, n. 9/50

  Hay Garden, 56-7, 79
  Heating of houses, 107, 109, 110, 120, 121, 134, 194
  Herbaceous Grounds, 56, 58
  Index Seminum (seed exchange lists), 179, 183, 237-8
  Insurance, 121
  Irish Garden, 81, 86, 88
  Labels, 31, 135, 219
  Lady Gardeners, 201-5, 219, 239, n. 11/81
  Lawns, 57, 203
  Lecture syllabus, 66-7
  Lectures, 58, 65, 68, 69, 78, 120, 137, 143, 148, 183
  Library, 46, 103, 120, 149-51, 195, 220, 232
  Maps, 54, 55, 86, 90, 112, 180
  Medicinal Garden, 79
  Messenger, 135
  Meteorological readings, 58
  Mill Field, 81, 88
  Nightwatchman, 134
  Nursery, 56, 150, 229
  Ornamental fountain, 80
  Paths, 53, 57, 63, 112
  Pinetum, 190
  Plant collector, 217
  Plants, numbers, 91, 94, 101, 127, 184, 187
    arrangement and classification of, 58, 65, 79. 85, 89-90, n. 5/25
  Pond, 56-7, 73, 88, n. 5/74
  Prams, 198-9
  Propagator, 135-6
  Refreshments, 200-201
  Rock garden, 58, 73, 88, 124, 185, 192-3, 220, 222, 231, 238
  Rock mound, 56, 79
  Rose garden, 58, 88, 230
  Sanitary arrangements, 197
  Scientific Superintendent, 177-9, 181
  Security, 134
  Shrubbery, 229 (see also Fruticetum)
  South Field, 88, 190, 229, 235
  Staff: see Appendix III
  Subsidiary staff: see under titles
  Sunday opening, 129-135
  Thefts, 134
  Title of Botanic Gardens, 181, n. Prologue 1
  Visitor numbers, 93, 128, 226
  Women as gardeners: see Lady Gardeners
  Willow Garden, 88
  Yew Walk, 40-1, 193
National Library, 149
National Museum, 149, 179, 182, 232, 237-8
Society: see under Societies
University College, 181, 203, 221, 225-6, 235
University, 7-20, 21, 24, 33, 69, 77-8, 97, 117, 120, 127, 140, 148, 174-5, 212, 234, n. 6/58
  Physic Gardens: first (1687-1722), 7-11
    second (1722-c. 1775), 12-17, 29
    Harold's Cross (1795-1803), 19-20, 97
    Ballsbridge (1806-1968), 77, 83, 97-99, 109, 112, 122, 127, 153-5, 158, 162, 174, 225, 234
    Palmerston Park (1968-), 234
Duffy, John, 71
Duffin, Mr, 58
Dun, Patrick, 17, 20
Dundee (Scotland), 97, 151, n. 8/1
Dyer, William Thiselton, 144, 148-9, 151, 179, 181, 188
Dynamite, 207

Easter Rising (1916), 211
Eccles, Miss, 189
Edgeworth, Richard Lovell, 69
Edinburgh (Scotland), 5, 78, 97, 99, 148-9

Edwards, Sydenham, 152
Egan, Daniel, 216
Ehret, Georg, 44
Ellis, John, 25-7
Elwes, Henry, 26, 188, 204, 220
Endlicher, Stefan, 117
Epacridaceae, 73, 121
Epidendrum crassifolium, 113
  elangatum, 113
  mooreanum, 184
Epimedium, 225
Equisetum moorei, 140
Erica (heathers), 49, 53, 56, 73, 96, 101-2, 110, 153, 179, 181
  cinerea 'Glasnevin Red', 238
  mackaiana, 91
Erinensis: see Green, Peter Hennis
Eriocaulon aquaticum (pipewort), 69
Escallonia, 209
  'Alice', 209, 238
  'C. F. Ball', 209, 238, Plate 14
  'Glasnevin Hybrid', 209, 238
Eucharistic Congress 1932, 225
Euphorbia, 194
Euryale amazonica, 122
Eustace, Clotilda (Mrs Thomas Tickell), 41, 45
Eustace, Colonel, 32
Eustace, Maurice, 41
Exhibitions:
  Exhibition of Manufactures 1850 (Dublin), 124
  Great Exhibition 1851 (London), 134
  Great Exhibition 1853 (Dublin), 95, 124, 128
  International Exhibition 1865 (Dublin), 95, 135

Fagus sylvatica (beech), 27, 53, 190
  sylvatica f. *purpurea* (copper beech), 27, 53
Fairburn, James, 49, 51
Falkland Islands, 140
Fanning, John, 227-8, 232, 236-7, n. 12/33
Farley, George, 217
Farrer, Reginald, 165, 171, 193, 203, 220
Fennessy, Patrick (nurseryman), 158
Fenzl, Edouard, 142
Ferdinand, King of Bulgaria, 209-210
Ferguson, Daniel, 123
Ferguson, Duncan, 105
Fermanagh (Ireland), 109, 114
Fernando Po (West Africa), 163-4, 166
Ferns (including tree-ferns), 49, 56, 91, 96, 101, 121, 146, 161-2, 167, 174, 194-5, 230-3, 235 (see also *Adiantum*, *Asplenium*, *Cyathea*, *Dicksonia*, *Hymenophyllum*, *Todea*, *Trichomanes*)
FitzRoy, Charles, 162
Florence (Italy): International Botanical Congress 1874, 142, 162, n. 9/54
Foley, Samuel, 8
Foot, Lundy Edward, 118
Foot, Simon, 118
Ford, Mr (engraver), 70
Forrest, George, 165, 171, 212, 220, 222
Foster, Anthony, 25
Foster, John, 18, 19, 25-31, 33-4, 46, 58, 62, 65, 72, 74-9, 81, 152-3, n. 3/20
Fox, Henry, 156
Francis, Richard, 37-8
Fraternity of Physicians (Dublin), 7
Fraxinus (ash), 190
Frazer, Charles, 154
Frost, John, 78
Fuchsia, 49, 83, 88
  coccinea, 56
  macrostemma 'Recurvata', 84-5
  magellanica, 56
Furneaux, Tobias, 152
Fusarium, 93

Gall, St, 3
Gallipoli, 210
Gardener, Master Jon, 6, 240

Gardens and houses:
  Aldenham House (England), 219
  Annes Grove (Cork), 223
  Arley Hall (England), 91
  Ballynure (Dublin), 138
  Belgrove (Cork), 176-7, 182, 197
  Belladrum House (Scotland), 83
  Bellevue (Fermanagh), 109
  Belvoir Castle (England), 217
  Birr Castle (Offaly), 223
  Bothwell Castle (Scotland), 83
  Camperdown House (Scotland), 97
  Chatsworth (England), 91
  Colebrook (Fermanagh), 109
  Collon (Oriel Temple, Louth), 25-7, 51-2, 75, 115, 153
  Delville (Dublin), 34, 36, 42-5
  Drumcondra (Garden Farm, Dublin), 93-6
  Dublin Society's at Summerhill, 15
    at Martin's Lane, 15
  Greenfields (Tipperary), 223
  Headfort (Meath), 217, 222-3, 228
  Holburn (England), 5
  Hoole (England), 91
  Howth Castle, 235
  Jenkinstown (Kilkenny), 229
  Kier House (Scotland), 83
  Kilakee (Dublin), n. 8/49
  Ladiston (Westmeath), 112
  Lakelands (Cork), 182
  Marlfield (Tipperary), 109
  Marlay (Dublin), 48
  Mitchelstown (Cork), 21
  Mount Stewart (Down), 72
  Mount Usher (Wicklow), 204, 213, 224-5
  Muckross (Kerry), 223
  Nutley (Dublin), 138
  Oriel Temple: see Collon
  Phoenix Park (Dublin) including Chief Secretary's Lodge, Vice-Regal Lodge (Áras an Uachtaráin), 34, 69, 83, 85, 103, 115, 138
  Rowallane (Down), 226
  St Dolough's (Dublin), 138
  Syon House (England), 123
  Trentham House (England), 184
  Wentworth (England), 91
  Westonbirt (England), 219
Gardiner, Douglas, 97
Gardiner, William, 97
Garrya × *issaquahensis* 'Glasnevin Wine', 238
Gasteria, 173, 193
Gentiana verna (spring gentian), 91
Geranium macrorhizum 209-10
Gerard, John, 5-6
Geum coccineum, 209
  bulgaricum, 209
Ghent (Belgium), 174
Ghini, Luca, 5
Gibbons, William, 83
Gibbs, Vicary, 219
Gibson, John, 154
Gilbourne, John, 15
Ginkgo biloba (maidenhair tree), 53
Glasgow (Scotland), 83, 99, 148, 151
Glasnevin, 35-45
  Cemetery, 134
  derivation of Irish name, n. 4/1
  St Mobhi's Parish Church, 37-8, 41, 44, 51, 220, n. 4/6, 5/15
  see also Delville (under Gardens and houses)
Glenelg, Earl of, 83
Gogarty, Dr A., 161
Goggins, Thomas, 206, 223, 230
Gore, Henry, 122
Goulbourn, Henry, 122
Graff, de (nurseryman, Belgium), 174
Grasses, 48, 57, 65, 72, 136-7
Graham, Robert, 88, 99-100
Graves, Mary, 201, n. 11/80
Gray, Asa, 148
Great Famine 1845-1848, 113-5

Green, Peter Hennis (*Erinensis*), 76, 78-9, 116, Appendix I
Greene, John, 37
Gregory, William, 130
Grevillea, 73
Grey, Earl, 162
Griffin, H., 80
Grimwood, John (nurseryman), 48
Grogan, Denis, 217
Guinness, Arthur, 175
Gumbleton, William Edward, 176-8, 182, 188, 190, 195, 209, 210, 232
Gwin, John (artist), 12

Haberlea rhodopensis, 209-10
Haenke, Thaddeus, 122
Hakea, 153
  victoriae, 146
Halley, Edmund, 8
Hamilton, Joseph, 81, 86
Hamilton, Letitia, 224
Hamilton, William, 153
Hammersmith Iron Works, Dublin, 208-9, 131
Hanover (Germany), 135, 173
Hardwicke, Lord, 65
Harlow, James, 8
Harold's Cross Botanic Garden: see under Dublin, University
Harrison, Celia, 224
Hart, Dr, 175
Hart, Henry Chichester, 175
Hartlib, Samuel, 7
Harvey, Joseph Massey, 118
Harvey, William, 19
Harvey, William Henry, 98, 117-22, 127, 139, 140, 148
Hastings, Marquis of, 154
Haughton, James, 129
Hayes, Revd F. C., 187
Hayes, Samuel, 19, 33
Headfort, Marquis of, 212, 217-8, 222-4, 225, 227, Plate 15
Heaton, Richard, 7
Hedychium, 190
Heer, Oswold, 140-1
Helsham, Richard, 39, 42
Hegarty, John, 104
Heliamphora nutans, 143
Helleborus, 183, 187, 219
  orientalis 'Dr Moore', Plate 9
Hely-Hutchinson, John, 17-19, 29
Hemphill, W. D. (photographer), 40, 56
Henry, Mrs Alice, 218, 220, 222, 226
Henry, Augustine, 26, 204-5, 214, 220, 223-5, 227
Henslow, John, 162
Herbaria, 4, 62, 91, 98-100, 116-7, 120-1, 137-9, 179, 181, 220-1, 226-9, 232, 236-7
Herbert, William, 112
Herrenhausen (Germany), 173
Higgins, William, 69
Hill, Arthur, 211, 224
Hill, Edward, 15, 17-21, 23, 29, 32-4, 97
Himalayas, 138, 159-160, 169, Plate 7
Hippeastrum, 190, 219
Hodgins, Edward (nurseryman), 48
Hodgkin, J. E., 224
Holford, Sir George, 219
Holland, Harry, 8
Holland House (London), 40, 42
Holly: see *Ilex*
Hone, Nathaniel, 138
Hooker, Joseph, 117, 135, 137-8, 142-3, 145, 151, 160, 190
Hooker, William Jackson, 83, 89, 98-9, 104-6, 109-10, 116-8, 122, 124, 145, 148, 152, 156-8, 160, 162
Hopkins, Revd, 15
Hopkirk, Thomas, 98
Hornibrook, Murray, 219, 222
Hospitals:
  Mercer's (Dublin), 15
  Royal (Kilmainham), 15, 103

  Dr Steeven's (Dublin), 14-5
How, William, 7
Howard, Mr, 184
Howth (Dublin), 53, 88, 192
Hu, Hsien-Hsu, 223
Humboldt, Alexander, 117
Hume, Mr, 58
Humphries, Josephine, 187, n. 11/61
Hunt, Sir Vere, 62
Hunter, Captain, 203
Huntingdon, Robert, 6, 8
Hutcheson, Dr, 17
Huxley, Thomas, 143
Hymenophyllum, 194
Hyoscyamus niger, 24

Impatiens, 164
India, 3, 158, 173, 218
Insectivorous plants, 7, 127, 142-3, 188-90, 216 (see also *Darlingtonia, Dionaea, Drosera, Drosophyllum, Heliamphora, Nepenthes, Sarracenia*)
Inula salicina (fleabane), 140, Plate 6
Ilex, 9, 56, 62, 190
  fragilis, 225
Iris, 219

Jack, John, 136
Jacob, Alice, 187, Plate 13, n. 11/57
Jamaica, 8, 53, 124, 161, 236
James, Robert Ernley, 229
Jasminum, 223
Jenkins, Abraham Lionel, 21
Johnston, Thomas, 181
Johnstown Castle (Wexford), 228, 235
Jones, John, 71
Jussieu, Adrien de, 117

Keit, Julius Wilhelm (William), 135-7, 139, 142, n. 9/31
Keith, William, 171
Kellerer, Herr, 209
Kemmis, Henry, 107
Kennedy, - (nurseryman): see Lee and Kennedy
Keogh, John, 21
Kerry (Ireland), 5, 181
Kew: see under Botanical Gardens
Kew Guild, 225
Kiernan, John, 36, 45
Kildare, Bishop of, 58
Kilkenny (Ireland), 21, 103, 158, 230
Killarney fern: see *Trichomanes speciosum*
Kilmore, Bishop of, 33
King, William (Archbishop of Dublin), 10
King, William (Professor), 124
King's Inns (Dublin), 107
Kingston, Earl of, 21
Kirkwood, J., 92-3
Kirwan, Richard, 69
Kniphofia, 190
Knox, George, 69

Lachenalia, 183, 187-9, 209, 212, Plate 10
  aureliana, 189
  quadricolor, 188
  rubra, 189
Lambert, Aylmer Bourke, 21, 69
Lambertia, 153
Lane-Poole, Charlotte, 218
Lankesteria barteri, 164, Plate 8
Lathraea, 142
Lathyrus palustris, 100
La Touche, David, 48
Lauche, Howard, 163-5
Lawrence, W., 62, 196, 232
Ledwich, Edward, 62
Lee and Kennedy (nurserymen), 49, 56, 64, 72, 152-3
Leiden (Netherlands), 4-5, 7, 8, 10-11, 174
Leinster, Duke of, 111
Leinster House (Dublin), 120, 124
Lemaire, Charles, 158
Lemon, John, 138-9
Lepidozamia peroffskyana, 179

INDEX

Leptospermum, 49
  lanigerum, 152, Plate 1
Leucogenes leontopodium, 167
Lichens, 97, 140
Lilies, 2-3
Lilium formosanum, 218
  jankae, 209
  wallichianum, 160
Lincoln, Lord, 109
Linden's Nursery (Belgium), 135, n. 9/31
Lindley, John, 99, 105, 114, 162
Lindsay, Mrs Alice, 175
Lindsay, Lady Mary, 156
Linen Board, 25-6
Linnaeus, Carl, 17, 21, 25, 27, 79
Lithops, 194
Littleton, Edward, 83
Litton, Edward, 78
Litton, Margaret: see Mrs Cuthbert
Litton, Samuel, 76, 78-81, 86, 88-9, 96, 101, 103-4, 106, 115-6, 118, 158, 238
Littonia modesta, 116
Liverworts, 98, 117, 139-40, 161, 182, 219
Livingstone, David, 161
Locke, Joseph, 8
Loddiges, Conrad (nurseryman), 61, 158
Loddiges, George (nurseryman), 161
Lombard, Peter (Archbishop of Armagh), 6
Londonderry (Ireland), 98-9, 140
Lonicera, 223
Loranthus europaeus, 142
Loudon, John Claudius, 86, 98, 104, 173
Lough Neagh (Ireland), 98
Lowe, Edward, 177
Lowe, J., 40
Lynch, N., 203, 206, 223
Lyon, John, 153
Lyons, John Charles, 109, 112, n. 8/49

McArdle, David, 181-2, 206, 216, 218-20
McArdle, Patrick, 85, 135, 218
McArdle, William, 104
McCalla, William, 91, 158, 162
McCann, James, 206, 222, 229
McCann, Sean, 230
McCarthy, Bucknall, 64, n. 5/58
McCarthy, Donagh, 188, 204
McClanaghan, H., 210-1
McCoy, Mrs Mary, 89, 101
McCoy, Timothy, 89
McDonagh, Thomas, 104
McGredy and Sons Ltd (nurserymen), 226, 230
Mackay, James Townsend, 77-9, 83, 91, 94, 97-9, 100, 153, 158
Mackie, Mrs Jane, 71
Mackie, John, 71, 74, 89
McNab, William, 99, 109, 149
McNab, William Ramsay, 148-51, 173, 177-9, 181, 224, n. 7/6
McNabb, Mrs Winifred, 169
Macrozamia denisonii, 179
Madden, Edward, 158-9, 160-1, 163, 173, Plate 7
Madden, Samuel, 44
Magee, Richard, 78
Mahonia, 209
Makon, Rhoda (Mrs Edward Litton), 78
Malcolm, William (nurseryman), 71
Malcomson, Anthony, 25, 29
Mallett and Ryder, Messrs, 79
Mandeville, H. J., 157
Mandevillea suaveolens, 157
Mansfield, Brendan, 164, 217, 225
Mansfield, Mrs Margaret, 215
Maple, William, 11-13
Masdevallia, 183
Mason, Maurice, 229
Mauritius (Indian Ocean), 124
Mayo (Ireland), 21
Melastomataceae, 164, 166
Melissa melissophyllum, 209-10
Mesembryanthemum, 53, 101, 194
Meylor, Anthony, 107
Michaux, François, 69

Micholitz, Wilhelm, 184
Middle Temple, London, 11, 25
Midland Great Western Railway Company, 198
Millard, Frederick William, 224
Miller, Elsie, 188, 203-4, 216, 218, 224
Miller, Philip, 27
Milne, Colin, 27
Milne, William Grant, 162-6, Plate 8
Miltown Malby (Clare), 120
Mitchell, Thomas, 161
Mobhi, St, 37, 219
Moir, Charles, 97, 162
Moir, David (see Moore, David), n.8/1
Molyneux, Thomas, 8-9, 11
Molyneux, William, 8-9, 13
Monceau, Duhamel du, 27
Monteagle, Lord, 108
Montpellier (France), 3
Moore, Alexander, 136
Moore, Charles, 97-8, 100, 146-6, 162-3, 179, n. 8/20
Moore, David, 57, 83, 96, 97-151 passim, 155, 162-4, 173-5, 178, 187-8, 190, 213, 225, 237, Plates 5, 6, n. 8/1 (portrait, 129, 139)
Moore, David (junior), 135, 173
Moore, Frederick William, 135, 151, 164, 167, 172-214 passim, 215, 223-5, 229, n. 11/112 (portrait, 177, 202, 213)
Moore, Mrs Hannah, 99, n. 8/56
Moore, Mrs Helen, 97-8
Moore, Miss Helen, 176
Moore, Hugh Armytage, 225
Moore, Mrs Isabella, 115, n. 8/56
Moore, Miss Isabella, 99, 115
Moore, Mrs Margaret, 173, 175, 202
Moore, Phylis, 188, 201-2, 213-4, n. 11/67
Moore, Thomas (poet), 229
Moorea argentea, 158
  irrorata (= *Neomoorea*), 184
More, Alexander, 140, 179, 237
Morgan, Isabella: see Mrs David Moore, 115, n. 8/56
Morley, Brian, 236
Mosses, 56, 91, 98, 101, 117, 137-40, 161, 182, 219 (see also *Buxbaumia*)
Mueller, Ferdinand von, 234
Muir, David: see Moore, David, n. 8/1
Mullay, Andrew, 96
Mullen, Alan, 8
Murphy, Edmund, 85
Murphy, Mr, 78
Murray, Andrew, 96
Murray, Richard, 19
Murray, Stewart, 83, 99-100
Myosotidium hortensia, 166

Narcissus poeticus, 145
  pseudo-narcissus, 145
Neill, Thomas, 147
Nelson, Ernest Charles, 236
Neomoorea wallisii, 184-5
Nepenthes, 143, 188-90
  distillatoria, 188
  × *dominiana*, 143
  rajah, 188-9
Nerine, 183, 188, 209
Neville, Dr, 175
Newcomen, William Gleadowe, 33
Newcommen, Mary (Mrs John Underwood), 51
Newgate Prison, Dublin, 65, 69
Newman, Edward, 140
New South Wales, 49, 61, 99, 136, 146, 154, 161-2 (see also Australia)
New Zealand, 53, 146, 158, 165-8, 194, 218
Nicholson, Edward, 10
Nicholson, Henry (Professor of Botany), 9-11
Nicholson, Henry (senior), 9
Nicholson, William (Bishop of Derry), 41
Niven, James, 83, 181
Niven, James (of Penicuik, Scotland), n. 7/2

Niven, Ninian, 57, 80-96, 97, 100-1, 103, 111, 122, 128, 134, 140, 152, 154-7, Plate 2, n. 7/2
Niven, Ninian (senior), 83
Nobel, Alfred, 207
Norbury, Lord: see Toler, John
Norris, Richard, 25
Northumberland, Duke of, 123
Noyer, George Victor du, 98-9
Nuttall, John, 65

Oaks: see *Quercus*
O'Connor, J. J., 232
O'Halloran, Patrick, 83
Olearia, 166, 210
  semidentata, 166-8, 238, n. 10/39
  traversii, 166
  virgata, 210
O'Mahony, The, 209
Orbigny, Alcide Dessalines de, 122
Orchids:
  see also *Catasetum, Coelogyne, Cymbidium, Cypripedium, Dactylorhiza, Epidendrum, Masdevallia, Neomoorea, Orchis, Phaius, Physosiphon, Spiranthes*
  collections at Glasnevin, 96, 112, 164, 182-7, 212, 215
  cultivation of, 98, 112, 126, 194, 226
  seed germination, 112-3
Orchis latifolia, 186
Ordnance Survey (Ireland), 96, 98-100, 140, 158, 162, n. 8/20
  Memoirs, 98
O'Reilly, Brendan, 228
O'Reilly, Richard, 161
Oriel, Lord: see Foster, John
  Temple: see Collon (under Gardens and houses)
Orobanche, 142
Orr, David, 138-9, n.9/44
Orrery, Lord, 42, 44
Ortgies, Edouard, 194
Osmunda, 226
O'Toole, Laurence, 37
Ouvirandra fenestralis, 143, n. 9/54
Owen, Miss C. M., 177
Owen, Jacob, 106, 124
Oxford (England), 5, 7, 8, 10, 40
Oxypetalum caeruleum, 152, 158

Paeonia, 183, 187, 219
  emodi, 187
  'Emodoff', 187
  officinalis, 187
Palms, 96, 111, 145, 161, 164, 173-4, 195
  (see also *Trachycarpus*)
Pampas grass: see *Cortaderia selloana*
Pancratium, 64
Paphiopedilum burbidgeanum, Plate 12
Paris (France), 7
Parke, Edward, 46, 51
Parkinson, John, 7
Parliament of Ireland, 15, 17-19, 23, 25, 27, 29-31, 33-4, 36, 46, 64, 69
  House of Commons, 17-19, 25, 27, 29-31, 33-4, 64
  House of Lords, 20
Parliament of United Kingdom, 64
  House of Commons, 64, 101, 104
Parnell, Frederick, 206, 223, 227, 230
Parnell, William, 135, 137-8, 205, 206, 214, 227
Pasley, Joshua, 60
Passiflora mooreana, 157
Paul, Kathleen, 188
Paul, Phylis (Lady Moore), 188, 198
Paul, Robert, 198
Paxton, Joseph, 105, 122
Pelargonium, 10, 121
Pelham, Lord, 65
Pemberton, Christopher, 230
Pendarves, Mrs Mary: see Delany, Mrs Mary
Perceval, Robert, 18-20, 33-4

Periodicals and journals:
   Botanical Register, 152
   Curtis's Botanical Magazine, 85, 145, 152, 157, 161, 164, 184, 190, 217, 220, 232, Plates 1, 2, 8, 15, Appendix II
   Dublin Chronicle, 23-4
   Dublin Penny Journal, 81
   Gardeners' Chronicle, 94-5, 105, 113-4, 125-6, 131, 135, 139, 142, 151, 164, 171, 176, 190, 194, 198, 210, 224
   Glasra, 237
   Illustrated London News, 108, 110
   Irish Farmer's and Gardener's Magazine, 85-6, 92, 94, 98, 104
   Irish Farmer's Gazette, 94-5
   Irish Farmer's Journal, 114
   Irish Gardening, 203-4, 207, 210-11, 224
   Journal of the Department of Agriculture, Ireland, 224
   Journal of the Royal Dublin Society, 140
   Journal of the Royal Horticultural Society (London), 224
   Orchid Review, 184
   Proceedings of the Royal Dublin Society, 79, 179
   Proceedings of the Royal Irish Academy, 142, 219
   Saunder's Newsletter, 106-7
   Scientific Proceedings of the Royal Dublin Society, 147
   The Garden, 150, 176, 190
   The Spectator, 43
   Transactions of the Royal Dublin Society, 52, 58-9, 65, 71
   Transactions of the Royal Irish Academy, 149
Perry, Amos, 210
Persoonia, 153
Petiver, James, 10-11
Petty, William, 7-8
Phaius albus, 113
Philadelphus, 223
Phlox drummondii, 154
Physic gardens: see Botanical gardens
Physosiphon moorei, 184
Phytophthora infestans (blight fungus), 114-5
Pine: see *Pinus*
   Norfolk Island: see *Araucaria*
Pinguicula grandiflora, 143
Pinus, 26-7
   cembra, 62
   coulteri, 154
   nigra, 56-7
   pinea, 192
   pungens, 103
   sylvestris, 62
Pisa (Italy), 4
Pleasants, Thomas, 61, 78, 103, 121
Pliny, Caius, 3, 104
Plumptre, Anne, 62
Poeppig, Eduard, 122
Pollock, Rosamund, 190, 203, 204
Pope, Alexander, 43
Pope, Patrick, 184, 206, 219
Pope, William, 134, 138, 139, 184, 218-9
Populus (poplar), 190, 205
   angulata, 205
   candicans, 205
   × *generosa*, 205
   'Robusta', 205
Porter, J. H., n. 9/23
Porter, Revd J. L., 143
Portlock, Joseph Ellison, 98-9
Potato:
   blight: see *Phytophthora infestans*
   cultivars, 96, 142-3
   diseases, 93-5, 113
   dry rot: see *Fusarium*
Potentilla erecta (tormentil), 12
Power, John, 37
Praeger, Robert Lloyd, 211, 213, 225
Prain, David, 211
Primula, 44, 188, 203
   deorum, 209
   minima, 209
Prior, Thomas, 13

Prochazka, Baroness Pauline, 171
Protea, 49, 73, 153
Proteaceae: 146, 163 (see also *Grevillea, Hakea, Persoonia, Protea, Telopea*)
Prunus, 223
Pseudotsuga menziesii (Douglas fir), 91, 204
Pseudowintera traversii, 167
Putland, John, 37, 39, 58, n. 4/9
Purdom, William, 165, 212, 214, 220
Pyracantha, 62

Quakers (Society of Friends), 21, 23, 118, 184
Queen's Colleges (Ireland — now University Colleges)
   Cork, 147
   Galway, 124, 147
Quercus (oaks), 26, 44, 53, 69
   cerris, 142
   petraea, 142

Ramsay, Allan, 50, n. 5/43
Rattray, Helen (Mrs Charles Moir), 97-8
Rawdon, Arthur, 8, 154
Regel, Edward von, 147
Reseda luteola, 15
Rhododendron, 56, 72, 121, 169-71, 219, 223-4
   arboreum, 155, 170
   burmanicum, 169, 171
   cuffeanum, 169-71
   maddenii, 160
   myrtifolium, 209
Ribes, 209
Richardson, Harold, 203
Richardson, Lionel J., 203
Richardson, Mr (artist), 70
Richmond, Duke of, 68, 176
Ridley, Henry, 184
Robertson, John, 85, 103, 138
Robinson, William, 126, 131, 135, 143, 146, 147, 150, 193, 212
Rock, Joseph, 222-4
Rodie, Dr, 122
Roe, George, 138
Roe, W., 73
Roebuck (Dublin), 34-5
Rogers, Charles, 171
Rolfe, Robert, 184
Romneya coulteri, 154-5, 190
Rosa (roses), 2-3, 7-8, 44, 99, 223
   'Glasnevin', 230
   "Last Rose of Summer" (= 'Old Blush'), 229
   cooperi, 171
   hibernica, 100
   laevigata, 171
Rose, Stephen, 211-2
Rose, William, 216
Roses: see *Rosa*
Rosse, Earl of, 222-4
Rothschild, Lionel de, 219
Royal Belfast Academical Institution, 99
Royal Botanic Gardens, Kew: see under Botanical Gardens
Royal Canal Company, 58, 198
Royal College of Physicians in Ireland, 7, 9, 15-9, 20, 23, 33
Royal College of Science, Dublin, 148-50, 204-5 (see also Dublin: University College)
Royal College of Surgeons of Ireland, 69, 147, 161
Royal Cork Institution, 78, 117
Royal Irish Academy, 201, 219
Royal University of Ireland, 201
Royle, Dr, 158
Rubus (brambles), 99
Russell, Thomas, 61
Rutland, Duke of, 218
Rutty, John, 21
Ryan, Daniel, 96

St Gall (Switzerland), 2
St Helena (Atlantic Ocean), 154

Salix (willows), 56, 62, 69, 88
   acutifolia, 70
   × *grahamii* var. *moorei*, 140
   herbacea, 91
   moorei, 140
Sanders Ltd (nurserymen), 184
Saracen Foundry, Dublin, 135
Sarracenia, 142-3, 145, 183, 187-8, 190, n. 9/54
   × *catesbaei*, 142, 190
   × *chelsonii*, 189
   drummondii, 142
   flava, 142-3, 188, 189
   leucophylla, 142, 188, 189
   × *moorei*, 142-3, 188, 190, n. 11/69
   × *patersonii*, 190
   × *popei*, 142, 189, n. 11/9
   purpurea, 142, 189
   rubra, 142, 189, Plate 11
   variolaris, 143
Saxifraga florulenta, 210
   rotundifolia, 209
Scannell, Maura Josephine Patricia, 237
School of Art, Dublin, 148, 184, 226
Schomburgk, Robert, 122
Scott, Robert, 69, 97
Scott, Thomas, 101
Scouler, John, 158
Seabrooks Ltd (nurserymen), 226
Seaweeds, 98-9, 120, 140, 158
Secrett, Frederick (nurseryman), 226
Sedum, 225
Sempervivum, 225
Shackleton, Joseph Fisher, 187
Shackleton, Lydia, 184, 187, 190, Plates 9, 10, 11, 12, n. 11/59
Shamrock, 5
Sheridan, James, 80
Sherrard, G. O., 203
Sherrard, Thomas, 35
Sibbald, Robert, 5
Silliard, Zanchi, 7
Sims, John, 152
Simpson and Sibbert (nurserymen), 48
Simpson and Sons (nurserymen), 153
Skirving (nurserymen, Liverpool), 91
Sligo, Marquis of, 27
Smith, James Edward, 34, 36, 45
Smyth, Edward, 8
Societies:
   Agricultural Society of Ireland, 93
   American Rock Garden Society, 212
   Belfast Botanic and Horticultural Society, 127
   Botanical Society of Edinburgh, 161
   Botanical Society of London, 122
   Caledonian Horticultural Society, 83
   Dublin Natural History Society, 114, 117
   Dublin Naturalists' Field Club, 225
   Dublin Philosophical Society, 8
   Dublin Society (founded 1684): see Dublin Philosophical Society
   Dublin Society (founded 1731), 8 (see also Royal Dublin Society)
   Experimental Society of Ireland for Promoting Natural Knowledge, 23
   Farming Society of Ireland, 27, 46, 64, 68
   Irish Gardeners' Association, 183
   Linnean Society, 21-2, 49, 51
   Medico-Botanical Society of London, 77
   Pharmaceutical Society (Dublin), 226
   Physico-Historical Society, 21
   Royal Botanic Society of London, 123
   Royal Dublin Society, 8, 13, 15, 18, 23, 29, 31, 33-4, 36, 39, 46-9, 53, 58, 64-150 passim, 225
      Committee of Agriculture, 46-7
      Committee of Botany, 58-60, 69, 71, 73-79, 81, 85, 88, 90, 96, 105-7, 115, 117, 124, Appendix III
   Royal Geographical Society, 122
   Royal Horticultural Society (London), 183, 186-8, 195, 203, 225, 238

INDEX

Royal Horticultural Society of Ireland, 83, 94, 122, 152, 162, 201, 203, 212, 225
Royal Horticultural Society of Tuscany, 142, n. 9/54
Royal Society (London), 8, 11, 49, 69, 122
Royal Zoological Society of Ireland, 161
Society of Apothecaries (London), 5
Thomas Moore Society, 230
Solanaceae, 115
Solander, Daniel, 44
Solanum crispum 'Glasnevin', 238
  tuberosum: see Potato
Sole, William, 103
Sonder, Otto, 121
Sophora, 223
  tetraptera, 53
South Africa, 145-6, 153, 179, 189, 225, 240, n. 7/2 (see also Cape of Good Hope)
Span, James, 15-6
Spartium junceum, 209
Spiranthes romanzoffiana, 142
Stearn, William T., 225
Stearne, John, 7, 9
Steele, Richard, 37
Steele, William Edward, 150-1, 176, 178
Stella (Ester Van Homrigh), 44
Stephens, Samuel, 19
Stephens, William, 11, 13-5
Stern, Frederick, 225
Steward, William, 78
Stewart, Richard, 39
Stone, Nicolas, 5
Strawberry tree: see *Arbutus unedo*
Strelitzia × *nivenii*, 85, n. 7/6
  reginae, n.7/6
Stronge, Lady Margaret, 175
Sunderland, Duke of, 184
Sundew: see *Drosera*
Suringar, Willem, 174, n. 11/6
Sweet, Robert, 173
Swift, Jonathan, 37, 39, 42,, 44-5, 239
Sylliard, Zanchi: see Silliard, Zanchi
Synnott, Donal Michael, 237

Tasmania, 49, Plate 1
Taxodium distichum, 56
Taylor, Thomas, 78, 117
Telopea speciossima, 61, 146
Templemore (Londonderry), 98-9
Templeton, John, 61
Theobroma cacao, 111, 126-7
Theophrastus, 3
Thompson, Charles Wyville, 148
Threlkeld, Caleb, 11, 21
Thuja plicata, 62
Tickell, Mrs Clotilda, 41, 45
Tickell, John, 45
Tickell, Thomas, 37, 39, 40-5, 58, 126
Tickell, Thomas (junior), 58

Todea, 194
  barbara, 234
Toler, John, 19, 25, 34
Tolka (river), 36-7, 42, 46, 73, 134, 190, 198, 231, n. 4/1, 5/24
Toole, Charles and Luke (nurserymen), 48, 83
Trachycarpus fortunei, 145
Trapp, Joseph, 40
Travers, Henry Hammersley, 165-8
Travers, William, 166
Tresco (Isles of Scilly), 168
Trevithick, William (senior), 217
Trevithick, William Edward, 192, 217, Plate 15, n. 11/9, Appendix II
Trichomanes speciosum, 146, 161, 182, 194, 230
Trinity College, Dublin: see Dublin: University
Trollius, 183
Tsuga chinensis, 218
Tubergen, Van (nurseryman), 174
Tulipa australis, Plate 15
Turner, Richard, 106-7, 109, 131-4, n. 8/49, 8/50
Turner, Timothy, 109
Turner, Revd William (botanist), 5
Turner, William (ironmaster), 131, 195
Tweedia caerulea = *Oxypetalum caeruleum*, 158
Tweedie, John, 91, 155-8, 161, 184, Plate 2
Tyndall, John, 143

Ulex downiensis, 72
  europaeus 'Strictus', 72, 152, 238, Plate 4
Underwood, John, 49, 51, 58, 61-2, 71, 79-81, 83, 89, 153-4, n. 5/11, 5/13
Underwood, Mrs Mary, 81, n. 5/15
Utrecht (Netherlands), 7
Utricularia vulgaris, 143-4

Vaccinium, 56
Vallency, Charles, 58-9, 68, 154
Van Houtte, Louis (nurseryman), 174, 226
Vartry Reservoir (Co. Wicklow), 198
Veitch, Harold (Harry), 177
Veitch, James, 136
Veitch and Sons (nurserymen), 136, 143, 184
Venus's fly-trap: see *Dionaea muscipula*
Verbena, 101, 156-7
  arraniana, 157
  phlogiflora, 156-7
  tweediana, 152, 156, Plate 2
Vernon, Edward, 78
Veterinary College, Dublin, 226
Victoria amazonica, 121-3
  regina, 122
Victoria, Queen, 101, 110, 122, 147, 201
Viola nummularieafolia, 210
  valderia, 210
Viscus album, 142

Wade, Charles, 23
Wade, John (apothecary), 23
Wade, John (senior), 23
Wade, Mrs Mary, 23
Wade, Dr Walter (Portugal), 23
Wade, Walter, 18-9, 21, 23-4, 46, 48-51, 65, 68-9, 71-2, 74-7, 118, 152-4, n. 3/6
  portrait, 50, n. 5/43
Wakefield, Edward, 62
Walker, J. C., 45
Wallace, John, 71
Wallich, Nathaniel, 154
Walpole, Edward, 224-5
Walsh, Robert, 62, 95
Walsh, Thomas James, 226-35
Walsh, Wendy, Plates 4, 6, 7, 14
Ward, Frank Kingdon, 165, 171, 212, 220, 222
Ward, Nathaniel, 161
Wardian cases, 161-2, 164, 171, 194
Watson, William, 205-6, 212, 214
Watt, Helen (Mrs John Besant), 216
Webb, Captain, 145
Webb, David A., 140
Webb, Gertrude, 201
Weeks, Messrs, 194
Wellesley, Richard (Marquis), 83
Welwitschia mirabilis, 124
Wendland, Herman, 173
Wentworth, Thomas, 7
Westminster Abbey (London), 41, 86
Westmoreland, Earl of, 25
Whalley (nurseryman, Liverpool), 91
Whately, Richard (Archbishop of Dublin), 142, 161
Wheeler Cuffe, Lady Charlotte, 168-171
Wheeler Cuffe, Sir Otway, 168, 171
White, Edward, 72
White, Miss Henrietta, 201
White, James (of Ballitore), 118, n. 9/12
White, James (son of John White), 72
White, John (undergardener), 51, 64, 72, 75-7, 79, 81, 85, 89, 135, Plates 3, 4, n. 5/15, 5/16, 9/12
Wilkinson, Miss, 201
Willmott, Ellen Ann, 188
Willows: see *Salix*
Wilson, Ernest, 165
Wilson, John, 105
Wisteria, 87
Woodall, Edward, 176-7
Woolward, Florence, 183
Worsley, Benjamin, 7-8
Wright, Edward Percival, 175, 180

Yeats family, 218
Young, Arthur, 25, 31
Yu Dejung (Tse-Tsun Yu), 223

Zelkova carpinifolia, 238
Zurich (Switzerland), 140-1